SYMMETRIES

IN

PARTICLE PHYSICS

Professor Feza Gürsey

S Y M M E T R I E S
IN
PARTICLE
PHYSICS

EDITED BY

Itzhak Bars

University of Southern California
Los Angeles, California

Alan Chodos

and

Chia-Hsiung Tze

Yale University
New Haven, Connecticut

Springer Science+Business Media, LLC

Library of Congress Cataloging in Publication Data

Main entry under title:

Symmetries in particle physics.
"Proceedings of a symposium celebrating Feza Gürsey's sixtieth birthday, held April 11,
1981, at Yale University, New Haven, Connecticut"—T.p. verso.
"List of publications of Feza Gürsey": p.
Bibliography: p.
Includes index.
1. Symmetry (Physics)—Congresses. 2. Particles (Nuclear physics)—Congresses. 3.
Gursey, Feza. I. Bars, Itzhak. II. Chodos, Alan. III. Tze, Chia. IV. Gürsey, Feza.
QC793.3.S9S93   1984   539.7   84-13418
 ISBN 978-1-4899-5315-5

ISBN 978-1-4899-5315-5   ISBN 978-1-4899-5313-1 (eBook)
DOI 10.1007/978-1-4899-5313-1

Proceedings of a symposium celebrating Feza Gürsey's sixtieth birthday,
  held April 11, 1981, at Yale University, New Haven, Connecticut

Editors' Foreword

On April 11, 1981, many friends of Feza and Suha Gürsey descended on the Yale Campus from near and far for a day-long celebration of Feza's sixtieth birthday. The present collection of articles is, in part, a memento of that joyful gathering. Besides the talks delivered at the Symposium that was part of the festivities, this volume includes additional invited papers from a representative sample of Feza's long list of close friends and colleagues.

We hope to have captured between the covers of this book some small part of Feza's remarkable breadth of knowledge and interests, as well as his unique style of doing physics. We dedicate it to him as a sincere expression of our affection and admiration.

The production of this book benefited greatly from the efforts of several members of the Yale Theory Group. We wish to thank Thomas Appelquist who was instrumental in making the Yale Symposium an immense success. We gratefully acknowledge the skills and labor of John Gipson, Mary LaRue, Sait Umar, Jeff Hersh and Roger Ove who spent interminable hours typesetting the text and equations. We warmly thank the Yale Printing Service without whose essential collaboration this book would not have been possible. Finally, we deeply appreciate the cooperation of Ellis Rosenberg of Plenum Press who shepherded this enterprise through to its completion.

Itzhak Bars
Alan Chodos
Hsiung Chia Tze
New Haven, April, 1984

Contents

Some Perspectives on the Nonlinear Sigma Model   1
T. Appelquist

The Turbulent Aether   9
Yoichiro Nambu

Does Quantum Chromodynamics Imply Confinement?   19
Gerard 't Hooft

Chaos and Cosmos   33
Luigi A. Radicati di Brozolo

Dynamic Symmetries in Nuclei, Atoms and Molecules   47
F. Iachello

The Symmetry and Renormalization Group Fixed Points of Quartic Hamiltonians   63
Louis Michel

Relativistic Heavy Ion Collisions and Future Physics   93
T. D. Lee

The Fermion Determinant in Massless Two-Dimensional QCD   105
Ralph Roskies

Dynamic Mass Generation for Fermions   115
Abdus Salam and J. Strathdee

Tomographic Representation of Quantized Fields   127
Charles M. Sommerfield

On Spontaneously Broken Supersymmetry   141
G. Domokos and S. Kövesi-Domokos

An Action in Superspace for $SO(N)$-Supergravity   159
S. W. MacDowell

Contents

Intrinsic Geometry of Supergravity   177
   V. I. Ogievetsky

On the Physics of Dimensional Reduction   191
   Peter G. O. Freund

Backlund Transformations and Deformations of Linear Differential   201
Equations with Applications to Diophantine Approximations
   G. Chudnovsky

Backlund Transformations and Geometric and Complex-Analytic Back-   221
ground for Construction of Completely Integrable Lattice Systems
   D. V. Chudnovsky

Unfashionable Pursuits   265
   Freeman J. Dyson

Epilogue   287
   Maurice Goldhaber

List of Publications of Feza Gürsey   291

Curriculum Vitae of Feza Gürsey   299

List of Contributors   301

Index   303

SOME PERSPECTIVES ON THE NONLINEAR SIGMA MODEL

T. Appelquist

J. W. Gibbs Physics Laboratory
Yale University
New Haven, CT 06511

Abstract

When the nonlinear sigma model was first written down in 1959 by Feza Gürsey [1], it was modestly offered as "an illustration of the possibility that the symmetries of the weak interactions may be essentially contained in those of the strong interactions, and certainly not as a definite proposal for a theory of elementary particles." In the more than twenty years since that paper appeared, the nonlinear sigma model has been extended in several ways, applied to a variety of low energy phenomena, and analyzed as both a classical and quantum field theory. While it has never gained the status of a fundamental theory of elementary particles, it has certainly played a key role in the development of particle physics. Its beautiful and realistic symmetry properties and its deep dynamical structure have made it much more than an illustration of possibilities.

Most of the important features of the nonlinear sigma model were already emphasized in Gürsey's original paper. He introduced a proton-neutron doublet and constructed the pion-nucleon interaction to be invariant under the chiral symmetry group $SU(2)_L \times SU(2)_R$. He pointed out that a nonlinear realization of this symmetry requires the presence of pion self-interactions of a definite form, accompanying the usual pion kinetic energy term in the Lagrangian. He noted that the nonlinear constraint would lead to a nucleon mass along with pion-nucleon interactions and, most importantly, that it would not allow the existence of a pion mass.

The symmetry group $SU(2)_L \times SU(2)_R$ is perhaps the simplest, and yet still the most physically relevant, on which to base a nonlinear theory. Nonlinear realizations of

1

larger groups, such as $SU(3)_L \times SU(3)_R$ are of less obvious physical importance and, furthermore, they lead to very little that is new in terms of field theoretic structure. In the $SU(2)_L \times SU(2)_R$ model, the fields are represented by the two-by-two matrix

$$M(x) \equiv \sigma(x) + i\, \vec{\tau} \cdot \vec{\pi} \tag{1}$$

which transforms from the left and right according to the (1/2, 1/2) representation of $SU(2)_L \times SU(2)_R$. The nonlinear constraint

$$M^\dagger M = M M^\dagger = f^2 \tag{2}$$

can be enforced by requiring

$$\sigma(x) = [f^2 - \vec{\pi}^2(x)]^{1/2} \tag{3}$$

The Lagrangian density for the purely pionic part of the theory is

$$\mathscr{L} = \frac{1}{4}\, \text{Tr}\, \partial_\mu M^\dagger \partial^\mu M = \frac{1}{2}(\partial_\mu \vec{\pi})^2 + \frac{1}{2}\frac{(\vec{\pi} \cdot \partial_\mu \vec{\pi})^2}{f^2 - \vec{\pi}^2} \tag{4}$$

and, as with any Lagrangian field theory, it can be cast into a variety of different forms by making changes of variables [2]. With the Lagrangian written in terms of the π fields, invariance under the isomorphic group $O(4) \cong SU(2)_L \times SU(2)_R$ is nearly manifest. The set $\phi = (\sigma, \vec{\pi})$ transforms as an $O(4)$ vector with $\vec{\pi}$ transforming linearly under the $O(3)$ subgroup. In addition, there is a set of three "boosts" under which π transforms nonlinearly:

$$\vec{\pi} \rightarrow \vec{\pi} + \vec{\epsilon}[f^2 - \vec{\pi}^2]^{1/2} \ . \tag{5}$$

The work of Nambu and Jona-Lasinio [3] and Goldstone and others [4] showed that many of the properties of the nonlinear sigma model are common to a larger class of models. The unifying feature is the spontaneous breakdown of continuous symmetries, leading necessarily to the presence of massless Goldstone bosons. The nonlinear realization of $SU(2)_L \times SU(2)_R$ is an example of spontaneous breakdown, with the massless π fields creating the Goldstone bosons. A simple way to enter the larger class of models is to remove the nonlinear constraint (Eq. 2). Then the σ field becomes an independent coordinate and invariant interactions of the form $\mu^2 M^\dagger M$, $\lambda(M^\dagger M)^2$, etc., can be added. This is the linear sigma model, and with an appropriate choice of μ^2 and λ the symmetry will again break spontaneously and the π field will be massless. The mass of the σ field will be μ.

The linear sigma model has an apparent advantage over the nonlinear model. It is renormalizable if the highest dimensional interaction is $(M^{\dagger}M)^2$, and it can, therefore, be reliably used at very large energies $(E \gg \mu)$. On the other hand, it has the same low energy $(E \ll \mu)$ behavior as the nonlinear model. The reason for this is that the nonlinear model can be viewed as a limiting form of the linear model in which λ and $\mu^2 \to \infty$ with the ratio $\mu^2/\lambda \equiv 4f^2$ fixed. While the nonlinear theory is nonrenormalizable, meaning that this limit does not really exist at the one-loop level and beyond, it is only the tree-level theory which is relevant for the leading low-energy behavior. It is by now well understood that the same low energy behavior is common to any theory with a spontaneously broken $(SU(2)_L \times SU(2)_R)$ symmetry [5]. Whether it is the linear model or a model in which the Goldstone bosons are bound states of fermionic constituents such as quarks, the most concise way of summarizing the low energy behavior is the nonlinear model used at the tree graph, or "phenomenological" level.

We still do not have a genuine understanding of how, and under what circumstances, the spontaneous breakdown of symmetry takes place. If QCD is indeed the correct theory of strong interactions, then it evidently does lead to spontaneous breakdown of $SU(2)_L \times SU(2)_R$ and to the Goldstone nature of the pion. The challenge is to show how this happens and how the nonlinear sigma model, which correctly describes the interactions of the pion at energies below a few hundred MeV, emerges out of QCD. If QCD, or some other theory of quark constituents, is eventually shown to be the correct high energy extension of the nonlinear $SU(2)_L \times SU(2)_R$ model, then there is little point in extrapolating to higher energies by means of the linear model. The lure of a simple renormalizable high energy extension may lead to a theory which has nothing to do with reality. The linear model should then be viewed simply as a convenient way of regularizing the worst divergences of the nonlinear model if its quantum properties are being studied.

This kind of study has, in fact, been going on for more than a decade [2]. A number of surprises have been encountered and our understanding of quantum field theories in the presence of nonlinear constraints has grown considerably. The most important fact to emerge from these studies, however, is one which surprises almost no one. The nonlinear model is indeed nonrenormalizable. At each order in the loop expansion, new counterterms must be introduced in order to render finite the physical predictions, in the form of the on-mass-shell S-matrix. The counterterms are higher dimensional operators to be added to the Lagrangian and they, therefore, produce only nonleading terms in the low energy expansion.

The form and cutoff-dependence of the counterterms have been computed through two loops [6]. The one-loop counterterms are

$$\mathcal{L}_1 = \left(\frac{1}{16\pi^2 f^2} \ln\mu\right)\left(\mathrm{Tr}\partial_\mu M^\dagger \partial^\mu M\right)^2$$

$$\mathcal{L}_2 = \left(\frac{2}{16\pi^2 f^2} \ln\mu\right)\left(\mathrm{Tr}\partial_\mu M^\dagger \partial_\nu M\right)^2$$

(6)

where I have used the linear model, with mass μ, to provide the cutoff. From the point of view of the nonlinear model, this cutoff dependence must be absorbed into two new free parameters which must be introduced at this stage. Then, although the interactions of the original Lagrangian, $\mathcal{L}$, involving only two derivatives, will surely dominate at sufficiently low energies, the scale for the low-energy expansion cannot be predicted. On the other hand, the mild, logarithmic cutoff dependence will amount to only a factor of order unity unless the cutoff is enormously large. Thus, a comparison of $\mathcal{L}_1$ and $\mathcal{L}_2$ with $\mathcal{L}$ would lead the pragmatist to the conclusion that the low energy expansion is set by the scale $4\pi^2 f^2$. In the parlance of the linear model, this is simply $\mu^2 (\lambda/\pi^2)^{-1}$, where λ/π^2 can be recognized as the loop expansion parameter. The scale μ^2, by itself, also enters the low energy expansion of the linear model (already at the tree level), but

$$\mu^2 \gtrsim \mu^2 (\lambda/\pi^2)^{-1}$$

(7)

as long as $\lambda/\pi^2 \lesssim 1$. It seems very natural to me that the cutoff μ^2 on the nonlinear model is of order $4\pi^2 f^2$, so that this inequality is saturated. This corresponds to the existence of strong self-interactions among the Goldstone bosons at energies of order $4\pi^2 f^2$ and higher.

At low energies $E \ll 2\pi f$, it is thus reasonable to conclude that the effective expansion parameter governing the use of the nonlinear sigma model is of order $E^2/4\pi^2 f^2$. Studies at two loops and beyond [6,7] reinforce this conclusion by showing that new higher dimensional operators with coefficients of order $\frac{1}{4}(4\pi^2 f^2)^n$ enter at each order n. In addition, each lower dimensional operator already generated at lower orders gets renormalized by an amount of order $\mu^2/4\pi^2 f^2 \cong \lambda/\pi^2$. While these are quadratic ultraviolet divergences from the point of view of the nonlinear model ($\mu \to \infty$), the natural cutoff size $\mu^2 \cong 4\pi^2 f^2$ makes them of order one. All of this reinforces the notion that $E^2/4\pi^2 f^2$ is the effective expansion parameter and that the nonlinear sigma model can be usefully employed up to energies at least of order $2\pi f$.

Although the above conclusions have been drawn by making use of the linear model as a cutoff, it is the nonlinear model which plays the essential role. The allowed counterterms are dictated by the symmetries of the nonlinear model and by a dimensional analysis of its loop expansion. The actual computations can be done either directly with the nonlinear model using, say, dimensional continuation or using the linear model as I have discussed. In either case, some subtleties must be dealt with in order to make the computed cutoff dependence correspond precisely to that expected from symmetry and dimensional analysis. If canonical perturbation theory is applied directly to the nonlinear model, it is found that some off-mass-shell divergences appear in addition to those corresponding to the allowed counterterms such as $\mathscr{L}_1$ and $\mathscr{L}_2$ [2, 6, 7]. Although these extra terms vanish on shell, their appearance is annoying and, at the very least, they complicate the analysis of higher-loop diagrams.

In 1972, Honerkamp showed [8] that these terms originate in the inappropriate use of canonical perturbation theory in the presence of a nonlinear constraint. The Goldstone fields live on a closed, curved manifold and the usual loop expansion, being an essentially linear procedure, simply doesn't know about this. Honerkamp suggested that a modified loop expansion be employed which takes into account the natural geometry of the manifold. He used a variant of the background-field method in order to do this, which leads to a modified set of Feynman rules. It has been checked that this procedure does indeed lead to one-loop Green's functions, whose divergent parts correspond precisely to the allowed counterterms both off shell and on shell. An elegant exposition of this approach has recently been given by Boulware and Brown [9].

While dealing directly with the nonlinear model has an undeniable geometric elegance, I have tried to suggest that using the linear model as a cutoff can help in understanding the scales in the low energy expansion. Canonical perturbation theory and the usual loop expansion can then, of course, be employed. After conventional renormalization, λ and μ^2 can be taken large, with their ratio fixed, and the effective, cutoff-dependent interactions (Eq. 6) can be identified. However, there will again be additional, off-shell interactions, now proportional to $ln\mu$, whose structure is the same as the divergent off-shell terms encountered in the canonical treatment of the nonlinear model. These connections between the nonlinear and linear models have been elucidated in Ref. 2, but many interesting questions are still only partially answered. In particular, is there a perturbative treatment of the linear model which takes into account, in a simple way, the fact that it is only a convenient regulation for the nonlinear model? One would want such a procedure to lead only to the allowed structures (Eq. 6) in the large μ limit and not to the additional off-shell structures. Somehow, information about the nonlinear constraint (Eq. 2) would have to be built into the linear theory.

Perhaps the most interesting use of the nonlinear sigma model is as a description of the Higgs sector of spontaneously broken gauge theories. In the Weinberg-Salam model, the scale associated with symmetry breakdown (the value of f in Eq. 2) is about 250 GeV. Thus, if the Higgs self-coupling is strong, then at energies small compared to $2\pi f \simeq 1.5$ TeV, it is most appropriate to view the model as a gauged nonlinear sigma model. A few more counterterms, in addition to those of Eq. 6, must be included at the one loop level, but all the cutoff dependence remains logarithmic [8]. It is possible to use this kind of analysis to, in effect, catalogue all possible physical sensitivity to a heavy, strongly interacting Higgs sector in an electroweak theory [10]. The results in the standard Weinberg-Salam model are not experimentally encouraging, but that may not be the case in more exotic models. A strongly interacting, dynamical origin for the Higgs mechanism may require larger symmetry groups and therefore more Goldstone bosons than can be absorbed by the $W^{\pm}$ and Z^0 .

As a theoretical laboratory, the nonlinear sigma model has proven its value again and again. It is renormalizable and asymptotically free in two space-time dimensions, making it an interesting prototype for four-dimensional gauge theories. It can give rise to topologically nontrivial classical solutions [11] and it can be analyzed nonperturbatively using $1/N$ expansions. Recently, nonlinear models which incorporate supersymmetry have been developed and studied [12]. If supersymmetry proves to be a chosen symmetry of nature, it is clear that it must appear broken at accessible energies. If the breaking is spontaneous, then supersymmetric nonlinear sigma models will, at the very least, play an important role as an effective low energy theory. Even solid state theorists have used the nonlinear sigma model to describe collective phenomena with Goldstone modes. It has, for example, proven to be useful in the study of surface physics [13].

It seems very likely to me that during the second twenty years of its existence, the nonlinear sigma model will loom even larger in the theory of elementary particles than it did during its first twenty years. It has taught us a great deal but there is surely much more for us to learn.

REFERENCES

[1]. F. Gürsey, Nuovo Cimento 230 (1960) 230.

[2]. T. Appelquist and C. Bernard, Phys. Rev. D23 (1981) 425 and references contained therein.

[3]. Y. Nambu and G. Jona-Lasinio, Phys. Rev. D122 (1961) 345.

[4]. J. Goldstone, Nuovo Cimento 19 (1961) 154; J. Goldstone, A. Salam and S. Weinberg, Phys. Rev. 127 (1962) 965.

[5]. For reviews, see: B. Lee, *Chiral Dynamics*, Gordon and Breach, S. Gasiorowicz and D. Geffen, Rev. Mod. Phys. 41 (1969) 531.

[6]. D. Kazakov and S. Pushkin, JINR Dubna preprint E2-10655 (1977).

[7]. T. Appelquist, *Broken Gauge Theories and Effective Lagrangians*, Proceedings of the 21st Scottish Universities Summer School in Physics, St Andrews, Scotland, 1980.

[8]. J. Honerkamp, Nucl. Phys. B36 (1972) 130.

[9]. D. Boulware and L. Brown, Univ. of Washington preprint, 1981.

[10]. T. Appelquist and C. Bernard, Phys. Rev. D22 (1980) 200 and A. Longhitano, Phys. Rev. D22 (1980) 1166.

[11]. J. Gipson and H. C. Tze, Nucl. Phys. B183 (1981) 524 and references contained therein.

[12]. See, for example, L. Alvarez-Gaumé and D. Freedman, MIT Dept. of Mathematics preprint, 1981; C. Zachos and T. Uematsu, Fermilab-DUB-81/76, Nov. 1981.

[13]. D. Wallace, *Field Theories of Surfaces*, Proceedings of the 21st Scottish Universities Summer School in Physics, St Andrews, Scotland, 1980.

THE TURBULENT AETHER

Yoichiro Nambu

The University of Chicago
The Enrico Fermi Institute
5630 Ellis Avenue
Chicago, Illinois 60637

1. MY ENCOUNTER WITH FEZA GURSEY

My first encounter with Feza occurred probably in the summer of 1959 at BNL. We were among the summer visitors there. Recognizing his name on the door of his office, I went in and introduced myself, because I knew his work on the so-called Pauli-Gürsey transformation, and also because I had heard a lot about him from K. Nishijima. In any case, I immediately felt close to him. I don't think he has changed much over the years, nor has my feeling toward him.

My association with him has been both professional and personal. I was invited twice to summer schools in Istanbul; I have taken on two Turkish students. But above all, I must say, I have learned a great deal from him, not only through his original work, but also because of his erudition in mathematics and physics. He is a good source of information and enlightenment for me, and my indebtedness to him often appears in the form of references and footnotes to my papers.

In a more direct and important way, his contributions to physics have influenced my thinking, too. The idea of an effective Lagrangian embodying certain symmetry properties first appears in his work on chiral symmetry. This was further developed by other people in the late 60's and made into a polished method of chiral Lagrangians. Most recently, the effective Lagrangian seems to be coming back again in a broader context in connection with the GUTs, or the grand unified theories. The Pauli-Gürsey transformation is also quite commonplace now. Feza has built a fine and unique school of theoretical physics at Yale. Every bit of it has the color and flavor of Feza. It also fits nicely into the dignified Yale tradition dating back to Willard Gibbs. If

Feza is not exactly a Connecticut Yankee, still he can pretend that King Arthur lives in New Haven.

2. REALITY OF AETHER

The subject I would like to discuss has at least partly grown out of my contact with the Yale school, as you will presently see. However, this is not a place to discuss the history of aether. Nor do I know much about the history anyway. So let me immediately come down to modern times.

After Maxwell and Einstein, the physicists of the 20th Century have learned to do away with any mechanical model to understand or explain electromagnetism and gravitation. The space-time itself is endowed with the basic properties that are necessary to account for these phenomena. However, this does not mean that the elimination of the concept of aether is altogether complete.

I think it is meaningful, even in physics, to distinguish between constitution and law. The intrinsic properties of space-time, as uncovered by Einstein, and the principles of quantum mechanics may be called the constitution of physics to which all the individual laws of physics have to conform. The former gives the basic framework for the latter, and has an appearance of being more permanent, pervasive and compelling, whereas the latter are more susceptible to evolution, modification and addition.

Today we know that there exist a large number of distinct kinds of particles and fields. This number has grown considerably in the last 50 years, and is most likely to increase even further. We do not know why they exist, how many there should be ultimately, or whether they can further be broken down to even more basic elements. In any case, all these diverse particles and fields are described according to the principles of quantum mechanics, and may be regarded as different types of excitations that propagate in space-time.

Thus the space-time looks as if it is occupied by a very complex medium, to which I will attach the old name aether. The complexity has become more and more pronounced in recent years not only because of the increase in the number of excitations, but also because of the conceived possibility that the types of excitations can be altered by the prevailing conditions of space-time itself, just as an ordinary material medium can change from gas to liquid to solid under changing temperature and pressure. Regrettably, we have not yet observed such a phase transition of the aether. It will be a remarkable event when the phenomenon actually is confirmed some day.

On the other hand, there have been persistent efforts by physicists starting with Einstein, to reduce all physical laws to a geometry of space-time itself. These efforts fall under the name of unified field theories. Although dormant for many years, we are currently seeing great upheaval and progress in this direction with the advance of gauge theories and the success of the unification of electromagnetic and weak interactions.

If this program can be carried out to the end, we may be able to say that the aether is indeed the space-time itself, which is endowed with a rich and complex geometry. But somehow we will have to be able to incorporate the mechanism for the above mentioned phase transition, or spontaneous breakdown of symmetry as it is usually called. This mechanism is the only known one to explain the fact that there exist massive particles in nature, like the electrons and the protons that form the basis of ordinary matter.

Massless fields, like gravity and electromagnetism, can be easily viewed as geometrical constructs. Mass is the stumbling block that spoils the elegance of a geometrical picture of physical laws. This point is epitomized in Einstein's remark about his field equation

$$R_{\mu\nu} - \frac{1}{2} g_{\mu\nu} = T_{\mu\nu} \, .$$

He observed, or complained, that the equation looks very uneven, like a house with two wings, of which the left one is made of marble and the right is made of wood. The marble wing, of course, represents the curvature of space-time, or the gravitational field, whereas the wooden wing represents the energy-momentum tensor of everything else present in space-time. We may take out the electromagnetic field from the latter and move it to the left side because it can be interpreted as curvature in a generalized geometry. But what about massive particles? If we somehow succeed in reducing them to geometry, we may bring them over to the left side, one by one, until nothing remains on the right. The house is now all made of marble. But will it really be beautiful? It will look lopsided: it will have lost symmetry. Maybe this is as it should be, because Nature does not have a left-right symmetry, either.

Currently people are trying to achieve this lofty goal in the form of grand unified theories and supersymmetric theories. They are conceptually very elegant and appealing. Certainly I like them up to a point, but something still seems missing, something that makes the right-hand side of Einstein's equation distinct from the left-hand side. The fundamental principle underlying this distinction should be simple and elegant, although the outcome is superficially ugly and complicated.

3. TURBULENT AETHER

I have no answer to the question I just posed to myself. Instead, I will now ask another. As I have remarked, massless fields, namely long range fields such as gravity and electromagnetism, are natural manifestations of a geometrical principle. This principle is nowadays called the gauge principle, a name coined by the mathematician Hermann Weyl in his attempt at a unified theory of gravity and electromagnetism.

Gravity and electromagnetism are the only long range fields we know. The rest are massive fields, which manifest themselves as short range forces or massive particles. One exception is the neutrinos, which have been thought to be massless, but this presumption is now becoming suspect. Lacking an obvious gauge principle, the neutrinos may in fact be massive, if very slightly.

Any other types of forces and objects? That is my question. In the past decade there have been interesting theoretical developments which actually have prompted me to ask this question. The forces and objects I am talking about are not quite like the ones we are used to. The objects are not particles, or pointlike concentrations of energy, but strings which are linearly extended. A string can join two particles and keep them together. Strings may break up or join, but always there is a particle attached at the end of a string, being created or destroyed in such a process. A closed string is an entity in itself, not requiring any particles associated with it.

Experimental indications are that the hadrons indeed behave like quarks joined by means of strings. Furthermore, quantum chromodynamics -the gauge theory of strong interactions -tells us rather convincingly that these strings are nothing but chromo-electric lines of force squeezed into thin tubes, like the paths of spark discharge between two electrodes. There are other ways to visualize the strings. One is to compare them to magnetic flux lines -the Abrikosov flux lines- trapped in a superconducting medium. In this analogy, you replace the electric field with a magnetic field, and the quarks with the magnetic monopoles - the hypothetical particles first introduced by Dirac 50 years ago.

These monopoles have not been found yet. But again the modern gauge theory shows that they should in general exist, though the existence depends on the details of the theory. Some of them may be able to exist temporarily in pairs and joined by magnetic strings. Such objects will look like the ordinary hadrons. Monopoles and monopole-strings, however, are much heavier than the hadrons. The unified gauge theories expect them to be upwards of 1 Tev. Although they are beyond experimentally accessible energies, their existence has important implications for the Big Bang cosmology.

Yet another analogy for the hadron strings is the hydrodynamic vortices. Vortices are a very common phenomenon observed in everyday life; they are characteristic of the motion of fluids in general. In an infinite vessel, the core of a vortex forms a closed loop, so we may compare it to a closed chromo-electric string which is also called a glueball. This analogy has a very intriguing speculative aspect. A line of vortex induces an organized rotational pattern of motion around it. Because of this, two vortex lines repel each other when they are parallel, and attract each other when they are anti-parallel, exactly like two electric currents except that the sign of the force is opposite.

Just as the electromagnetic field exists in our space-time and exerts forces between the electric charges and electric currents, shouldn't a hydrodynamics field also be an intrinsic property of the aether, and exert forces between vortex cores?

I am here paraphrasing the question first posed by Pierre Ramond, a former member of Yale. The electromagnetic field is a natural one associated with mass points: the vector potential couples to the tangent vector of the world line of a mass point, and this interaction is unchanged if the potential is altered by the gradient of a scalar.

Ramond and his student Kalb asked the question: what is the natural field associated with strings rather than mass points? A string sweeps out a 2-surface in space-time, so an anti-symmetric tensor field, not a vector field, may couple to the tangential 2-plane at each point on the surface. The consequences of the ensuing mathematical formalism are what I have described above using hydrodynamic terminology. It should be pointed out that this Kalb-Ramond theory predicts the existence of massless fields propagating in vacuo (or in the aether), just as the Maxwell theory predicted the electromagnetic waves, except that the hydrodynamic waves are longitudinal.

The Kalb-Ramond theory is a geometrical theory like electromagnetism and gravity. It is based on a new gauge principle appropriate to strings rather than mass points. Physics remains invariant if the potential is altered by the curl of a vector.

For some time I have been wondering if Mother Nature has adopted this geometrical principle along with the other known ones. If so, the aether must be turbulent, filled with vortices of all sizes. Actually, such a view is quite old. Whittaker's book on the history of the aether tells us about the activities of the 19th Century physicists along this line. In particular, W. Thomson (Lord Kelvin) proposed that the atoms are vortex rings. He also envisaged an aether filled with vortex rings, and called this state "vortex sponge". Of course they were on the wrong track. What they were after was not electromagnetism. And yet they might still turn out to be right after all.

I have been searching for clues or evidence for the existence of vortices and their interactions. How big is a typical vortex ring? Is it of atomic scale, or human scale, or perhaps astronomical scale? The hadrons and glueballs of course behave like vortices in many ways, but they do not produce long range fields. Open strings break the Kalb-Ramond gauge principle, and turn the field into a massive one.

I now think that the most promising place for vortices is cosmology. The relevant vortex lines in this context will be of astronomical size. Their interaction with ordinary matter may be very weak, but at least they must couple to gravity. Being intrinsically extended, they can influence the evolution of the universe in a manner different from ordinary matter. Such a possibility was in fact recently suggested by Zeldovich in order to account for the inhomogeneities of the universe that are found in the form of galaxies. He was talking about the magnetic or electric flux lines expected in the GUTs. But it might be interesting to contemplate macroscopic vortices of the Kalb-Ramond type.

4. MATHEMATICAL DESCRIPTION OF STRINGS

I will close with a bit of mathematical formalism. Classical dynamics of mass points is a well developed branch of physics. It can be reduced to the Hamiltonian system of equations of motion, or the Hamilton-Jacobi form of partial differential equations. These also serve as the basis for going over to quantum mechanics. The whole formalism is a very natural one for dealing with mass points. It is true that a string may be viewed as a continuous collection of mass points, and as such, can be handled by this formalism, but it is not a natural connection. The point is that a mass point evolves with one internal clock - the proper time - into a world line, whereas a string may be said to have two internal clocks, one timelike and one spacelike, as it evolves into a world sheet. When the "times" are treated symmetrically, we arrive at a generalized form of Hamiltonian systems which determines how a patch of sheet should evolve in two directions. A patch of sheet should actually evolve into a minimal surface (like the surface of a soap bubble), just as a mass point should evolve into a geodesic. So the formalism amounts to a new way of handling minimal surfaces.

A further interesting situation emerges when we consider the analog of the Hamilton-Jacobi system. For mass points it describes a family of geodesics which fill up the space without intersecting each other, and is represented by a scalar field, or Hamilton function. The gradient of this scalar is the tangent to the geodesic, representing the momentum vector, at that point. In quantum mechanics, the scalar field gets translated into the Schrödinger wave function.

What is the corresponding Hamilton-Jacobi system for strings? It will be a family of non-overlapping minimal surfaces. Such a family can be represented by a vector field. The curl of this vector, formally looking like the electromagnetic field, determines the tangential 2-plane to a minimal surface. The analogy with electromagnetism is rather striking, although there are important differences. I will illustrate this by writing down, side-by-side, the system of equations for 1) a mass point, 2) string, and 3) electromagnetic field.

Mass Point

$$\frac{dx_\mu}{d\tau} = \frac{\partial H}{\partial P_\mu} \quad , \quad H = - L = \frac{1}{2}(P_\mu - eA_\mu)^2 \quad , \quad P_\mu = \partial_\mu \mathcal{S}$$

$$\frac{dP_\mu}{d\tau} = -\frac{\partial H}{\partial x_\mu} = constant \quad , \quad (\partial_\mu \mathcal{S} - eA_\mu)^2 = constant$$

String

$$\frac{\partial(x_\mu, x_\nu)}{\partial(\tau,\sigma)} = \frac{\partial H}{\partial P_{\mu\nu}} \quad , \quad H = - L = \frac{1}{4}(P_{\mu\nu} - \Omega_{\mu\nu})^2 \ , \ P_{\mu\nu} = \partial_\mu B_\nu - \partial_\nu B_\mu \equiv F_{\mu\nu}(x)$$

$$\frac{\partial(P_{\mu\nu}, x_\nu)}{\partial(\tau,\sigma)} = -\frac{\partial H}{\partial x_\mu}$$

$$\partial_\mu F^{*}_{\mu\nu} = 0 \quad , \tag{4.1.a}$$

h_ν: *end point current*

$$\partial_\mu (\phi F_{\mu\nu}) = h_\nu \quad , \tag{4.1.b}$$

ϕ: *a scalar field*

$$F^{*}_{\mu\nu} F_{\mu\nu} = 0 \quad , \tag{4.1.c}$$

$\Omega_{\mu\nu}$: *Kalb–Ramond field*

$$(F_{\mu\nu} - \Omega_{\mu\nu})^2 = constant \quad , \tag{4.1.d}$$

Electromagnetic Field

$$- L = \frac{1}{4} F_{\mu\nu}^2 \quad , \qquad\qquad F_{\mu\nu} = \partial_\mu A_\nu - \partial_\nu A_\mu$$

$$\partial_\mu F^{*}_{\mu\nu} = 0 \quad , \tag{4.2.a}$$

$$\partial_\mu F_{\mu\nu} = j_\nu \quad . \tag{4.2.b}$$

The string system is more restricted than electromagnetism in one sense, because of eqn.(4.1.c) and eqn.(4.1.d). But it is more general in another sense, because of the appearance of the "dielectric coefficient" ϕ in eqn.(4.1.b).

The emergence of an Abelian gauge field in the description of strings is very suggestive. In fact it can be put to good use. We can reduce the non-Abelian gauge theory of QCD to an effective Abelian one for a dielectric medium. Its dielectric constant is related to the running coupling constant of QCD considered as a function of the field itself. This approach is not entirely new, but in the present formalism the effective gauge field is Abelian, and the equations for it reduce to those of the string system in the strong coupling limit. The effective Lagrangian takes the following form:

$$L = \frac{1}{4} g^2 S_{\mu\nu} S_{\mu\nu} + \frac{1}{4} f S_{\mu\nu} S_{\mu\nu}^{*} - \frac{1}{2} S_{\mu\nu} F_{\mu\nu} - j_{\mu} B_{\mu} \ ,$$

$$F_{\mu\nu} = \partial_{\mu} B_{\nu} - \partial_{\nu} B_{\mu} \ .$$

Here j_{μ} is the external source current. B_{μ} and $S_{\mu\nu}$ are independent Abelian variables. g^2 is the running coupling constant of the original QCD, and f yet another parameter, both being regarded as functions of S^2 and $(S \cdot S^{*})^2$ in general. If $S \cdot S^{*} = 0$ and $f = 0$, the equations of motion can be reduced to

$$\partial_{\mu}(\epsilon F_{\mu\nu}) = j_{\nu} \quad ; \quad S_{\mu\nu} = \epsilon F_{\mu\nu} \quad ; \quad \frac{1}{\epsilon} = \frac{d(g^2 S^2)}{dS^2}.$$

The function ϵ plays the role of a dielectric coefficient of the medium. According to the strong coupling theory, $(g^2 S^2)$ is expected to go like $1/S$ for small S, i.e., $\epsilon \sim S$. In this limit F approaches a constant, which is precisely the property of a string system. The f term in the Lagrangian seems improper and unnecessary as it violates T and P. But I am keeping it because it allows a more general description of string systems even if $S \cdot S^{*} = 0$ for them, and also because it may be related to the problem of the Θ-vacuum in QCD, arising from the presence of instantons.

Clearly, this effective Abelian theory stands midway between classical string dynamics and quantum chromodynamics. It is like the WKB approximation to quantum mechanics. A point to be noted here is that the question of what constitutes quantum string dynamics is avoided. Quantization of string dynamics as such is still plagued with mathematical problems, and may in fact be meaningless in a strict sense.

The properties of the effective Lagrangian have not yet been studied in detail. For the typical problem of two static point sources, it seems that the solution is a bag-like object, as is to be expected. The electric displacement D is confined within the bag, and vanishes toward the boundary, where ϵ becomes zero, and the strong coupling limit is reached. An important question is what lies outside the bag. Although the D and H fields are zero outside, the E and B fields must remain finite. It has been argued that the vacuum state in QCD is not a state of no gluon fields, but corresponds to a nonzero density of fields, $<B^2 - E^2> \neq 0$. This seems to be the case in the present Abelian theory, too. The direction of these fields will not be fixed however. So we are again talking about a turbulent kind of vacuum.

According to the recent developments in lattice gauge theory, it is claimed that QCD has only one phase, i.e., there is no sharp boundary like the bag surface separating the inside from the outside. This may well be so, as our theory is something like mean field theory, and a local approximation at that.

To summarize my theme, the aether is a far more complex medium than has been thought until recently. It might in fact be turbulent in many ways, although it is not dissipative. The scales of turbulence could range from cosmological to subatomic. In the subatomic phenomena, the running coupling of gauge fields is their Reynolds number.

DOES QUANTUM CHROMODYNAMICS IMPLY CONFINEMENT?

Gerard 't Hooft

University of Utrecht
Institute for Theoretical Physics
Postbox 80.006
3508 TA Utrecht
The Netherlands

Abstract

The mechanism of permanent quark confinement in Quantum Chromodynamics is cast in a language that might enable one to perform more detailed dynamical calculations. The choice of gauge condition and the topological features added to the theory by fixing the gauge are crucial in this description. However, it also becomes clear that in spite of its popularity Quantum Chromodynamics has an "identity problem": the theory still has no sound mathematical basis. We speculate on remedies that one can envisage in the future.

1. INTRODUCTION

It is a long-standing problem how to devise a reliable method for computing physically observable quantities accurately in non-abelian gauge theories with strong interactions. One crucial step in solving this problem is to isolate the relevant dynamical variables at the critical distance scale (in quantum chromodynamics that is, of course, the hadronic mass scale). Approximation methods that are popular at present sometimes ignore some of these variables. In ordinary perturbation theory the compactness of the gauge group is not reflected, so that topologically non-trivial effects such as instantons literally slip through the meshes. Theories based on instanton gases in their turn do not admit magnetic vortices, which we know to be crucial for understanding confinement. We do not as yet have a remedy for this situation. Nor do we have a clear insight on how seriously this drawback should be taken. Experiments suggest without much doubt that with "QCD" we are on the right track. But how well do we understand this theory?

In essence there are two problems. We list the two in the order that they are likely to be solved. The first is to obtain a good and reliable *qualitative* understanding. The observation that topological considerations are crucial here has been made a number of times [1,2]. In this paper we emphasize this and elucidate the present state of mind of the author (sects 2-5). Indeed these qualitative topological considerations yield a picture that could explain the experimental observations neatly. Very neatly, were it not that experimenters seem to be turning the clock backwards by observing some fractionally charged "quarks" [3]. To the best of my own understanding theory is unable to follow experiment here, in spite of some very ingenious attempts [4].

It is unlikely that the second problem on my list has anything to do with this fractional charge problem. It is the more subtle question of mathematical definition. Careful consideration of the perturbative theory, with its beautiful property of asymptotic freedom, strongly suggests that short-distance perturbation theory determines the dynamics completely, including the stronger interactions at long distances, because interaction at long distances could be viewed as an interplay of many weaker interactions at short distances. Renormalization group arguments suggest that one should choose as a bare coupling constant

$$g_B^2 \xrightarrow[M^2 \to \infty]{} \frac{16\pi^2}{\beta_0 \log (M^2/\Lambda^2) + (\beta_1/\beta_0)\log \log (M^2/\Lambda^2)} \tag{1.1}$$

where M is the cutoff with dimension of a mass and Λ is the QCD Λ parameter. β_0 and β_1 are the first two coefficients of the β-function, computed in diagrams with one and two loops, respectively. The other β-coefficients need not be known! They would only give corrections to g_B^2 that are so small that they can be absorbed in finite shifts of the Λ parameter. It would seem then that two-loop perturbation theory at small distances, together with the topologically defined instanton Θ parameter determine the dynamics of the theory. My problem is this: is that true? It is true up to any finite order in the perturbation expansion. But this expansion is known to diverge badly. This question has received relatively little attention in the literature. We will come back to it in section 6.

2. TOPOLOGY IN THE GAUGE CONDITION

The space of all vector potential configurations described by the set of functions $\{A_\mu^a(x)\}$ is essentially "flat" if all elements in this space are considered to be distinct and the usual topology for function spaces is introduced. A linear superposition of two non-vanishing vector potentials then gives a different vector potential.

However, if we introduce gauge invariance and all points on a gauge trajectory in our vector potential space are identified then we get an altogether different and highly non-trivial topology [5]. How is this reflected in the *physical* degrees of freedom? Can we localize topologically non-trivial features in space-time? This is the first problem of the two mentioned in the introduction.

We are interested in the question of what the *physical* degrees of freedom are. Therefore it is necessary to "fix the gauge", that is, find a subset (section) of the space $\{A_\mu^a(x)\}$ that uniquely represents all gauge equivalence classes. Traditionally this is done by writing down a gauge condition such as

$$\partial_\mu A_\mu^a = 0 \tag{2.1}$$

(Lorentz gauge). Gauges such as these minimize the values of the vector potential fields (more precisely, the Lorentz gauge minimizes

$$W_1 = \int A_\mu^2 \, d^4x \tag{2.2}$$

under gauge rotations in Euclidean space). Therefore the Lorentz gauge is important to obtain ultra-violet convergent ("renormalizable") perturbation expansions.

However, the trouble with the Lorentz gauge (2.1) is that at every point x the gauge function $\Omega(x)$ that transforms an arbitrary vector potential A_μ into one that satisfies (2.1) depends on field values at all other points. Thus spurious modes are created that propagate from one point to any other point, without having any direct physical interpretation. In perturbation theory these spurious modes ("ghosts") can be dealt with completely [6]. Non-perturbatively, these modes obscure the topological structure of the theory. We will show this by choosing a "non-propagating" gauge; no ghost propagates, at the expense of a more violent ultraviolet behavior. Later one could choose to tame the ultraviolet divergences by allowing some very heavy ghost to occur (thereby introducing only some very short range propagation of non-physical gauge effects).

Since it is of foremost importance to understand pure quantum chromodynamics (i.e. without additional scalar fields), the gauge must be fixed by the vector field itself.

And, since we wish to have the relevant gauge transformations $\Omega(x)$ determined by fields at the same point $A_\mu(x)$ only, we should use a field or field combination that transforms under Ω non-trivially, but without the derivative term (the term containing $\partial_\mu \Omega$ or $\partial_\mu \Omega^{-1}$). Such a field is the covariant curvature $G_{\mu\nu}^a$ or $(G_{\mu\nu}^{ij} \equiv \lambda_{ij}^a G_{\mu\nu}^a)$. So a good candidate for a gauge condition is the gauge that brings one of the components, say G_{12}^{ij}, into diagonal form:

$$X^{ij} = \lambda^{(i)} \delta_{ij}, \tag{2.3}$$

where

$$X^{ij} = G_{12}^{ij}. \tag{2.4}$$

We will choose the λ^i to be ordered:

$$\lambda^{(1)} > \lambda^{(2)} > \ldots >. \lambda^{(N)}. \tag{2.5}$$

But because this X is not Lorentz-invariant, one breaks Lorentz-invariance in the same process, which is usually not very desirable.

So instead of (2.4) we could substitute for X in (2.3) some other field combination, such as

$$X^{ij} = (G_{\mu\nu} G_{\mu\nu})^{ij} \tag{2.6}$$

or

$$X^{ij} = (G_{\mu\nu} D_\alpha^2 G_{\mu\nu})^{ij}. \tag{2.7}$$

The problem with Eq. (2.6) is that it does not fix the gauge if the gauge group is $SU(2)$, because then Eq. (2.3) is trivially satisfied, and Eq. (2.7), though Lorentz-invariant, may cause problems because of its complicated form (D_α is the covariant derivative).

An intermediate gauge choice could be

$$\alpha D_\mu^o A_\mu^{ch} + i\beta [X^{ch}, X^o] = 0, \tag{2.8}$$

where the superscript "o" stands for the diagonal part of X only, and the superscript "ch" for the off diagonal part of X only. D_μ^o is the covariant derivative with respect to the diagonal part of A_μ only. α and β are (positive) gauge parameters. Eq. (2.8) corresponds approximately to extremizing

$$W_2 = \int d^2x \, \mathrm{Tr}\{ \, g\alpha \, (A_\mu^{ch})^2 + \beta(X^{ch})^2 \, \} \qquad\qquad (2.9)$$

under gauge rotations in Euclidean space.

The gauges Eq. (2.3) and Eq. (2.8) have in common that they leave the largest abelian subgroup unbroken. This is the group of $\Omega(x)$ that commute with X and therefore in general with itself (if the λ do not coincide). Indeed, in any generalization of the above procedure (for other gauge groups, say) one should choose the gauge condition such that one of the largest abelian subgroups remain unbroken.

3. TOPOLOGICAL GAUGE ARTIFACTS

In the gauge Eq. (2.3) but also in the hybrid gauge Eq. (2.8) one finds that the Euclidean functional integral gets contributions from topologically stable singular field configurations. These singularities occur where and when two eigenvalues of λ^i of X coincide. At such points the gauge is not properly fixed. Let us demonstrate this explicitly of the gauge Eq. (2.3) when the gauge group is $SU(2)$.

The gauge quantum numbers of X^{ij} are those of an isovector (X^1, X^2, X^3): $X^{ij} \equiv X^a \tau_{ij}^a$, and the gauge Eq. (2.3), together with Eq. (2.5) corresponds to

$$X^a = (0, 0, X) \; ; X > 0, \qquad\qquad (3.1)$$

except when $\lambda^{(1)} \to \lambda^{(2)}$. Then X^{ij} is an isoscalar and in Eq. (3.1) $X^a \to 0$; $X \to 0$.

The unbroken subgroup mentioned in the end of the previous section is $U(1)$, the set of rotations about the X-axis. However at those points in space-time where the two eigenvalues coincide we have $X = 0$, and there our gauge condition leaves an invariance group $SU(2)$. Now prior to fixing the gauge, our X-vector was just an arbitrarily chosen field, whose zeros had no special significance. The location of those zeros is determined by the three equations $X^a = 0$; therefore they are in general one dimensional subspaces of $\mathcal{R}^4$, or pointlike in $\mathcal{R}^3$. After fixing the gauge, these point-particle-like structures correspond to singularities in our fields. Singularities, because the gauge rotation $\Omega(x)$ that restores Eq. (3.1) in their vicinity gets an infinite gradient at those points. So, the vector fields $A_\mu(x)$ that transform with $\partial_\mu \Omega$ become singular.

The significance of these special points with respect to the $U(1)$ ("electromagnetic") subgroup of the gauge group is easy to deduce from our study of their cousins in the 1972 Georgi-Glashow model [7]. That is a model in which the vacuum expectation value $<X^a> = F\{0, 0, 1\} \neq 0$, which implies that the $U(1)$ invariance of the X-gauge is also an invariance of the vacuum, so that the $U(1)$ photons are physical. It has been established [8], and we do not repeat those arguments here, that the zeros of X in that model carry two Dirac units of pure magnetic monopole charge. In our system also, the zeros of X are to be identified with magnetic monopoles. However they might undergo Bose-condensation which gives the system a long distance structure entirely different from that of the 1972 Georgi-Glashow model.

We can readily generalize our arguments for the case that the original gauge group was $SU(3)$. Our X field is an $SU(3)$ octet, and its diagonalization in the generic case leaves a subgroup $U(1) \times U(1)$ invariant. Indeed this is again the largest abelian subgroup. It will be easier to represent this as

$$U(1)^3/U(1). \tag{3.2}$$

Each gluon carries one positive $U(1)$ charge and one negative $U(1)$ charge:

$$\begin{aligned}
Q^{(i)} &= \pm (g, -g, 0), &\text{(3.3.}a\text{)}\\
or &\pm (g, 0, -g), &\text{(3.3.}b\text{)}\\
or &\pm (0, g, -g). &\text{(3.3.}c\text{)}
\end{aligned}$$

The two neutral gluons correspond to $U(1)$ photons. Notice that $\sum_i Q^{(i)} = 0$, so that one $U(1)$ drops out, as indicated in Eq. (3.2).

Again, we have singularities if two eigenvalues of X coincide. That could either be the first two or the last two but not the first and last because the λ 's were ordered. These singularities again turn out to be particle-like and carry the following magnetic charges.

$$M^{(i)} = \pm (\frac{2\pi}{g}, -\frac{2\pi}{g}, 0), \tag{3.4.a}$$

$$or \pm (0, \frac{2\pi}{g}, -\frac{2\pi}{g}). \tag{3.4.b}$$

The combination analogous to Eq. (3.3.b) does not occur. Again, $\sum_i M^{(i)} = 0$.

Our claim is that magnetic monopole singularities will always arise whenever one attempts to fix the gauge locally, using only the fields occurring in the pure gauge theories. They even persist in smoother gauges such as Eq. (2.8). See [9] where also the effects of instantons are taken into account.

4. PHANTOM SHEETS

Let us once again consider the intermediate gauge Eq. (2.8). What kind of ghost should one expect ? If β is large enough then according to Eq. (2.9), X^{ch} tends to become small. After an infinitesimal gauge transformation we have approximately:

$$X^{ch} \to i\,[\Lambda^{ch}, X^{o}], \tag{4.1}$$

and

$$A_{\mu}^{ch} \to -\,D_{\mu}\Lambda^{ch}. \tag{4.2}$$

Substituting this in Eq. (2.9) we get a Lagrangian for the ghosts:

$$\mathcal{L} = g\alpha\,(D_{\mu}\,\Phi^{ch})^2 + \beta|[\Phi^{ch}, X^{o}]|^2 + \textit{interaction terms}. \tag{4.3}$$

From this we read off that the mass of the ghost will be

$$m_{gh} = \mathcal{O}[\sqrt{g\beta/\alpha}\ |\lambda^{(i)} - \lambda^{(j)}|], \tag{4.4}$$

where the $\lambda^{(i)}$ are the eigenvalues of X. As long as the coupling constant $g \neq 0$, we expect that this mass will tend to infinity if β tends to infinity. In the limit $\alpha \to 0$ the non-propagating gauge should be reached.

This gauge is unstable against formation of monopole singularities because at such singularities W_2, Eq. (2.9) still converges. However, a new topological structure can now appear: "phantom sheets". This is because the system will not always keep the eigenvalues $\lambda^{(i)}$ of X^o ordered. We get various regions of space-time where their order will be different. These regions will be separated by domain walls much like Bloch walls. Let us take a closer look at these walls.

Consider for simplicity again the case $SU(2)$. We write

$$X = \lambda + a_k\sigma_k. \tag{4.5}$$

Here σ_k are the Pauli matrices. We are far away from any singular points, so

$$|\vec{a}| = a \neq 0. \tag{4.6}$$

We take m_{gh} to be large with respect to the scale in which the physical fields vary. Then we are close to a pure gauge transformation of the approximately constant field configuration,

$$X' = \lambda + a'_3\sigma_3 \ ; \ A'_\mu = 0, \tag{4.7}$$

where a'_3 and A'_μ are space time independent. This configuration is transformed by a gauge transformation $\Omega(x)$ such that at $x_3 \gg 0$ we have

$$X \equiv \Omega X' \Omega^{-1} = \lambda + a'_3\sigma_3 \ ; \qquad\qquad A_\mu \equiv ig^{-1} \Omega \partial_\mu \Omega^{-1} = 0 \tag{4.8}$$

and at $x_3 \ll 0$ we have

$$X = \lambda - a'_3\sigma_3 \quad ; \quad A_\mu = 0. \tag{4.9}$$

Ω must be of the form

$$\Omega = \exp(i\omega\sigma_1), \tag{4.10}$$

where ω depends on x_3, and

$$W_2 = \int d^2x \ \mathrm{Tr}[\ -g^{-1}\alpha\{(\Omega\partial_\mu\Omega^{-1})^{ch}\}^2 + \beta\{(\Omega X'\Omega^{-1})^{ch}\}^2 \ ]$$

$$= 2\int d^2x \ \{ \ \beta a^2 \ sin^2 2\omega + g^{-1}\alpha(\partial_\mu\omega)^2 \ \}. \tag{4.11}$$

This is the Lagrangian for a sine-Gordon equation, and the boundary conditions Eq. (4.8) and Eq. (4.9) are those of the sine-Gordon soliton. The contribution of the sheet to W_2 is

$$W_2 = 4a\sqrt{\alpha/\beta} \ g^{-3/2} \int d^2x, \tag{4.12}$$

where the integral is over the sheet. The thickness of the sheet is of order

$$a^{-1} \sqrt{\alpha/\beta} \ g^{-1/2} = \mathcal{O}(m_{gh}^{-1}). \tag{4.13}$$

We may safely assume that the phantom sheets contribute to W_2 by an amount proportional to their area. Since they have no natural boundary they only come in the form of bubbles, and minimizing W_2 will correspond to minimizing the area of these bubbles, which will therefore never grow to substantial sizes. So, in the bulk of space-time, with only small exceptional regions, we may assume that the Λ's are ordered in the same way as in the ghost-free gauge Eq. (2.3), see Eq. (2.5). Monopole singularities will occur in this gauge as much as in Eq. (2.3). Because of the above, in the case of larger gauge groups, the magnetic charges will be consecutive (in the sense of Eq. (3.4.a)) in the "regular" regions of space time, but non-consecutive charges might show up occasionally, inside a bubble. We see that although gauges of the type Eq. (2.8) might be easier to handle in perturbation theory, they give rise to a more complicated topology.

We conclude that perhaps the most crucial physical dynamical variables that govern the strong-interaction region of a non-Abelian theory may be obtained by first fixing the non-Abelian part of the gauge symmetry. We get an Abelian theory with magnetic monopoles. The short range non-electromagnetic interactions between the electrically charged particles and the monopoles may perhaps be computable by "ordinary" perturbative techniques.

5. CONFINEMENT

Let us now concentrate on the quantized theory of electromagnetic fields, electrically charged particles and magnetically charged particles. In the simplest case Dirac's condition [10] for the charge quanta is:

$$QM = 2\pi n, \tag{5.1}$$

with n an integer. More precisely, and in the case that there are more than one kind of electromagnetisms (in the case that the original gauge group was $SU(3)$ or larger),

$$\sum_i (Q_i^{(1)}M_i^{(2)} - Q_i^{(2)}M_i^{(1)}) = 2\pi n^{(1,2)} \tag{5.2}$$

for any pair of particles (1) and (2). So either the $Q's$, or the $M's$, or both, must be large. This makes quantization extremely difficult.

We now notice that the *classical* theory has an invariance under rotations of the type

$$\begin{aligned} Q_i &\to Q_i \cos\Theta_i + M_i \sin\Theta_i, \\ Q_i &\to M_i \cos\Theta_i - Q_i \sin\Theta_i \end{aligned} \tag{5.3}$$

as far as the electromagnetic interactions are concerned. But we also know that the quantized interactions of the electromagnetic fields are completely determined by the classical theory. So the invariance Eq. (5.3) is an exact invariance of the quantized Maxwell system as well. Notice, indeed, that Eq. (5.2) is invariant.

Now Eq. (5.2) is not exactly the same as Eq. (5.1) if there is only one kind of charge. Eq. (5.2) allows monopoles themselves to carry fractional electric charge. Our monopoles will carry fractional electric charge if the so-called instanton angle differs from zero. Let us only concentrate on the case that this angle equals zero. Then we will choose a transformation Eq. (5.3) with $\Theta_i = 90^0$. After that we will do ordinary field theory. This implies that, at least to some extent, the original electric charges, now magnetic charges, are ignored. The fundamental assumption in quantum-chromodynamics is that that is reasonable.

Another assumption goes even farther than that. We assume that the vacuum becomes a superconductor with respect to the new "electric" charges. That is not so unreasonable. Perturbation theory tells us that the Q_i become large at the hadronic distance scale. According to Eq. (5.2) then, the magnetic charges become weak. One can imagine very well that this will bring their chemical potential to the level of Bose condensation. The mechanism for Bose condensation is in no way different from the one in ordinary superconductors, or, in the language of particle physicists, the Higgs model. However, whether or not Bose condensation does take place in our case, depends on the dynamics. We still have great difficulties in doing reliable computations here: at small distance scales the Q_i are small, so that perturbation expansions with respect to the Q_i must be performed, and at large distance scales the M_i are small. It is in the transition region where both are fairly large that we do not know how to extract reliable results.

There is no doubt that *if* the original magnetic charges become superconducting, quarks, or any other objects with color, will be permanently confined. The argument is simple. After our transformation Eq. (5.3), with $\Theta = 90^0$, these objects become magnetically charged. But their magnetic flux is expelled from the superconductor by the Meissner effect: The flux lines are squeezed into narrow tubes (the Abrikosov fluxes in a superconductor). These tubes are the strings that tie quarks together permanently. We conclude that we are perfectly able to understand permanent quark confinement qualitatively. It is "reasonable" for the system to condense into a "quark confining mode". What is still lacking is a convincing proof that the dynamics is indeed such that this condensation takes place. Model dependent calculations, such as the Monte Carlo computations on lattice gauge theories give strong evidence in favor of the above picture. However the next section should be a warning against too rapid conclusions from those results.

6. IS QUANTUM CHROMODYNAMICS A THEORY?

All we really understand of quantum chromodynamics is its perturbation expansion. For calculations at small distance scales the expansion parameter is small, so the small distance structure of the theory seems to be well understood. Since we expect that the large distance interactions can be seen as an interplay of small distance interactions, one might conclude that such an asymptotically free theory has a sound and unique mathematical basis. This however is not true, as yet. Sound "constructive field theories" have been discovered in two and three space-time dimensions, but not in four. One fundamental obstacle seems to be that in four space-time dimensions a-symptotic freedom is only logarithmic, which is too slow. Another is a divergence in the Borel resummation procedure due to instantons. There are three different attitudes a physicist can have regarding these problems, besides ignoring them. One, obvious-ly, is assuming that although our present day mathematics is not yet capable of pro-ducing accurate numbers in QCD, improvement will be possible, so that, eventually, one will be able to prove that all or some asymptotically free field theories in four dimensions are well defined and unique. Numbers such as the ratio between the proton mass and the string constant can perhaps not in practice but in principle be computed up to any desired accuracy.

What if this is not so? One then must realize that since the floating coupling con-stant depends only logarithmically on the distance scale, one might sooner or later have to deal with quantum gravity effects. That is possibility number two, and if this is inevitable then there seems to be little chance that we will be able to handle QCD within our lifetime.

A third possible attitude is that including quantum gravity can be avoided, but quan-tum chromodynamics all by itself is not good enough. We must replace it by a theory with perhaps more additional fields, in such a way that the diseases can be cured. Indeed we have in mind that such a procedure might exist. In the following section we sketch the argument. The assumption that quantum chromodynamics all by itself is not good enough implies for instance that the presently popular lattice approximations will eventually run into trouble. Physical quantities computed this way will not become independent of the lattice artifacts of the models, even in the limit (lattice par-ameter) $\rightarrow$ 0. Probably computations at present are not sensitive enough to reveal this. Replacement of QCD by the model(s) in the next section might be comparable to the replacement of non-renormalizable theories by renormalizable ones: the latter are more valuable because computations can be done more accurately, even if the former are not evidently wrong. The price is often the introduction of new degrees of free-dom.

7. ASYMPTOTICALLY FREE 1/N EXPANSIONS

Let us try to change the theory such that the coupling constant goes to zero *faster* than logarithmically. Thus we imagine that the β-function increases with energy. The first coefficient is the only one we have to look at. It is proportional to N, for $SU(N)$

theories. So it is natural to suggest that N increases indefinitely with energy. Does this indeed improve the theory? This is not known, but there is a fair chance that such models will be much better from a mathematical point of view than ordinary QCD with N=3.

What we can say with some confidence is that the $N \rightarrow \infty$ limit of $SU(N)$ gauge theories is unique. As is well known [11], this limit is governed by planar Feynman diagrams. These can still become infinitely complicated, and no analytic solution of this limit theory is known. But the diseases of the finite N theory do not arise here. In the limit, g^2N is kept fixed whereas instantons carry an action proportional to $1/g^2$, without any further N-dependence. Consequently, bothersome instanton singularities will disappear from the Borel sum. In terms of g^2N, the theory is still only logarithmically asymptotically free, but on the other hand the expansion in terms of g^2N is much better behaved. The total number of different planar diagrams with L loops and E external lines is of order C^L, where C depends on details of the theory [12]. At finite N this number goes as $C^L L!$, a fundamentally faster divergence. What we claim is now that, in the presence of a sufficiently effective infrared cut-off, the Dyson-Schwinger equations determine all Green functions of the planar theory uniquely. A proof, or at least arguments that favor this claim strongly, will be published elsewhere.

The infrared cut-off can be realized by enclosing the system in a sufficiently small box. The complete theory is then obtained by gluing boxes together. Planarity seems to be crucial for the argument. An important lemma that we can prove states that the subset of all ultraviolet *convergent* planar diagrams in Euclidean space converges with a finite radius of convergence when summed. That lemma even holds without infrared cut-off.

Given that the $N \rightarrow \infty$ theory is fine, what remains to be done is to invent a "tumbling mechanism" enabling us to go gradually from $N = 3$ to $N \rightarrow \infty$. What first springs to mind is that the $SU(N)$ symmetry breaks down spontaneously, step by step into $SU(3)$. Of course the symmetry breaking mechanism could be dynamical in nature, but a model can be designed where ordinary scalar Higgs fields are responsible. This is a super-symmetric model with an $SU(N)$ gauge super-multiplet and roughly 2N super-fields in the elementary representation of $SU(N)$. Due to super-symmetry it is still asymptotically free. Then mass terms are added that ignore super-symmetry entirely. They vary wildly in order of magnitude. One obtains an infinite series of thresholds $M_3, M_4, \ldots M_N, \ldots$ where the apparent symmetry is raised from $SU(3)$ to $SU(4)$ to $\ldots$ to $SU(N)$, etc.

Past experience gives us reason to hope that the spontaneous symmetry breaking will not upset the superior convergence properties of this theory, but a lot of work still must be done to sort this out.

REFERENCES

[1]. H. B. Nielsen and P. Olesen, Nucl. Phys. B61 (1973) 45; B. Zumino, in *Renormalization and Invariance in Quantum Field Theory*, p 367. Edited by E. R. Caianiello, Plenum, New York, 1974.

[2]. G. 't Hooft, Nucl. Phys. B138 (178) 1; B153 (1979) 141.

[3]. G. S. La Rue, W. M. Fairbank and A. F. Hebard, Phys. Rev. Letters 38 (1977) 1011; 42 (1979) 142, 1019E; 46 (1981) 967.

[4]. R. Slansky et al, Los Alamos preprint LA-UR-81-1378 (1981).

[5]. M. F. Atiyah, in *Recent Develpments in Gauge Theories*, Cargese 1979, p. 1. Edited by G. 't Hooft et al, Plenum, New York and London. R. Bott, ibid., p. 7. M. F. Atiyah and J. D. S. Jones, Commun. Math. Phys. 61 (1978) 97.

[6]. G. 't Hooft and M. Veltman, Nucl. Phys. B50 (1972) 318; DIAGRAMMAR, Cern report 73/9 (1973)

[7]. M. Georgi and S. L. Glashow, Phys. Rev. Lett. 28 (1972) 1494.

[8]. A. M. Polyakov, JETP Lett. 20 (1974) 194. G. 't Hooft, Nucl. Phys. B79 (1974) 194. G. 't Hooft, Nucl. Phys. B190 [FS3] (1981) 455.

[9]. P. A. M. Dirac, Proc. Roy. Soc. A133 (1931) 60; Phys. Rev. 74 (1948) 817.

[10]. G. 't Hooft, Nucl. Phys. B72 (1974) 461.

[11]. E. Berezin et al, Commun. Math. Phys. 59 (1978) 35.

CHAOS AND COSMOS

Luigi A. Radicati di Brozolo

Scuola Normale Superiore
Pisa, Italy

Felix Klein [1] once said that the concept of group is perhaps the most characteristic concept of 19th century mathematics. However, in his Erlangen Program (1872), an extraordinarily effective advertisement for the group concept, Klein does not make any attempt to "sell" it to the natural scientists. At that time he was probably too absorbed with geometry to think of other applications, but already a few years later his friend and one-time collaborator, Sophus Lie, in the preface to the third part of his celebrated "Theorie der Transformationsgruppen" wrote "The principles of mechanics have a group theoretic origin". Soon afterwards, in 1904, this was explicitly proved by G. Hamel [2]. Little by little under the influence of Klein and of his school it began to be recognized that invariance under the group of the automorphisms of space-time was the true criterion of objectivity for the physical laws. Already in 1910 Klein [3] explicitly stated that relativity was synonymous with invariance under a group and Klein's school systematically investigated the physical consequences of invariance under both finite and infinite dimensional Lie groups [4].

With the exception of Einstein, physicists were much slower than mathematicians to recognize the importance of the group concept. They only accepted it, and even then with considerable reluctance, after Wigner [5] and Weyl [6] showed that with its help the complicated phenomenology of atomic spectroscopy could be deduced from a few general symmetry principles. Quantum mechanics proved to be a much easier testing ground for the application of group theory than classical mechanics with its intrinsically non-linear structure. Partly because of the extraordinary success of quantum mechanics, and partly due to the greater simplicity of the theory of linear representations, most physicists, with the notable exception of a few farsighted people [7], tended to overlook the implications of symmetry in non-linear phenomena. In fact, until some twenty years ago most of the applications of group theory to physics were simply ex-

33

tensions of the methods pioneered by Wigner and Weyl, with results that were often quite spectacular. As time went by, order was brought into nuclear spectroscopy by the same group theoretical methods [8] used for atomic spectroscopy. It was then the turn of what used to be called elementary particle spectroscopy, and once more it was found that all new particles could be classified by using the representations of various unitary groups, successors of the old $SU(2)$ of isospin symmetry which in turn had its roots in atomic spectroscopy [9]. The symmetry groups of the fifties and sixties had of course nothing to do with transformations in space-time, but acted on some internal space that was progressively enlarged to accommodate the new degrees of freedom that became excited as the energy of the accelerators increased. Besides being introduced in a somewhat unimaginative way, these symmetry groups could not claim to have universal validity. They did not represent fundamental symmetries of nature as the Poincaré group, but only the symmetry of one part of nature on the basis of its pre-dominance in the energy range accessible to experiments. These were therefore not exact symmetries but only broken, and often badly broken, symmetries. The art at that time was to select some character a set of particles had in common and then find a group with a representation that could accommodate the whole set. As for the differ-ences between the particles of the set, they were blamed on some symmetry violating force often of unknown origin.

During the period that extends from the mid-thirties, when Wigner successfully ap-plied group theory to nuclear physics [8], to the end of the sixties, when the unifi-cation of the electromagnetic and weak interactions based on a gauge group was achieved [10], the really hard question about symmetry was seldom asked: how could approximate symmetries arise out of a truly exact symmetry without any external a-symmetrical cause? The reason it was not asked is, I think, that the linear framework of orthodox theories could not even allow one to formulate it. There is no such thing as spontaneous symmetry breaking in linear theories. To break a symmetry in a linear theory one needs to add an external perturbation and indeed the example of the Zeeman effect was often quoted to explain the lack of degeneracy that exact symmetry would have required.

For quite a while the fact that different fundamental interactions showed different symmetry properties did not seem to have worried people unduly and few attempts to develop an evolutionary theory of symmetry were consistently pursued. That different interactions had different symmetries was taken more or less for granted and it was accepted without much discussion that besides the Poincaré group there was in nature no other general exact symmetry. The obvious question: how does the asymmetry and the order that we observe in the actual world arise from the symmetrical, unstructured framework provided by the fundamental laws, was usually answered by some vague references to the asymmetry of the initial conditions. For people who believed in the fundamental role of symmetry in physics this could not be the final answer and sooner or later the explanation of the lack of symmetry of our universe was bound to become the most pressing problem.

The notion of spontaneous symmetry breaking emerged painfully in the late fifties. It is perhaps worth noticing that such a fundamental notion did not arise from the study of the so called *fundamental physics* but rather from that of condensed matter, in particular of superconductivity [11]. Today we finally understand, at least in a broad sense, how states with different symmetries can arise spontaneously from a symmetrical theory or, to put it in a different, and perhaps too general way, how the perfect order and arrangement of κοσμοζ can evolve from the formless void of primordial matter, i.e., from χαοζ. Let me stress that formless does not mean devoid of symmetry: indeed in the Hesiodic theogony [12] chaos is synonymous with infinite empty space and therefore possesses, by definition, the highest symmetry. It is only by breaking this symmetrical uniformity that the opposite principles become separated so that forms can emerge and order is progressively established.

To illustrate the complementary role of symmetry and order, let us consider two equal empty vessels connected by a pipe as illustrated in Fig. 1. The geometry is such that the system is invariant under the group $\mathfrak{D}_2$ whose elements are the identity and the rotation by π around the vertical axis. By filling the vessels with two different monatomic gases at the same pressure and temperature and with the same number of molecules, the $\mathfrak{D}_2$ symmetry is broken and an order is established which allows one to distinguish the two vessels. If the two gases are now allowed to mix, the symmetry is restored while the disorder, as measured by the entropy per particle, increases by ln 2.

As a second example, consider a system of free magnetic dipoles with spin 1/2 in an external magnetic field. At infinite temperature the entropy per dipole is maximum (equal to ln 2) and the system is invariant with respect to the reversal of each spin. As the temperature is lowered this symmetry is broken until at zero temperature the system reaches a completely ordered state with all spins oriented parallel to the magnetic field. In this idealized case, which neglects the interaction between the dipoles, the order is enforced by an external cause, the magnetic field, which distinguishes the parallel from the anti-parallel orientation. However, even without any symmetry breaking external field, order spontaneously arises at low temperatures, provided we take into account the (symmetry preserving) interaction between dipoles. This example shows that a symmetrical system can, when the values of a scalar parameter (the temperature in this case) reach a critical point, go over from a symmetrical, disordered

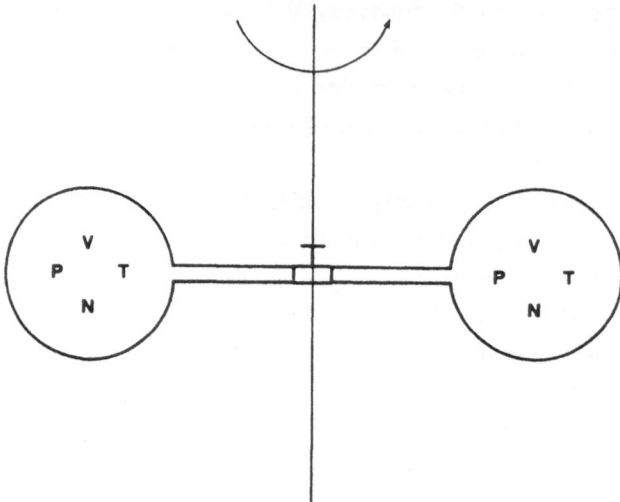

Figure 1:   Mixing of two different monatomic gases.

state, into a less symmetrical but more ordered one without the intervention of any symmetry breaking cause. The entropy released in this ordering process is taken over by the heat reservoir.

The concept of spontaneous symmetry breaking is one of the most significant new ideas that has emerged in physics in recent years. Whereas it certainly arose in connection with quantum phenomena, it is my contention that many problems of classical physics already show that spontaneous symmetry breaking is a common feature of most non-linear theories, quantum mechanical as well as classical. The oldest example of a non-linear theory having both symmetrical and asymmetrical stable solutions is probably Euler's elastica where the rotational symmetry of a loaded rod is broken in the buckled solution which obtains when the load exceeds a critical value. Another famous historical example is the problem of determining the equilibrium shape of a self-gravitating fluid mass rigidly rotating around an axis [13]. The hydrostatic equilibrium equation is invariant under the group $D_{\infty h}$, i.e., the symmetry group of a biaxial ellipsoid. One finds that, when the angular momentum reaches a critical value, the equilibrium shape changes from a biaxial to a triaxial ellipsoid thereby breaking the $D_{\infty h}$ symmetry of the problem. When the angular momentum increases beyond another critical value, even the D_{2h} symmetry of the triaxial ellipsoids is lost and pear-shaped figures acquire stability. Similar phenomena occur in fast rotating nuclei where the equilibrium shape depends not only on the interparticle interaction, but on the surface tension as well.

Even more striking is the onset of convection in a fluid with a temperature, or density (or both) gradient directed opposite to the gravitational field [14]. In spherical geometry the problem has obvious astrophysical and geophysical implications of the greatest importance. Laboratory experiments done under widely different conditions [15], but usually in a plane geometry, have in general confirmed the original Bénard results, namely, the abruptness of the onset of convection and consequent breaking of Euclidean invariance in the horizontal plane as soon as the buoyancy acceleration exceeds a critical value proportional to the product of the kinematical viscosity times the thermal conductivity. This critical value can be predicted with reasonable accuracy by linearizing the Boussinesq approximate equations, but the prediction of the convection pattern or, in other words, the way symmetry is broken, is a much more difficult problem. One thing is now known for sure: the convection pattern, i.e., the way the Euclidean symmetry is broken, depends critically on the details of the equations. Allowing certain parameters to vary with temperature, or including some new (symmetry conserving) effects, may give rise to widely different symmetry breaking patterns [16]. Let me just quote in passing another striking example of spontaneous symmetry breaking, namely, the so called dynamo problem, i.e., the generation of a magnetic field by convection currents in an electrically conducting fluid. The problem is of the greatest importance for both astrophysics and geophysics and poses even more formidable mathematical problems than the convection problem. These are only a few examples, chosen because of the simplicity of their mathematical structure. Many others could be quoted, like pattern formation and chemical waves exemplified by the Belousov reaction where, above a critical point, the concentration varies with a period of about 30 seconds over several orders of magnitude, giving rise to elegant patterns.

Though most of these phenomena were known for a long time, until very recently they were not discussed *sub specie symmetriae* partly, I believe, because the applications of group theory were usually limited to linear field theories. Non-linearity is indeed the common feature of all these phenomena, a feature they share with the fundamental theories where spontaneous symmetry breaking is more often discussed. In practically all the examples I have mentioned we are concerned with non-linear functional equations invariant under a group G and depending upon a real parameter or a set of parameters λ called control parameters:

$$F(\lambda, u) = 0. \tag{1}$$

Here u is a vector in some functional space $\mathscr{E}$ and F is a G-invariant mapping from $\mathscr{R} \times \mathscr{E}$ to some other functional space $\mathscr{F}$. The mapping F represents not only the equations of motion but also the boundary conditions, and the invariance group G is thus the symmetry group of the complete physical problem described by F. For Euler's elastica the control parameter λ is proportional to the applied load; for the rotating ellipsoids λ is the square of the angular momentum; for Bénard's convection λ is the Rayleigh number, i.e., the ratio of the buoyancy acceleration to the acceleration related to viscosity and thermal conduction. All of these parameters are scalars and varying them does not introduce any asymmetry.

It is usually relatively easy to find a solution u_0 invariant under G, valid for every λ, which we shall call the symmetrical solution. Without loss of generality we can set $u_0 = 0$. In all the examples mentioned above, when λ increases above a critical value, the symmetrical solution loses its stability and a new family of asymmetrical stable solutions appear. Therefore by varying the scalar parameter λ we pass from the symmetrical regime to an asymmetrical one. The points $\lambda = \lambda_c$ where the new solution appears is a bifurcation point of F: it has the property that in every neighborhood of it there exist vectors u different from u_0 such that $F(\lambda, u) = 0$.

The possibility of symmetry breaking bifurcations [17] rests on the simple observation that the covariance of F under G does not imply that all solutions of $F(\lambda, u) = 0$ be G-invariant. It only requires that if u is a solution, $g \cdot u$ also be a solution for all $g \in G$. It often happens that the symmetry is not completely lost at a bifurcation but the bifurcating solution u is still invariant under a subgroup of G, $G_u = \{g \mid g \cdot u = u\}$ called the isotropy group of u. If $G_u \neq G$, the solution of Eq. (1) is not unique but we have instead a multiple eigenvalue branch point, i.e., a whole orbit of solutions $\Omega(u) = \{g \mid g \in G\}$, a phenomenon entirely equivalent to the degeneracy of multiple eigenvalues in linear theories. Notice that for $\lambda > \lambda_c$ the orbit of asymmetrical solutions is stable and thus corresponds to the lowest energy: we can therefore call it a degenerate vacuum invariant under G. The set of all solutions of Eq. (1) is separated by the action of G into strata, i.e., sets of orbits with conjugate isotropy groups. If $\Theta_1, \Theta_2 \ldots \Theta_n$ is the set of all independent invariants for the action of the group on the set of solutions, it is possible, by a suitable choice of the Θ's, to label the strata in such a way that the stratum containing the symmetrical solution corresponds to $\Theta_1 = \Theta_2 = \ldots = \Theta_n = 0$, the one with the largest isotropy group $G_n \supset G$ to $\Theta_1 \neq 0, \Theta_2 = \ldots = \Theta_n = 0$ and so on. The Θ's are the order parameters of the problem: as the order increases (i.e., the isotropy group becomes smaller) more and more Θ's become different from zero.

We may at this point wonder where symmetry breaking has gone since the orbit $\Omega(n)$ of bifurcating solutions is G-invariant. The answer is that only one solution will be actually realized and one particular solution on the orbit is only invariant under a group conjugated to G_n, not under the full group G. Which of the elements of $\Omega(u)$

will occur cannot be predicted on the basis of a G-invariant equation since it depends upon the effect of thermal or quantum fluctuations.

This brief summary should be sufficient to show that most of the concepts that are now currently used in fundamental physics are already present in these classical non-linear problems: the onset of asymmetry without the intervention of asymmetrical causes (*pace* Pierre Curie who stated [18] that "les éléments de symétrie des causes doivent être présents dans les effets qu'elles produisent"); the degeneracy of the vacuum; the zero-frequency modes at a symmetry changing bifurcation; and the non-vanishing values of symmetry breaking quantities in the asymmetrical ground state.

I mentioned earlier that spontaneous symmetry breaking can occur only in invariant non-linear problems. To see the importance of non-linearity let us consider a special case of Eq. (1), namely

$$F(\lambda, u) = (L - \lambda)u, \tag{2}$$

where L is a linear operator. This equation clearly admits the trivial symmetrical solution $u = 0$ for any λ. At each eigenvalue λ_i of Eq.(2) a new solution, or actually a new set of solutions $\{u_{im}\}$ appear which, however, are not uniquely determined by Eq.(2). Indeed, if u_{im} is a solution, so is cu_{im}, where c is an arbitrary complex number. The response diagram, i.e., a plot of the invariant norm $\|u\|$ against the control parameter λ, thus takes the form represented in Fig. 2. The symmetrical solution $\|u\| = 0$ exists for every λ. At each eigenvalue, new solutions appear whose norm is independent of the control parameter and must be fixed by some (non-linear) normalization condition to be added externally.

If on the other hand the problem is nonlinear the response diagram takes the form of Fig. 3 (or a more complicated one) and the norm $\|u\|$ of the solution bifurcating from the symmetrical solution $\|u\| = 0$ *is* determined (at least in principle) by λ.

Figure 4 shows the response diagram for the equilibrium of the rotating ellipsoids. The symmetrical (i.e., $D_{\infty h}$ invariant) solutions are the points of the $\eta^2 = 0$ axis which represent the stratum of the one-dimensional ellipsoids. These solutions are stable for $J^2 < J_j^2$ but become unstable for $J^2 \geq J_j^2$. The curve originating at $J^2 = J_j^2$ represents the stratum of the infinite dimensional orbits of triaxial ellipsoids whose isotropy groups are conjugated to D_{2h}. For $J^2 \geq J_p^2$ these ellipsoids become unstable against pear-shaped deformations.

If we have a functional equation of the type considered before, $F(\lambda, u) = 0$, covariant under a group G, we would like to know how G-invariance is broken as we change the values of the parameters. To my knowledge, only a few concrete examples

Luigi A. Radicati di Brozolo

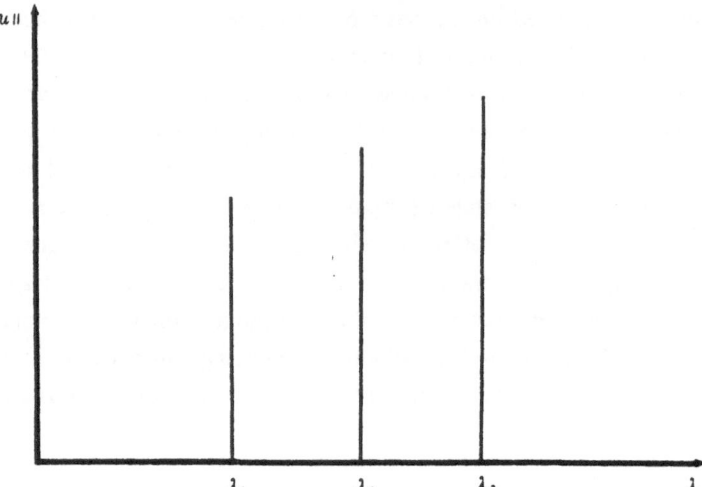

Figure 2:   Response diagram for a linear problem.

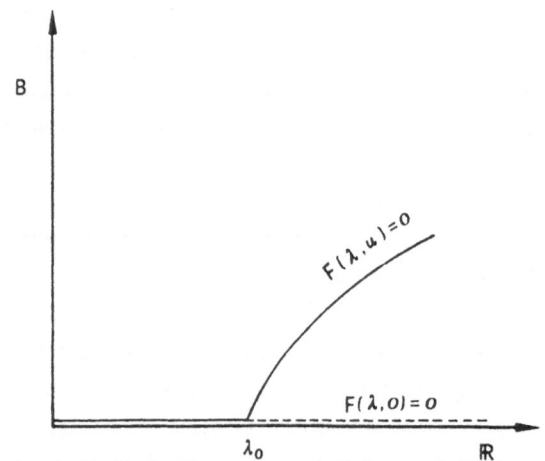

Figure 3:   Typical response diagram for a nonlinear problem.

have been analyzed in detail and they already show that the chain of symmetry breaking depends critically on the explicit form of F, i.e., on the dynamics. In the case of
rotating ellipsoids, Constantinescu, Michel and myself [19] have proved that, if we
assume that the solutions of the linearized problem belong to an irreducible representation of the symmetry group of that equation, then several selection rules can be
deduced by group theoretical arguments which eliminate a number of *a priori* possible
solutions. It is interesting to note that in this problem the forbidden isotropy groups
are always smaller than the allowed ones. In other words, the absence of accidental

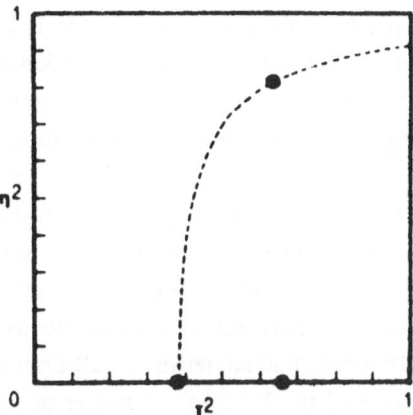

Figure 4: Response diagram for the rotating ellipsoids; $\eta^2 = 1 - a_1^2 / a_2^2$; where a_1 and a_2 are the semiaxes in the equatorial plane.

degeneracy, which is a different way of phrasing our assumption, leads in a natural way to minimal symmetry breaking.

Things are not as simple in the Bénard problem whose group theoretical aspects have been analyzed by Sattinger [20]. In this case the linearized equation which is used to determine the bifurcation point depends only one parameter, the Rayleigh number, the dependence on the other parameter, the Prandtl number, being lost in the linear approximation. As a result, several convection patterns corresponding to hexagons and to rolls are degenerate, and to decide which will actually occur one has to discuss their stability and take into account other nonlinear symmetry conserving effects, such as the temperature dependence of the viscosity and the conductivity, the surface tension, etc. The dynamo problem is of the same nature, only somewhat more intricate.

These examples show that to predict how the original G-invariance of a non-linear theory gets progressively broken as we change the value of the parameters, we need to know all the details of the theory. Even small G symmetrical nonlinear terms may remove the degeneracy between different sets of orbits with conjugate isotropy groups (strata) and thus lead to a definite choice between different types of ordering. In contrast with linear theories where the group theoretical structure and the dynamics can be treated independently, these two aspects become inseparable in non-linear theories. This makes them at the same time much more difficult to analyze and much more rich and predictive. Indeed, they allow one in principle to construct an evolutionary theory of the progressive establishment of order through a succession of symmetry breaking bifurcations. Are bifurcations always accompanied by symmetry breaking? The answer is almost certainly no, but to my knowledge little is known on the conditions which are necessary for symmetry breaking. Long ago Michel [21] and I investigated a special case of non-linear equations with spontaneous symmetry breaking and found some conditions which seem to be generally satisfied by the orbits where the phenomenon occurs. Much more work, however, is necessary to generalize these results [17].

One lesson to be drawn from the evolution of physics in these past twenty years is that we are only beginning to understand the richness of non-linear symmetrical theories. It is not unreasonable to think that the symmetrical laws which govern the unstructured chaos already contain *in potentia* all the complexity of the ordered cosmos that will develop from it. In more concrete terms, a full knowledge of the grand symmetrical equation which described the behavior of matter at the beginning of the universe should allow us to predict the way symmetry breaks down as the temperature cools off, allowing it to realize all the order potentially contained in it. At present this is little more than wishful thinking: we do not know that grand non-linear equation, and even if we knew it our knowledge of symmetry breaking at successive bifurcations is still too primitive to implement such a grandiose program.

The way physics advances has of course very little to do with this program: We know, or better, we painfully discover, bit by bit, the laws that govern the ordered, unsymmetrical state of matter at our low temperatures, and at the somewhat higher temperatures that can be artificially produced, and from this partial knowledge we have to work our way up to the symmetrical laws of the primordial, fiery chaos where order has given place to symmetry. In this completely symmetrical state all the differences of our familiar world have disappeared: we cannot tell a weak from a strong force, a lepton from a quark, a particle from an anti-particle; all order has merged in the amorphous symmetry of chaos.

The symmetry of this primordial state of matter goes much beyond the simple symmetries of the elementary examples I have mentioned: it is not a global symmetry but a local one, and can only be discussed in the framework of Riemannian geometry. In that framework the concepts of space, time and matter can no longer be artificially separated and even symmetry, as Cartan has shown [22], becomes a fundamental element of the structure. This new bold point of view, which, as Weyl said long ago "seals the doom of the idea that a geometry may exist independently of physics in the traditional sense" [23] is the latest and most promising development of the stream of ideas originating from that most characteristic concept of 19th century mathematics, the group concept. Another instance of the "unreasonable" [24] applicability of mathematics which astonished one of the founding fathers of the applications of group theory to physics.

REFERENCES

[1]. Quoted by H. Weyl; *Gruppentheorie und Quantenmechanik*, Leipzig 1931, 2nd ed., p. 2; see also H. Weyl, *Philosophy of Mathematics and Natural Science*, Princeton, 1952, p.28.

[2]. G. Hamel, *Die Lagrange-Eulerschen Gleichungen der Mechanik*, Zeit. f. Math. u. Physik., 50 (1904) 1.

[3]. F. Klein, *Über die geometrischen Gründlagen der Lorentzgruppe*, Jahresberichte der Deutschen Mathematikervereinigung, 19 (1910); reprinted in F. Klein, Gesammelte Mathematische Abhandlungen I, p. 553, Berlin, 1921.

[4]. B. Herglotz, *Mechanik der deformierbaren Körpers vom Standpunkte der Relativitätstheorie*, Ann. d. Physik, 36 (1911) 493; F. Engel, *Uber die zehn allgemeine Integrale der klassischen Mechanik*, Nachrichten. d. kön. Gesellschaft der Wissenschaften Göttingen 70 (1916); F. Klein, *Uber die Differentialgesetze für die Erhaltung von Impuls und Energie in der Einsteinschen Gravitationstheorie*, ibid., 171 (1918); *Uber der Integralform der Erhaltungssätze und die Theorie der Räumlich-geschlossen Welt*, ibid., 394 (1919); E. Noether, *Invariante beliebiger Differentialausdrücke*, ibid., 1918 p. 37.

[5]. E. Wigner, *Gruppentheorie und ihre Anwendung auf die Quantenmechanik der Atomspektren*, Braunschweig 1931.

[6]. H. Weyl, *Gruppentheorie und Quantenmechanik*, Leipzig 1928.

[7]. C. N. Yang and R. L. Mills, *Conservation of isotopic spin and isotopic gauge invariance*, Phys. Rev. 96 (1954) 191; J. Goldstone, *Field Theories with superconductor solutions*, N. Cimento, 19 (1961) 154; Y. Nambu and G. Jona-Lasinio, *Dynamical model of elementary particles based on analogy with superconductivity I. and II.*, Phys. Rev. 122 (1961) 345 and 129 (1961) 246; W. Heisenberg, *Introduction to the unified field theory of elementary particles*, London, 1966, where references to early papers can be found; F. Gürsey, *On the symmetries of strong and weak interactions*, N. Cimento, 16 (1960) 230.

[8]. E. Wigner, *On the consequencees of the symmetry of the Nuclear Hamiltonian on the spectroscopy of Nuclei*, Phys. Rev. 51 (1937) 106; B. H. Flowers, *Studies in j-j coupling, I. Classification of nuclear and atomic states*, Proc. Roy. Soc. A, 212, (1952) 248; J. P. Elliot, *Collective motions in the nuclear shell model. I. Classification schemes for states of mixed configurations*, ibid., 245 (1958) 129.

[9]. M. Gell-Mann, *Symmetries of Baryons and Mesons*, Phys. Rev. 125 (1962) 1067, 1084; Y. Ne'eman, *Derivation of Strong Interactions from Gauge Invariance*, Nuclear Physics, 26 (1961) 222; F. Gürsey and L. A. Radicati, *Spin and Unitary Spin Independence of Strong Interactions*, Phys. Rev. Lett. 13 (1964) 173; B. Sakita, *Supermultiplets of Elementary Particles*, Phys. Rev. B1756 (1964) 136.

[10]. S. Weinberg, *A model of leptons*, Phys. Rev. Lett. 19 (1967) 1264; Abdus Salam, in: *Elementary Particle Theory*, N. Svartholm Ed.,

p. 367, Stockholm 1968; J. Bjorken and S. L. Glashow, *Baryon Resonances in W_3 symmetry*, Phys. Lett. 11 (1964) 84; S. L. Glashow, J. Iliopoulos, and L. Maiani, *Weak Interactions with Lepton-Hadron Symmetry*, Phys. Rev. D2 (1970) 1285.

[11]. P. W. Anderson, *Random Phase Approximation in the Theory of* Superconductivity, Phys Rev. 112 (1958) 1900; see also P. W. Anderson, *Concepts in Solids*, New York, 1963, 1975; J. Goldstone, *Field Theories with Superconductor Solutions*, N. Cimento, 19 (1961) 154; see also B. D. Josephson, *The discovery of tunneling Superconductivity*, Rev. Mod. Phys. 46 (1974) 251. The earliest application to the fundamental interactions is probably: J. Goldstone, Abdus Salam, S. Weinberg, *Broken Symmetries*, Phys. Rev. 127 (1962) 965. The application to gauge theories was first given by: P. H. Higgs, *Broken Symmetries, Massless Particles and Gauge Fields*, Phys. Letters 12 (1964) 132.

[12]. Hesiodus, *Theogonia*, 115, 123.

[13]. S. Chandrasekhar, *Ellipsoidal Figures of Equilibrium*, New Haven, 1969. This book contains extensive references to previous works on the subject.

[14]. H. Bénard, *Les Tourbillons cellulaires dans une nappe liquide*, Revue Générale des Sciences Pures et Appliquées, 11 (1900) 1261 and 1309. See also Lord Rayleigh, *On Convective Currents in a Horizontal Layer of Fluid when the higher Temperature is on the Under Side*, Phil. Mag. 32 (1916) 529.

[15]. E. L. Koschmieder, *On convection on a uniformly heated plane*, Beitr. Phys. Atmos., 39 (1965) 1. J. Wesfreid, Y. Pomeau, M. Dubois, C. Normand, P. Bergé, *Critical Effects in Rayleigh-Bénard convection*, J. de Phys., 7 (1978) 725.

[16]. F. Busse, *The Stability of finite amplitude cellular convection and its relation to an extremum principle*, J. Fluid Mech., 30 (1967) 625.

[17]. D. H. Sattinger, *Bifurcations and Symmetry Breaking in Applied* Mathematics, Bull. Math. Soc. 3 (1980) 779.

[18]. P. Curie, *Sur la Symétrie dans les phénomènes physiques. Symétrie d'un champ électrique et d'un champ magnétique*, Journal de Physique 3me Série, 3 (1894) 393.

[19]. D. H. Constantinescu, L. Michel and L. A. Radicati, *Spontaneous Symmetry Breaking from the Maclaurin and the Jacobi sequences*, Journal de Physique, 40 (1979) 147.

[20]. D. H. Sattinger, *Group representaion theory, bifurcation theory and pattern formation*, J. Functional Anal., 28 (1978) 58.

[21]. L. Michel and L. A. Radicati, *The geometry of the octet*, Ann. Inst. Henry Poincaré Sect. A, 18 (1973) 185. *Properties of the Breaking of Hadronic Internal Symmetry*, Ann. of Physics, 66 (1971) 758.

[22]. E. Cartan, *Lecons sur la géométrie des espaces de Riemann*, Paris 1946; see also M. F. Atiyah, *Geometry of Yang-Mills Fields*, Pisa 1979.

[23]. H. Weyl, *Space, Time and Matter*, New York, 1952, p. 220.

[24]. E. Wigner, *The unreasonable effectiveness of Mathematics in the Natural Sciences*, Communications in Pure and Applied Mathematics, 13 1960 1.

DYNAMIC SYMMETRIES IN NUCLEI, ATOMS AND MOLECULES

F. Iachello

Wright Nuclear Structure Laboratory
Yale University
New Haven, CT 06520

and

Kernfysisch Versneller Instituut
Rijksuniversiteit Groningen
The Netherlands

Abstract

I summarize recent applications of symmetry considerations to the study of complex spectra of nuclei, atoms and molecules.

1. INTRODUCTION

In recent years, the idea that dynamic symmetries may be useful in describing physical phenomena has received considerable attention. Applications to elementary particle physics are very well known [1]. Feza himself has contributed much to this field [2]. In this contribution to his sixtieth birthday I will summarize recent applications of this idea to nuclear, atomic and molecular physics.

A simple definition of a dynamic symmetry can be given as follows: Consider a physical system described by a Hamiltonian (or mass operator) H, and let this Hamiltonian have group structure G. In general H will be written in terms of all the generators of G and the eigenvalue problem for H cannot be solved in closed, analytic form.

However, suppose that H contains only invariant (Casimir) operators of a complete chain of subgroups of G

$$G \supset G' \supset G'' \supset \ldots \tag{1.1}$$

Then the eigenvalue problem for H can be solved in closed form and one obtains energy (or mass) formulas. These are very useful in classifying data and in providing a simple understanding of the experimental situation.

A familiar example of a dynamic symmetry is provided by Gell-Mann-Ne'eman $SU(3)$ [1]. Here, the group chain Eq. (1.1) is

$$
\begin{array}{cccc}
G \supset & G' & \supset & G'' \\
\downarrow & \downarrow & & \downarrow \\
SU(3) \supset & SU(2) \otimes U(1) & \supset & SO(2) \otimes U(1).
\end{array}
\tag{1.2}
$$

Introducing the labels characterizing the irreducible representations of the groups G', G'', . . .

$$
\left| \begin{array}{ccc}
SU(3) \supset SU(2) \otimes U(1) \supset SO(2) \otimes U(1) \\
\downarrow \quad\quad \downarrow \quad\quad \downarrow \\
I \quad\quad Y \quad\quad I_3
\end{array} \right\rangle
\tag{1.3}
$$

and writing the mass operator in terms of some of the Casimir invariants of the group chain, Eq. (1.2) leads to the mass formula [3]

$$E(I, I_3, Y) = E_0 + bY + c \left[I(I+1) - \tfrac{1}{4} Y^2 \right]. \tag{1.4}$$

This formula describes the low-lying experimental spectra of hadrons quite accurately (Fig. 1-1). A generalization of this mass formula to more complex situations was provided by Feza himself and Radicati [2]. By combining ordinary spin with Gell-Mann-Ne'eman $SU(3)$, one can form the group chain

$$
\left| \begin{array}{c}
SU(6) \supset SU(3) \otimes SU(2) \supset SU(2) \otimes U(1) \otimes SU(2) \supset SO(2) \otimes U(1) \otimes SU(2) \\
\downarrow \quad\quad\quad \downarrow \quad \downarrow \quad\quad\quad\quad \downarrow \\
J \quad\quad\quad I \quad Y \quad\quad\quad\quad I_3
\end{array} \right\rangle .
\tag{1.5}
$$

Writing the mass operator in terms of some of the Casimir invariants of this chain leads to the Gürsey-Radicati mass formula

$$E(J, I, I_3, Y) = E_0 + bY + c \left[I(I+1) - \tfrac{1}{4} Y^2 \right] + a J(J+1). \tag{1.6}$$

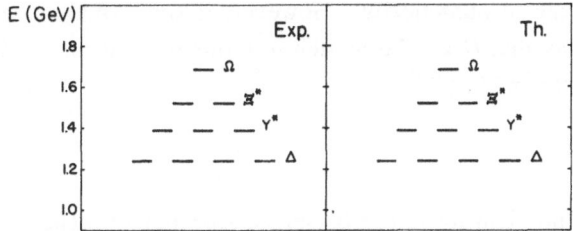

Figure 1-1: An example of a dynamic symmetry in particle physics: the spectrum of the baryon decuplet. The theoretical energies are calculated using Eq. (1.4).

2. DYNAMIC SYMMETRY IN NUCLEI

Most applications of dynamic symmetries in nuclei are based on the interacting boson model [4]. In this model, one describes low-lying collective states of even-even nuclei in terms of six dynamical bosons divided into a scalar boson with $J = 0$ (called s) and a quadrupole boson with $J = 2$ (called d). The six dynamical bosons span a six-dimensional space with group structure $G \equiv U(6)$. The bosons have a dual interpretation. On one side, they can be thought of as correlated pairs of nucleons with angular momentum $J = 0$ and $J = 2$ respectively [5], similar to the Cooper pairs of the electron gas [6]. On the other side, they can be thought of as a quantization of the surface oscillations of a liquid drop with quadrupole shape [7].

Starting from the group $U(6)$, one can form several chains of subgroups. However, since states in nuclei are characterized by a good value of the angular momentum, L, one must require that the rotation group, $O(3)$, be contained in the chain. There are then three and only three possible chains [8,9,10],

$$
U(6)
\begin{cases}
U(5) \supset O(5) \supset O(3) \supset O(2) & (I) \\
U(3) \supset SU(3) \supset O(3) \supset O(2) & (II) \\
O(6) \supset O(5) \supset O(3) \supset O(2) & (III).
\end{cases}
\tag{2.1}
$$

In discussing the algebraic structure of this model it has been found convenient to introduce boson creation ($b_\alpha^\dagger$, $\alpha = 1, \ldots, 6$) and annihilation (b_α, $\alpha = 1, \ldots, 6$) operators. The Hamiltonian describing the system of interacting bosons will, in general, contain all the generators, $G_{\alpha\alpha'} = b_\alpha^\dagger b_{\alpha'}$, of $U(6)$

$$
H = \sum_{\alpha,\beta} \epsilon_{\alpha\beta} G_{\alpha\beta} + \frac{1}{2} \sum_{\alpha,\beta,\gamma,\delta} u_{\alpha\beta\gamma\delta} G_{\alpha\beta} G_{\gamma\delta}
\tag{2.2}
$$

F. Iachello

and must be diagonalized numerically. However, if some of the coefficients in (2.2) vanish, in such a way that H can be written in terms only of the Casimir operators $\mathscr{C}_i$ of one of the group chains in (2.1),

$$H = \sum_i \alpha_i \, \mathscr{C}_i, \tag{2.3}$$

then the eigenvalue problem for H can be solved analytically. The eigenvalues depend on the labels characterizing the irreducible representations of the groups appearing in (2.1). For example, for the chain III the labels are [10]

$$\left| \begin{array}{ccccc} U(6) \supset O(6) \supset O(5) \supset O(3) \supset O(2) \\ \downarrow \quad \downarrow \quad \downarrow \quad \downarrow \quad \downarrow \\ N \quad \sigma \quad \tau(\nu_\Delta) \quad L \quad M \end{array} \right\rangle . \tag{2.4}$$

The label ν_Δ is related to the fact that the group $O(5)$ is not fully reducible with respect to $O(3)$.

Eq. (2.3), when written for chain III, becomes

$$H = A' \, \mathscr{C}_2(O6) + B'' \, \mathscr{C}_2(O5) + C' \, \mathscr{C}_2(O3) \tag{2.5}$$

where $\mathscr{C}_2(O6)$ denotes the quadratic Casimir operator of $O(6)$, etc., and A', B' and C' are arbitrary constants. The eigenvalues of H in the basis (2.4) are then

$$E(N,\sigma,\tau,\nu_\Delta,L,M) = A' \, \sigma(\sigma + 4) + B' \, \tau(\tau + 3) + C' \, L(L + 1). \tag{2.6}$$

The technique of writing H in terms of Casimir operators of a group chain can be used to find energy formulas for all three chains in (2.1). The complete results are [8,9,10]

$$E(N,n_d,v,n_\Delta,L,M) = \epsilon \, n_d + \alpha \, \tfrac{1}{2} \, n_d(n_d - 1) + \beta(n_d - v)(n_d + v + 3) \tag{2.7}$$

$$+ \gamma \, [L(L + 1) - 6 \, n_d] \tag{I}$$

$$E(N,\lambda,\mu,K,L,M) = \left(\tfrac{3}{4}\kappa - \kappa'\right) L(L + 1) - \kappa \, [\lambda^2 + \mu^2 + \lambda \mu + 3 \lambda + 3 \mu] \tag{II}$$

$$E(N,\sigma,\tau,\nu_\Delta,L,M) = \tfrac{A}{4} \, (N - \sigma)(N + \sigma + 4) + \tfrac{B}{6} \, \tau \, (\tau + 3) + C \, L(L + 1) \tag{III}$$

where I have used a notation slightly different from that of Eq. (2.6) in order to conform with the notation in the original articles [8, 9 and 10]. Several nuclei whose spectra can be approximately described by one of the three chains have been found. Three examples are shown in Figs. 2-1, 2-2 and 2-3 respectively.

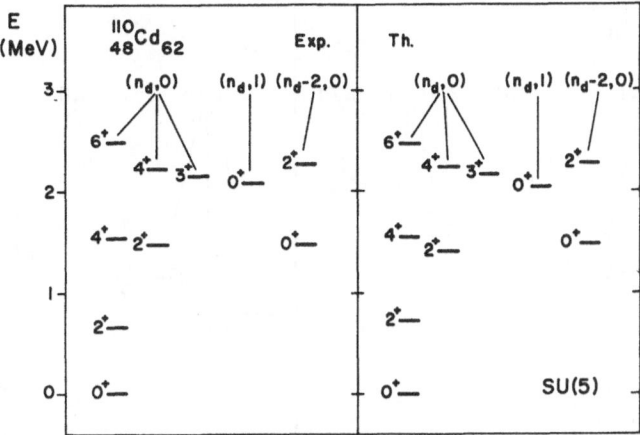

Figure 2-1: An example of a spectrum with $U(5)$ symmetry in nuclear physics: ^{110}Cd. The theoretical energies are calculated using Eq. (2.7), I.

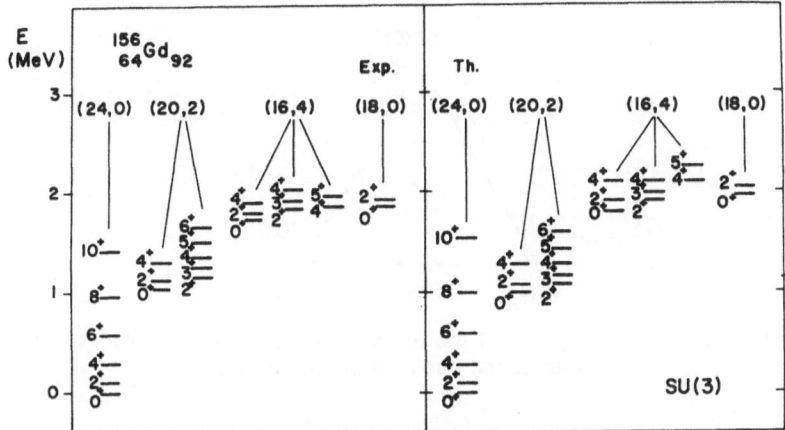

Figure 2-2: An example of a spectrum with $SU(3)$ symmetry in nuclear physics: ^{156}Gd. The theoretical energies are calculated using Eq. (2.7), II.

The dynamic symmetries of Eq. (2.7) are special cases of the more general Hamiltonian (2.2). It has become customary to represent the situation by a phase diagram [11], Fig. 2-4.

Here the three dynamic symmetries are placed at the vertices of a triangle. Each nucleus is represented by a dot in this diagram. If the dot is close to one of the vertices, the nucleus displays the corresponding symmetry. The microscopic theory of the interacting boson model can also explain why a particular type of dynamic symmetry appears in a particular nucleus [4]. I will not, however, discuss this subject here.

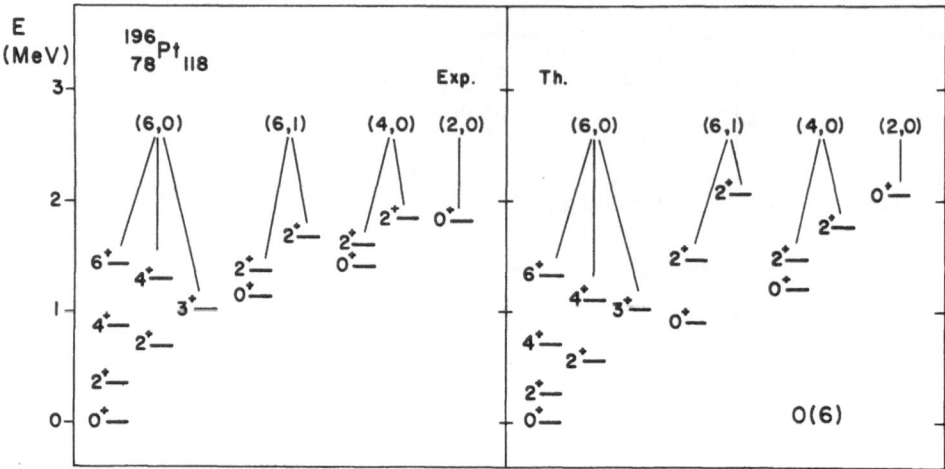

Figure 2-3:   An example of a spectrum with $O(6)$ symmetry in nuclear physics: ^{196}Pt.   The theoretical energies are calculated using Eq. (2.7), III.

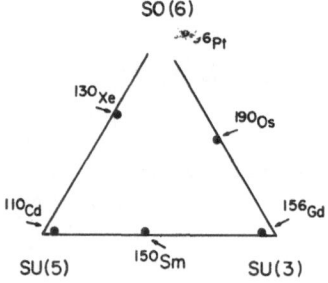

Figure 2-4:   Phase diagram of the interacting boson model.   The location of some of the nuclei is shown by dots.

Instead, I will return briefly to the boson operators b_α and divide them explicitly into a scalar (s) and quadrupole $(d_\mu, \mu = \pm 2, \pm 1, 0)$ operator.   It is then possible to study the properties of the geometric space associated with this problem by introducing a coherent state [11,12]

$$| N, \alpha_\mu > = \left\{ s^\dagger + \sum_\mu \alpha_\mu \, d_\mu^\dagger \right\} | 0>. \tag{2.8}$$

The c-number variables α_μ are variables in the coset space $U(6)/(U(5) \otimes U(1))$ [13]. The use of coset variables α_μ allows one to associate a shape to each of the three dynamic symmetries discussed above.   For purposes of visualization, it has been found convenient to associate to each set of values α_μ an ellipsoid of radius [7]

$$R = R_0 \left\{ 1 + \sum_\mu \alpha_\mu Y_{2\mu}(\Theta, \Phi) \right\} \tag{2.9}$$

Symmetry I then corresponds to a spherical shape, symmetry II to a deformed shape with axial symmetry and symmetry III to a deformed shape without axial symmetry [12]. The dynamic symmetry of the interacting boson model may thus be viewed as a way of classifying the possible shapes of nuclei, much in the same way in which the point symmetries provide a way of classifying the possible shapes of crystals and molecules. The seven symmetry classes (triclinic, monoclinic, orthorhombic, trigonal, tetragonal, hexagonal and cubic) are replaced here by the three symmetry classes I, II and III.

3. DYNAMIC SYMMETRIES IN MOLECULES

The success of the dynamic symmetry approach to nuclear spectra has stimulated applications of this idea to the study of molecular spectra. This study is being based on a model, whose algebraic structure is very similar to that of the interacting boson model of nuclei, but whose physical interpretation is very different. In this model, called vibron model [14,15], states of diatomic molecules are built in terms of four dynamical bosons divided into a scalar boson with angular momentum and parity $J^P = 0^+$ (σ-boson) and vector boson with $J^P = 1^-$ (π-boson). The bosons here represent oscillations of the relative distance between the two atoms. They are therefore called vibrons. The reason why vector (π) bosons appear in molecular spectra while quadrupole (d) bosons appear in nuclear spectra is that molecular spectra are characterized by a *dipole* deformation, while nuclear spectra are characterized by a quadrupole deformation, Fig. 3-1.

' The group structure of the molecular problem is thus $G \equiv U(4)$ and an analysis similar to that of Sect. 2 can be performed here. When $G \equiv U(4)$ there are only two possible chains of subgroups containing the rotation group, $O(3)$ [14, 15].

$$U(4) \Bigg\langle \begin{array}{l} U(3) \supset O(3) \supset O(2) \qquad (I) \\[2mm] O(4) \supset O(3) \supset O(2) \qquad (II). \end{array} \tag{3.1}$$

The Hamiltonian describing the system of interacting vibrons which contain in general all the generators $G_{\alpha\alpha'}$ ($\alpha, \alpha' = 1, \ldots, 4$) of $U(4)$ and will have the same structure as Eq. (2.2). In the case in which it can be written in terms only of the Casimir operators of one of the group chains in (3.1), it can be diagonalized analytically. This leads to the energy formulas [14,15]

F. Iachello

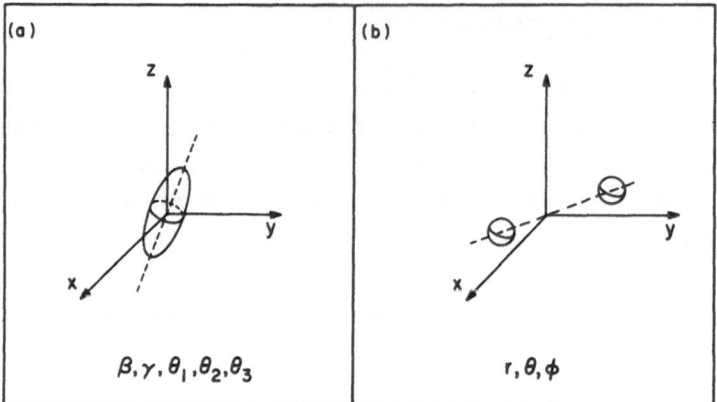

Figure 3-1: Schematic representation of the geometric structure of nuclei (a) and diatomic molecules (b).

$$E(N,n_\pi,J,M) = \epsilon\, n_\pi + \frac{1}{2}\, \alpha\, n_\pi\, (n_\pi - 1) + \beta\, J\, (J + 1) \qquad (I)$$

$$E(N,\omega,J,M) = A\, \omega\, (\omega + 2) + B\, J(J + 1) \qquad\qquad (II) \qquad (3.2)$$

where again N, ω, . . . label the representations of the groups appearing in Eq. (3.1). For example,

$$\left| \begin{array}{cccc} U(4) \supset O(4) \supset O(3) \supset O(2) \\ \downarrow\quad \downarrow\quad \downarrow\quad \downarrow \\ N\quad\ \omega\quad\ J\quad\ M \end{array} \right\rangle. \qquad (3.3)$$

It should be noted that both here and in the previous section, the representations of U(4) and U(6) that appear are the totally symmetric representations characterized by the Young tableau

$$[N] = \square\,\square\,\ldots\,\square, \quad (\,N\ boxes\,) \qquad (3.4)$$

where N is the total number of bosons (vibrons). The low-lying energy spectra of many diatomic molecules are well reproduced by the energy formula II. An example is shown in Fig. 3-2. There does not seem to be, at present, any example of an energy spectrum of a diatomic molecule that can be described by the energy formula I.

An analysis of the geometric properties of the two dynamic symmetries of the vibron model can also be done by introducing [10] variables r_μ ($\mu = \pm 1, 0$) in the coset space $U(4)/(U(3)\otimes U(1))$. This analysis shows that the dynamic symmetry II corresponds to rigid molecules, while the dynamic symmetry I corresponds to non-rigid

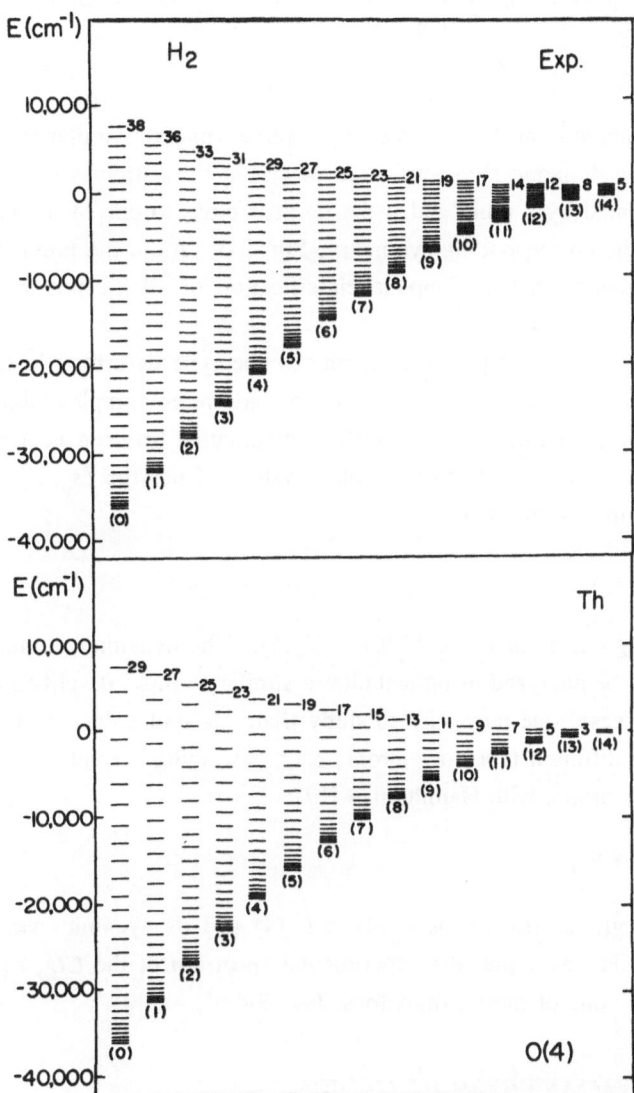

Figure 3-2: An example of a spectrum with $O(4)$ symmetry in molecular physics: $H_2(^1\Sigma_g^+)$. The theoretical energies are calculated using Eq. (3.2) II.

Figure 3-3:   Phase diagram of the vibron model. The location of the H_2
molecule is shown by a dot.

molecules. The situation can be described by a phase diagram similar to that of Fig.
2-4. where the two dynamic symmetries are placed at the extremes of a line. Each
molecule is represented by a point and if the point is close to one of the extremes the
molecule displays the corresponding symmetry, Fig. 3-3. As in the previous case, the
group approach provides then a complete classification of all the possible shapes of
diatomic molecules.

The applications of symmetry considerations of the type described in this and the
preceding section can be extended further to encompass more complex situations. For
example, in the nuclear case, one may wish to distinguish between neutron (ν) pairs
and proton (π) pairs. This leads to a coupled system of neutron (s_ν, d_ν) and proton
(s_π, d_π) bosons, with Hamiltonian

$$H = H_\pi + H_\nu + V_{\nu\pi} \ . \tag{3.5}$$

The corresponding group structure is $U_\nu(6) \otimes U_\pi(6)$. The dynamic symmetries of the
coupled system can be analyzed using techniques similar to those described above.

In the molecular case, one may wish to study triatomic molecules. In this case one
must introduce two different types of vibrons, each describing a bond. This leads to a
coupled system of vibrons, with Hamiltonian [17]

$$H = H_1 + H_2 + V_{12} \ . \tag{3.6}$$

The corresponding group structure is $U_1(4) \otimes U_2(4)$ and its dynamic symmetries can
again be analyzed. For example, the experimental spectrum of the CO_2 molecule can
be well described by one of these symmetries, Fig. 3-4.

4. DYNAMIC SYMMETRIES IN ATOMS

The dynamic symmetries described in Sects. 2 and 3 are symmetries of systems of
bosons. One can also have dynamic symmetries of systems of fermions. Gell-Mann-

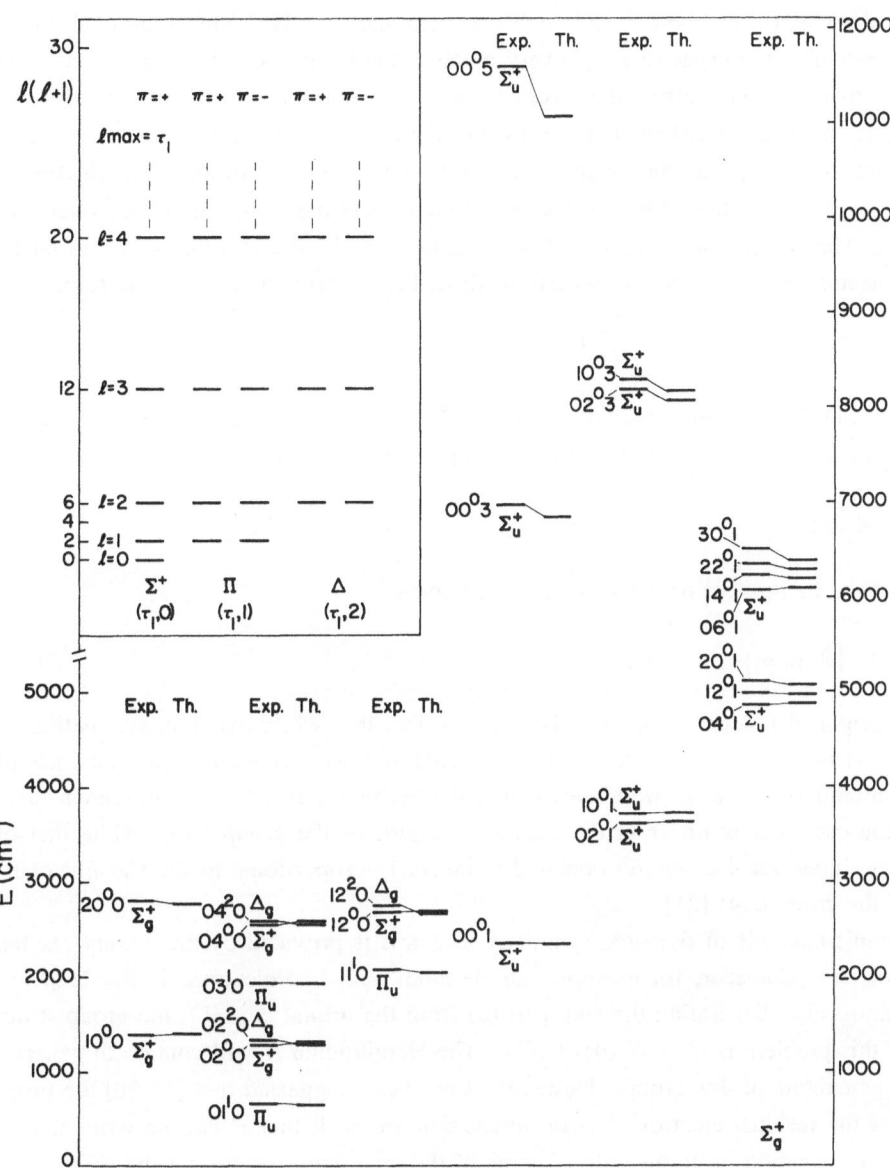

Figure 3-4: An example of a dynamic symmetry in coupled systems: The molecule CO_2.

Ne'eman $SU(3)$ is one such case (the fermions here are the quarks). Herrick and Kellman [18] have suggested that another of these symmetries occurs in highly excited states of atoms, for example in the doubly excited states of He. The study of dynamic symmetries in atomic spectra is still very much at the beginning. It is, however, clear that the dynamic symmetries that seem to occur in atoms are different, in many respects, from those described in the previous sections. As remarked above, the first difference is that symmetries in atoms apply to a system of fermions (the electrons), while symmetries in the collective spectra of nuclei and molecules apply to systems of bosons. The second, and major, difference appears to be the fact that while in nuclei and molecules dynamic symmetries can be described by Hamiltonians of the form

$$H = \sum_i \alpha_i \, \mathscr{C}_i,$$ (4.1)

i.e. linear in the invariant operators $\mathscr{C}_i$ in atoms dynamic symmetries appear to require more complex expansions, either of the form [19,20]

$$H = \sum_i \beta_i \, \mathscr{C}_i^x$$ (4.2)

where x can be positive or negative, or of the form

$$H = \left\{ \sum_i \beta_i \, \mathscr{C}_i \right\}^{-x} \quad x > 0.$$ (4.3)

The origin of this difference is related to the fact that while dynamic symmetries in nuclei and molecules are of the harmonic oscillator type, symmetries in atoms are of the Coulomb type. It is well known that the Hamiltonian of the three-dimensional harmonic oscillator is linear in the Casimir operator of the group $U(3)$, while that of the three-dimensional Coulomb potential is inversely proportional to the Casimir operator of the group $O(4)$ [21].

A simple example of dynamic symmetry in atoms is provided by the doubly excited states of He. Consider, for example, an He atom with two electrons in the $2s2p$ hydrogenic levels. Separating the spin part (S) from the orbital part (L), the group structure of this problem is $G \equiv U_L(4) \otimes U_S(2)$. The Hamiltonian H will contain in general, all the generators of this group. However, it has been suggested that [19,20] the properties of the residual electron-electron interaction are such that H can be written, to a good approximation, in terms only of some of the invariant operators of the chain

$$U_L(4) \otimes U_S(2) \supset O_L(4) \otimes SU_S(2) \supset O_L(3) \otimes SU_S(2) \supset SU(2) \supset SO(2) \ , \qquad (I) \qquad (4.4)$$

i.e.

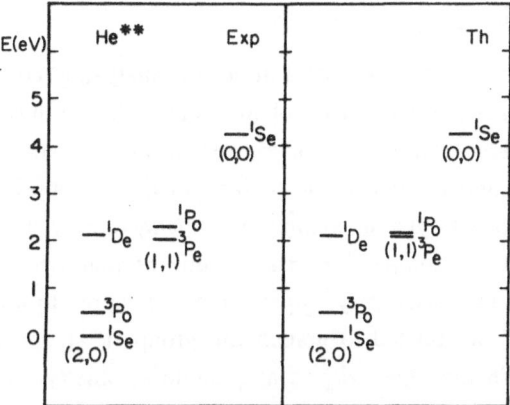

Figure 4-1: An example of a dynamic symmetry in atomic physics: the doubly excited states of He ($n=2$). The theoretical energies are calculated using Eq. (4.7)

$$H = \frac{1}{[\alpha + \beta \mathscr{C}_2(O_L 4) + \gamma \mathscr{C}_2(O_L 3)]^{1/2}} \cdot \qquad (4.5)$$

Introducing the labels that characterize the irreducible representations of the chain (4.4)

$$\left| \begin{array}{c} U_L(4) \otimes U_S(2) \supset O_L(4) \otimes SU_S(2) \supset O_L(3) \otimes SU_S(2) \supset SU(2) \supset SO(2) \\ \downarrow \qquad\qquad \downarrow \qquad\qquad \downarrow \qquad\qquad\qquad \downarrow \qquad\quad \downarrow \\ (\omega_1,\omega_2) \qquad S \qquad\quad L \qquad\qquad\qquad J \qquad\quad M_J \end{array} \right\rangle, \qquad (4.6)$$

the eigenvalue problem for H can be solved analytically, yielding

$$E\{(\omega_1,\omega_2),L,S,J,M_J)\} = \frac{1}{[\alpha + \beta \{\omega_1(\omega_1 + 2) + \omega_2^2\} + \gamma L(L + 1)]^{1/2}} \cdot \qquad (4.7)$$

These energy eigenvalues are compared with experiment in Fig. 4-1. The comparison here is not so striking as in the cases discussed previously since there are in Eq. (4.7) three free parameters and only five experimental energies. However, it is conceivable that the excitation spectrum of He^{**} with the two electrons in the $3s3p3d$ levels and/or the excitation spectrum of the triply excited He^{-**} atom with the three electrons in the $2s2p$ levels can be measured soon. These new measurements will provide a more stringent test of the occurrence of dynamic symmetries in atoms.

5. CONCLUSIONS

Dynamic symmetries provide a useful tool for analyzing complex spectroscopic problems. The systematic use of this tool in nuclear physics has provided a way to classify all collective low-lying spectra of even-even nuclei. The same techniques are beginning to have an impact also in molecular and atomic physics. In addition, an even more complex type of dynamic symmetry, supersymmetry, has begun to make an appearance in the study of complex spectra. While normal symmetries apply to systems of bosons or fermions separately, supersymmetries apply to mixed systems of bosons and fermions and are related to graded Lie groups [22]. A dynamic supersymmetry can be defined in the same way as a dynamic symmetry. Let the Hamiltonian (or mass operator), H, of the mixed system of bosons and fermions be described by a supergroup G^*. If H can be written in terms of invariant operators of a complete chain of groups (or supergroups) of G^*

$$G^* \supset G'^* \supset G''^* \supset \ldots , \tag{5.1}$$

then the corresponding eigenvalue problem can be solved analytically.

The idea of dynamic supersymmetries has already found useful applications in nuclear physics, where it has been used to study the low-lying collective spectra of odd-even nuclei [23,24]. It is conceivable that other applications could be found in molecular and atomic physics.

It is a pleasure for me to dedicate this paper to the sixtieth birthday of Feza Gürsey, whose inspiration and frequent discussions have stimulated much of my work. In particular, his view that symmetry considerations permeate the physical world has been a crucial guiding principle in most of my investigations of complex spectra.

This work was supported in part by the USDOE Contract EY-76-C-02-3074 and in part by the Stichting voor Fundamenteel Onderzoek der Materie (FOM).

REFERENCES

[1]. M. Gell-Mann, Phys. Rev. 125 (1962) 1067; Y. Ne'eman, Nucl. Phys. 26 (1961) 222.

[2]. F. Gürsey and L. A. Radicati, Phys. Rev. Lett. 13 (1964) 173.

[3]. S. Okubo, Prog. Theor. Phys. 27 (1962) 949.

[4]. For a review of this model, see A. Arima and ·F. Iachello, Ann. Rev. Nucl. Part. Sci. 31 (1981) 75.

[5]. A. Arima, T. Otsuka, F. Iachello and I. Talmi, Phys. Lett. 66B (1977) 205.

[6]. L. N. Cooper, Phys. Rev. 104 (1956) 1189.

[7]. A. Bohr, Mat. Fys. Medd. Dan. Vid. Selsk. 26, No. 14 (1952).

[8]. A. Arima and F. Iachello, Ann. Phys. (N.Y.) 99 (1976) 253.

[9]. A. Arima and F. Iachello, Ann. Phys. (N.Y.) 111 (1978) 201.

[10]. A. Arima and F. Iachello, Ann. Phys. (N.Y.) 123 (1979) 468.

[11]. D. H. Feng, R. Gilmore and J. R. Deans, Phys. Rev. C23 (1981) 1254.

[12]. A. E. L. Dieperink, O. Scholten and F. Iachello, Phys. Rev. Lett. 44 (1980) 1747.

[13]. R. Gilmore, *Lie Groups, Lie Algebras and Some of Their Applications*, J. Wiley and Sons, New York (1974).

[14]. F. Iachello, Chem. Phys. Lett. 78 (1981) 581.

[15]. F. Iachello and R. D. Levine, J. Chem. Phys., to be published.

[16]. O. S. van Roosmalen and A. E. L. Dieperink, Ann. Phys. (N.Y.) 139 (1982) 198.

[17]. O. S. van Roosmalen, A. E. L. Dieperink and F. Iachello, Chem. Phys. Lett. 85 (1982) 32.

[18]. D. R. Herrick and M. E. Kellman, Phys. Rev. A21 (1980) 418; M. E. Kellman and D. R. Herrick, J. Phys. B11 (1978) L755.

[19]. C. Wulfman, Chem. Phys. Lett. 23 (1973) 370.

[20]. F. Iachello and A. R. P. Rau, to be published.

[21]. See, for example, B. G. Wybourne, *Classical Groups for Physicists*, J. Wiley, New York (1974).

[22]. F. A. Berezin and G. I. Kac, Math. USSR Sbornik 11 (1970) 311.

[23]. A. B. Balantekin, I. Bars and F. Iachello, Nucl. Phys. A370 (1981) 284; Phys. Rev. Lett. 47 (1981) 19.

[24]. A. B. Balantekin, I. Bars and F. Iachello, Yale Preprint YNT 82-02, to be published.

THE SYMMETRY
AND
RENORMALIZATION GROUP
FIXED POINTS OF QUARTIC HAMILTONIANS

Louis Michel

Institut des Hautes Etudes
Scientifiques
35, route de Chartres
91440 Bures-Sur-Yvette, France

à Feza Gürsey, pour son soixantième anniversaire,
en témoignage d'amitié et d'admiration.

Abstract

This paper studies the number and the nature of the fixed points of the renormalization group for the ϕ^4 model, as used for instance in the Landau theory of second order phase transitions. It is shown that when it exists the stable fixed point is unique and a condition on its symmetry is given: it is often larger than the initial symmetry. Finally counter examples, with ν arbitrarily large, are given to the Dzyaloshinskii conjecture that there exist no stable fixed points when the Landau potential depends on more than $\nu = 3$ parameters.

INTRODUCTION

A great interest has risen from the study of the general n-vector model. Following the notations of Brézin's lectures [1] and with summation over repeated indices, the Hamiltonian density of this model is:

$$\mathcal{H}_n(x) = \frac{1}{2}\, \vec{\nabla}\phi_i(x)\cdot\vec{\nabla}\phi_i(x) + \frac{a}{2}\,\phi_i(x)\,\phi_i(x) + \frac{n^{4-d}}{4!}\, g_{ijkl}\,\phi_i(x)\,\phi_j(x)\,\phi_k(x)\,\phi_l(x)\,, \tag{1}$$

where the scalar field $\phi_i(x)$ has n components (the n values of its index) corresponding to inner degrees of freedom, while d is the space dimension. Brézin, Le Guillou and Zinn Justin [2] have written the renormalization group equations for this model and reached some general conclusion on the properties of stable fixed points, e. g. for $n < 4$ there is only one stable fixed point, that which is $O(n)$ invariant (see Eq. (4.18)).

Such a Hamiltonian density appears in the Landau theory [3] of second order phase transitions: the field values are the n component Landau order parameter. As a mean field theory one obtains good selection rules for the symmetry change in the transition but the wrong critical exponents; so one has to take account of the fluctuations at the critical points and applications to the Landau theory of the renormalization group techniques have been proposed [4-7], and many have been performed since.

The aim of this paper is different. Very few studies have been made up to now of the group covariance properties of the renormalization group techniques [8-10]. Here a more systematic study is made. As a result it is shown that, when it exists and for $n \neq 4$, the stable fixed point is unique; it often has a greater symmetry group than the starting polynomial: some sufficient conditions are given for this phenomenon to occur. In the applications studied in the literature few stable fixed points were found; so there is a conjecture (e.g., Dzyaloshinskii [11]) that there is a topological obstruction to their existence when the quartic part of the Hamiltonian depends on a number v of parameters larger than three. This is wrong. The stable fixed point of each of a family of Hamiltonians with increasing, and arbitrarily large, values of v are given explicitly: of course n has to be increasing with v; in our example $n = 2^{v-2}\times 3$. The invariance groups corresponding to these examples form a family of groups which may have some interesting properties.

This paper studies a new example of the equivariant symmetric non-associative algebra obtained from group representations with a third degree invariant (e.g., the "d" algebra introduced by Gell-Mann [12] in his $SU(3)$ paper). Radicati and I [13] have already studied several families of these algebras relevant to physics and shown that directions of spontaneous symmetry breaking are given by idempotents of these algebras. Sattinger [14] has extended this result to bifurcation theory. Here fixed points are also determined by idempotents of the algebra. It seems also that this algebraic method may be more efficient for computing solutions in the case of practical applications.

1. THE RENORMALIZATION GROUP EQUATIONS

In Eq. (1) the g_{ijkl} are coupling constants of the term quartic in Φ. They are completely symmetrical in the four indices so their number is $N = \binom{n+3}{4}$. Their set can be considered as a vector in the N dimensional real vector space $\mathcal{T}_4$ of quartic polynomials with n variables. This vector g generally depends on several parameters. Let λ be the renormalization parameter. The renormalization equations are:

$$\frac{\lambda \, dg_{ijkl}(\lambda)}{d\lambda} = \beta_{ijkl}(g_{i'j'k'l'}) \cdot \tag{1.1}$$

To simplify notations we will often use a multi-index α taking N values; then Eq. (1.1) reads:

$$\frac{\lambda \, dg_{\alpha}(\lambda)}{d\lambda} = \beta_{\alpha}(g_{\beta}) \cdot \tag{1.2}$$

The fixed points of g^{*} of these equations satisfy:

$$\beta_{\alpha}(g^{*}) = 0 \cdot \tag{1.3}$$

It can be shown that the matrix $\frac{\partial \beta_{\alpha}}{\partial g_{\beta}}$ has real eigenvalues which are related to the value of the critical exponents when the fixed point g^{*} is stable, that is when the $\frac{\partial \beta_{\alpha}}{\partial g_{\beta}}$ satisfy some partial positivity condition that we will make precise below.

The functions β_{α} are not known exactly, but the first few terms of an expansion in $\epsilon = 4 - d$ have been computed by Brézin et. al. [2]. Imposing the irreducibility condition:

$$g_{iikl} = \gamma(g) \, \delta_{kl} \, , \tag{1.4}$$

they have obtained

$$\beta_{ijkl}(\epsilon, g) = - \epsilon \, g_{ijkl} + \frac{1}{2}(1 + \frac{\epsilon}{2}) \left\{ g_{ijpq} \, g_{pqkl} + g_{ikpq} \, g_{pqjl} + g_{ilpq} \, g_{pqjk} \right\} \tag{1.5}$$

$$- \frac{1}{4} \left\{ g_{ipqr} \, g_{jprs} \, g_{klqs} + 5 \text{ other terms obtained by permutations of } ijkl \right\}$$

$$- \frac{1}{48} (1 + \frac{5\epsilon}{4}) \left\{ g_{ipqr} \, g_{apqr} \, g_{ajkl} + 3 \text{ other terms obtained by} \right.$$

$$\left. \text{permutations of } ijkl \right\} \, .$$

This can be written pictorially in a condensed notation: each g is "tetravalent", internal bonds express saturation of indices and average on the permutation of the indices corresponding to the four external bonds is assumed:

$$\beta(\epsilon,g) = -\epsilon(=g=) + \frac{3}{2}\left(1 + \frac{\epsilon}{2}\right)(=g=g=) - \frac{3}{2}\left(=g\!\!<\begin{smallmatrix}g & - \\ \| \\ g & -\end{smallmatrix}\right)$$
$$-\left(\frac{1}{12}\right)\left(1 + \frac{5\epsilon}{4}\right)\left(-g\equiv g - g\equiv\right). \tag{1.6}$$

For instance (and we introduce an even more condensed notation for this expression):

$$(=g=g=)_{ijkl} = \frac{1}{3}\{g_{ijpq}\,g_{pqkl} + g_{ikpq}\,g_{pqjl} + g_{ilpq}\,g_{pqjk}\} = (g_\vee g)_{ijkl}. \tag{1.7}$$

Wallace and Zia [15] have shown that, to this order β_{ijkl} is a gradient. Indeed:

$$\beta_{ijkl}(\epsilon,g) = \frac{d\Phi(\epsilon,g)}{dg_{ijkl}},$$

with:

$$\Phi(\epsilon,g) = -\frac{1}{2}\,\epsilon\, g_{ijkl}g_{ijkl} + \frac{1}{2}\left(1 + \frac{\epsilon}{2}\right)g_{ijkl}g_{klpq}g_{pqij} - \frac{3}{8}\,g_{ijkl}g_{ipqr}g_{jpqs}g_{rskl}$$
$$-\frac{1}{48}\left(1 + \frac{5\epsilon}{4}\right)g_{ijkl}g_{ipqr}g_{apqr}g_{ajkl}. \tag{1.8}$$

Equivalently:

$$\Phi(\epsilon,g) = -\frac{1}{2}\,\epsilon\big(g\equiv g\big) + \frac{1}{2}\left(1 + \frac{\epsilon}{2}\right)\left(g\!\!<\!\!\begin{smallmatrix}g \\ \| \\ g\end{smallmatrix}\right)$$
$$-\frac{3}{8}\left(\begin{smallmatrix}g \\ \| \\ g\end{smallmatrix}\!\!>g=g\right) - \frac{1}{48}\left(1 + \frac{5\epsilon}{4}\right)\left(\begin{smallmatrix}g\equiv g \\ | & | \\ g\equiv g\end{smallmatrix}\right). \tag{1.9}$$

The equation for stable fixed points becomes:

$$\frac{d\Phi(g^*)}{dg} = 0, \tag{1.10.a}$$

$$\frac{d^2\Phi(g^*)}{dg^2} \geq 0. \tag{1.10.b}$$

The first one expresses that g^* is an extremum of Φ; it is stable when it satisfies the second one, i.e., when it is minimum. However, the second condition, as it is written in Eq. (1.10.b) is too strong. We will make precise the exact condition in Section 3, after explaining in section 2 some properties of the action of the orthogonal group $O(n)$ on the vector space $\mathcal{T}_4$ of quartic polynomials. In Section 3 we also study when the fixed points can be found by an expansion in ϵ from the solutions in the first order in ϵ. These solutions are extrema of the polynomial:

$$\Phi^{(1)}(g) = -\frac{\epsilon}{2}\,(g,\,g) + \frac{1}{2}(g_\vee g,\,g), \tag{1.11}$$

where

$$(g, g) = g_{ijkl} g_{ijkl} ,\tag{1.12}$$

and the symbol $g \bigvee g$ has been defined in Eq. (1.7). The symbol $\bigvee$ defines an abelian algebra on $\mathcal{T}^4$ by:

$$g \bigvee h = \frac{1}{2} \left[(g + h) \bigvee (g + h) - g \bigvee g - h \bigvee h \right] .\tag{1.13}$$

It has $O(n)$ as the group of automorphisms. This algebra will be studied in Section 4. As said in the introduction, the extrema of $\Phi^{(1)}$ are idempotents of this algebra. Indeed:

$$\frac{d\Phi^{(1)}(g^*)}{dg} = 0 \quad \longleftrightarrow \quad g^* \bigvee g^* = \frac{2}{3} \epsilon g^* .\tag{1.14}$$

In Section 5 we will study general properties of the extrema of $\Phi^{(1)}$. The last section, Section 6, will study the family $G_{s_1, \ldots, s_k}$ of irreducible discrete subgroups of $O(n)$, their invariants, and give the counter examples to the Dzyaloshinskii conjecture.

2. GROUP ACTIONS -- THE ACTION OF O(N) ON REAL N-VARIABLE POLYNOMIALS.

We first recall some basic concepts and results on group action. When a group G acts on a set M, the set of transforms of $m \in M$ is denoted by $G \cdot m$ and is called the *G-orbit* of m. The elements of g which leave m invariant: $g \cdot m = m$, form a subgroup G_m of G which is called the *isotropy* group of m. Note that:

$$G_{g \cdot m} = g \, G_m \, g^{-1} .\tag{2.1}$$

So the set of isotropy groups of an orbit form a conjugation class of subgroups of G. We denote by $[H]$ the conjugation class of $H \subseteq G$. G orbits with the same set $[H]$ of isotropy groups form an equivalence class that we denote by $[G:H]$. The set of all points with isotropy groups in a given conjugation class form a *stratum*: this is also the union of all orbits of an equivalence class.

We denote by M^H the subset of M whose elements are invariant under H, i. e.

$$M^H = \{ m \in M, G_m \supseteq H \} .\tag{2.2}$$

The action of G on the set M defines an action on $\mathcal{T}(M)$, the set of subsets of M. Let $A \subset M$, i. e., $A \in \mathcal{T}(M)$. The *centralizer* $C_G(A)$ of A is the set of elements of G which leave fixed every element of A; it is a subgroup $C_G(A) \subseteq G$:

$$C_G(A) = \bigcap_{m \in A} G_m \ . \tag{2.3}$$

The *normalizer* $N_G(A)$ of A is the set of elements of G which transform A into itself. It is a subgroup of G which contains the centralizer $C_G(A)$. For instance in the action of G on itself by conjugation, $x \rightarrow g\ x\ g^{-1}$, $C_G(A)$ is the subset of elements of G which commute with every element of A; e. g. $C_G(G)$ is the center of G. If $H \subseteq G$ and $N_G(H) = G$, H is called an invariant subgroup of G and we denote it by $H \lhd G$. More generally, $N_G(H)$, the normalizer of H in G, is the largest subgroup of G which has H as an invariant subgroup. For a general G-action, we remark that:

$$C_G(A) \lhd N_G(A) \ . \tag{2.4}$$

(Indeed, let $c \in C_G(A)$, $n \in N_G(A)$; for any $a \in A$, $n(c(n^{-1} \cdot a)) = n\ n^{-1}\ a = a$). So from the definition of the normalizer of $C_G(A)$,

$$N_G(A) \subseteq N_G(C_G(A)) \ . \tag{2.5}$$

We study a situation when the equality holds. We first remark from Eq. (2.1) and Eq. (2.2) that:

$$H \subseteq C_G(M^H) \ , \tag{2.6}$$

but the equality is not necessary. In any case:

$$M^H = M^{C_G(M^H)} \ . \tag{2.7}$$

We now prove that Eq. (2.5) is always an equality when A is of the type M^H:

$$N_G(M^H) = N_G(C_G(M^H)) \ . \tag{2.8}$$

With Eq. (2.5) we need only to prove the inequality $\supseteq$. For every $c \in C_G(M^H)$, $n \in N_G(C_G(M^H))$ and $m \in M^H$, we have $n^{-1}cn \cdot m = m$, i.e., $cn \cdot m = n \cdot m$ so $n \cdot m \in M^{C_G(M^H)} = M^H$ and $n \in N_G(M^H)$.

Finally we recall a non-trivial result. If G is compact, there is a partial order on the set of conjugation classes of closed subgroups: $[H_1] \leq [H_2]$ if a group of $[H_1]$ is a subgroup of a group of $[H_2]$. (This does not define an order relation for general groups). Then Montgomery and Yang [16] have proven for smooth (infinitely differentiable) actions of compact groups that: Among the set of conjugation classes of isotropy groups there is a smallest element and the corresponding stratum is open dense.

We now study the linear action of $O(n)$, the n dimensional orthogonal group on the vector space $\mathcal{T}$ of n variable polynomials. $O(n)$ is the group of $n \times n$ orthogonal matrices, $u^T = u^{-1}$ acting on the real n-dimensional vector space $\mathcal{V}_n$. We denote by ϕ one of its vectors, and by ϕ_i, $1 \leq i \leq n$ its coordinates. $O(n)$ also acts on the functions defined on $\mathcal{V}_n$:

$$[u \cdot f](\phi) = f(u^{-1}\phi) . \tag{2.9}$$

This action transforms polynomials into polynomials, preserving the degree of homogeneous polynomials. Hence $\mathcal{T}$ is a direct sum of the $O(n)$ invariant vector spaces $\mathcal{T}_k$ containing all homogeneous n-variable polynomials of degree k.

$$\mathcal{T} = \bigoplus_{k=0}^{\infty} \mathcal{T}_k \quad , \quad \dim \mathcal{T}_k = \binom{n+k-1}{k} \tag{2.10}$$

Note that $\mathcal{T}_0$ is the set of real numbers and that $\mathcal{T}_1 = \mathcal{V}_n$. For $k \geq 2$, the representation of $O(n)$ on the space $\mathcal{T}_k$ is reducible. Indeed, in $\mathcal{T}_2$ there is, up to a factor, an invariant polynomial, the orthogonal product

$$(\phi, \phi) = \phi_i \phi_i . \tag{2.11}$$

So every quadratic form can be decomposed into two irreducible components.

$$q(\phi) = q_{ij} \phi_i \phi_j = \left\{ q_{ij} - \frac{q_{kk}\delta_{ij}}{n} \right\} \phi_i \phi_j + \frac{q_{kk}(\phi,\phi)}{n} . \tag{2.12}$$

Using the Laplacian

$$\Delta = \frac{\partial}{\partial \phi_i} \frac{\partial}{\partial \phi_i} = \nabla_i \nabla_i , \tag{2.13}$$

we note that

$$\Delta q(\phi) = 2 q_{ii} = 2 \operatorname{Tr} q . \tag{2.14}$$

More generally all harmonic polynomials, $\Delta p(\phi) = 0$, homogeneous of degree k form the space $\mathcal{T}_k^{(k)}$ of an irreducible representation of $O(n)$. In Dynkin labeling this representation is denoted by $(k, 0, 0, 0, \ldots, 0)$ with $\rho - 1$ zeros, where ρ is the rank of $O(n) = E(n/2)$, where $E(n/2)$ is the largest integer $\leq n/2$. We shall simply denote this representation abstractly by (k) and its carrier space by $\mathcal{E}^{(k)}$.

For the readers who still prefer indices, the polynomials $g(\phi) = g_{ijkl\ldots} \phi_i \phi_j \phi_k \phi_l$ $\ldots$ of $\mathcal{T}_k^{(k)}$ are tensors of $O(n)$ completely symmetrical in their k indices with partial trace $g_{iiklm\ldots} = 0$.

We can also say that $\triangle$ is a surjective linear map

$$\mathcal{T}_k \overset{\triangle}{\twoheadrightarrow} \mathcal{T}_{k-2} ,\tag{2.15}$$

whose kernel is $\mathcal{T}_k^{(k)}$. So, from Eq. (2.8) and Eq. (2.13) we deduce:

$$\dim \mathscr{E}^{(k)} = \binom{n+k-1}{k} - \binom{n+k-3}{k-2}\tag{2.16}$$

For instance:

$$\dim \mathscr{E}^{(2)} = (n+2)(n-1)/2 \quad , \dim \mathscr{E}^{(4)} = (n+6)(n+1)n(n-1)/24 .\tag{2.17}$$

We are especially interested in the case $k = 4$. We denote by $U(u)$ the operator representing the action of $u \in O(n)$ on $\mathcal{T}_4$. The decomposition of U into irreducible representations yields for the carrier space:

$$\mathcal{T}_4 = \mathcal{T}_4^{(4)} \oplus \mathcal{T}_4^{(2)} \oplus \mathcal{T}_4^{(0)} \quad ; \qquad\qquad g = g^{(4)} + g^{(2)} + g^{(0)} .\tag{2.18}$$

To compute the irreducible components of $g \in \mathcal{T}_4$, we use:

$$\triangle\, a\, b = (\triangle\, a\,)\, b + 2\,(\nabla_i\, a\,)\,(\nabla_i\, b\,) + a \triangle b ,\tag{2.19}$$

and, for a homogeneous polynomial of degree k,

$$\phi_i \nabla_i p_k(\phi) = k\, p_k(\phi) .\tag{2.20}$$

Moreover, $g^{(0)}$ is proportional to $(\phi, \phi)^2$, $g^{(2)}$ contains (ϕ, ϕ) as a factor and $\triangle \{(\phi,\phi)^{-1} g^{(2)} \} = 0$. So we find, computing $\triangle g$ and $\triangle\triangle g$,

$$g^{(0)} = \frac{1}{8n(n+2)}\,(\phi,\, \phi)^2\, \triangle\triangle g \;\; , \quad g^{(2)} = \frac{1}{2(n+4)}\,(\phi,\, \phi)\, \Big\{\triangle g - \frac{(\phi,\, \phi)}{2n}\, \triangle\triangle g\Big\}.\tag{2.21}$$

We shall set:

$$(\phi,\, \phi)^2 = s = s_{ijkl}\phi_i\, \phi_j\, \phi_k\, \phi_l\tag{2.22.a}$$

with

$$s_{ijkl} = \frac{1}{3}\Big\{\delta_{ij}\delta_{kl} + \delta_{ik}\delta_{jl} + \delta_{il}\delta_{jk}\Big\} .\tag{2.22.b}$$

We have already introduced in Eq. (1.12) the $O(n)$ invariant scalar product

$(g, h) = g_{ijkl} \, h_{ijkl}$.

Then

$$(s, s) = \frac{n(n + 2)}{3} , \qquad (2.23)$$

and from Eq. (2.19) and $g^{(0)} = s \, (s, g) \, (s, s)^{-1}$ one obtains

$$(s, g) = \frac{1}{24} \, \triangle \, \triangle \, g . \qquad (2.24)$$

We will find it convenient to use the shorthand notation

$$\gamma(g) = \frac{1}{n} \, (s, g) \qquad (2.25.a)$$

so

$$g^{(0)} = \frac{3 \, \gamma(g)}{n + 2} \, s . \qquad (2.25.b)$$

We will be led to study the polynomials of $\mathcal{T}_4$ whose isotropy groups are irreducible; i.e., their n dimensional representation defined as subgroups of $O(n)$ is irreducible. From the remark that for such a group of the form

$$G = O(n)_g = O(n)_{g^{(4)}} \cap O(n)_{g^{(2)}}$$

and that all isotropy groups of non-vanishing polynomials of $\mathscr{E}^{(2)}$ must be reducible we deduce the

Lemma 2.1: $G = O(n)_g$ irreducible implies $g^{(2)} = 0$, that is $g \in \mathcal{T}_4^{(4)} \oplus \mathcal{T}_4^{(0)}$; With the shorthand introduced in Eq. (2.25.a), and from Eq. (2.21) and Eq. (2.25.a) we get:

$$g^{(2)} = 0 \longleftrightarrow \frac{1}{12} \triangle(g) = \gamma(g) \, (\phi, \phi) \qquad (2.26)$$

or with indices, this is identical to Eq. (1.4).

Let us consider an example (corresponding to the Ising model)

$$c = \sum_{i=1}^{n} \phi_i^{\,4} \qquad (2.27.a)$$

so

$$c_{ijkl} = \begin{array}{ll} 1 & for \;\; i=j=k=l \\ 0 & otherwise \end{array} , \qquad (2.27.b)$$

then

$$(c, c) = (s, c) = n \ . \tag{2.28}$$

The isotropy group of c has been labelled B_n by Coxeter (see e. g. [17]). It is the semi-direct product of the abelian group of $(n \times n)$ diagonal matrices with ± 1 as entries and $(n \times n)$ permutation matrices (i.e., zero everywhere except one element in each row and column which is one). The order of this group (i. e. its number of elements) is

$$|B_n| = 2^n \, n! \ . \tag{2.29}$$

The group B_n is the symmetry group of the hypercube whose vertices have for coordinates ± 1. In Schönflies notation $B_2 = C_{4v}$, $B_3 = O_h$. For every n, $\mathcal{T}_4^{B_n}$ is generated by s and c:

$$\mathcal{T}_4^{B_n} = \left\{ \alpha' s + \beta' c \right\} \ . \tag{2.30}$$

Moreover, for $n = 2$ and $n = 3$, every polynomial $g \in \mathcal{T}_4$ whose isotropy group is irreducible belongs to the stratum of $[B_n]$ or is a multiple of s ; so by an orthogonal transformation it can be brought into a linear combination:

$$\alpha s + \beta c \in \mathcal{T}_4^{B_n} \ .$$

It happens that for $n = 2$ the condition (2.26) is also sufficient for the irreducibility of the isotropy group. Indeed:

$$n = 2 \ , \ \mathcal{T}_4^{(4)} = \left\{ \alpha \ \text{Re}(\phi_1 + i \, \phi_2)^4 + \beta \ \text{Im}(\phi_1 + i \, \phi_2)^4 \right\} \tag{2.31}$$

and the isotropy group of any non-zero polynomial of $\mathcal{T}_4^{(4)}$ is conjugated of $C_{4v} = B_2$.

The irreducibility of $G = O(n)_g$ requires that any quadratic form one can form with g and $\triangle$ (equivalently from g_{ijkl} by contraction of indices) is a multiple of $(\phi, \phi) = \delta_{ij} \, \phi_i \, \phi_j$; e. g. , using Eq. (2.22.a)

$$G \ \text{irreducible} \ \nrightarrow \ g_{ipqr} \, g_{jpqr} = \frac{1}{n} \, (g, g) \delta_{ij} \ . \tag{2.32.a}$$

Note that:

$$g_{ipqr} \, g_{jpqr} \, \phi_i \, \phi_j = \frac{1}{2^9 \, 3^2} \left\{ \triangle^3 \, g^2 - 6 \, (\triangle \, g) \, (\triangle^2 \, g) - 12 \, (\nabla_i \, \triangle \, g) \, (\nabla_i \, \triangle \, g) \right.$$
$$\left. - 24 \, (\nabla_i \, \nabla_j \, g) \, (\nabla_i \, \nabla_j \triangle \, g) \right\} \ . \tag{2.32.b}$$

Both of the necessary conditions Eq. (2.26) and Eq. (2.32.a) for the irreducibility of G are not sufficient and they are inequivalent as shown by the two following examples taken for $n = 3$; see [18] Table 2 for the determination of their isotropy group, which is irreducible:

$$G = D_{3d} , \qquad g(\phi) = \left(3\phi_1^2 - \phi_2^2\right)\phi_2\phi_3 , \qquad \triangle g = 0 = g_{iikl} , \qquad (2.33.a)$$

$$g_{ipqr}g_{jpqr}\, \phi_i\,\phi_j = \tfrac{1}{8}\left(3\phi_1^2 + 3\phi_2^2 + 2\phi_3^2\right) . \qquad (2.33.b)$$

$$G = C_{4h} , \qquad g(\phi) = 2\left(\phi_1^2 - \phi_2^2\right)\phi_1\phi_2 + \phi_3^4 , \quad g_{iikl}\phi_k\phi_l = \phi_3^2 = \tfrac{1}{12}\triangle g , \qquad (2.33.c)$$
$$g_{ipqr}\, g_{jpqr}\, \phi_i\,\phi_j = (\phi,\,\phi) . \qquad (2.33.d)$$

It would be interesting to have a simple set of sufficient conditions on g for the irreducibility of the isotropy group $G = O(n)_g$. As we have seen Eq. (2.26) is sufficient for $n = 2$ and I conjecture that Eq. (2.26) and Eq. (2.32.a) are sufficient for $n = 3$. More generally, for harmonic polynomials $\triangle g = 0$, the set of necessary conditions $\triangle^{2k-1}g^k = 0$ for $1 \le k \le n-1$ might be sufficient. For $n = 4$ an exhaustive study has been done in [19]: including that of $O(4)$, there are 15 strata corresponding to irreducible isotropy groups, instead of 2 for $n = 2$ or 3. For $n = 4$, the maximal dimension of a subspace $\mathcal{T}_4^G$ for G irreducible is 11; all these subspaces are included in $\mathcal{T}_4^{(4)} \oplus \mathcal{T}_4^{(0)}$ of dimension 26. No similar results are known for higher n.

We end this section with some very important remarks: The isotropy groups of polynomials $g \in \mathcal{T}_4$ are closed subgroups of $O(n)$ and for any mathematical discussion such as those in this paper, one has to consider the isotropy group of $O(n)_g$, i.e., the exact invariance group of g. However this is generally not $\mathcal{G}$, the physical symmetry group of the Hamiltonian defined by g. Indeed $\mathcal{G}$ acts on $\mathcal{V}_n$ (the space of ϕ_i) through an orthogonal representation $\mathcal{G} \xrightarrow{V} O(n)$ of image $V(\mathcal{G})$. For instance, in the Landau theory of second order phase transition $\mathcal{G}$ is the space group (the crystallographic group) of the crystal and either the image $V(\mathcal{G})$ is a finite subgroup of $O(n)$, or it is an infinite discrete subgroup of $O(n)$, so its closure is a compact Lie subgroup of $O(n)$ of positive dimension. In general, because we consider only polynomials of degree 4, any $V(\mathcal{G})$ invariant polynomial of $\mathcal{T}_4$ will have an isotropy group G larger than $V(\mathcal{G})$.

To conclude we emphasize that the symmetry group $\mathcal{G}$ and its image $V(\mathcal{G})$ are given by the physics of the problem. They determine the general quartic term $g(\phi)$ of the Hamiltonian. In general $g(\phi)$ is a function of several parameters and $g(\phi) \in \mathcal{T}_4^G$ where, in general $G \supseteq V(\mathcal{G})$.

3. THE DEFINITION OF STABLE FIXED POINTS

As we have seen, the isotropy group $O(n)_g = G$ of the quartic term $g(\phi)$ of the Hamiltonian density (1) is the symmetry group of the Hamiltonian. It is a closed subgroup of $O(n)$, hence it is a Lie group of dimension $m < \binom{n}{2}$; in the particular case $m = 0$, G is a finite group.

Physically we need only to consider the case of irreducible subgroups $G \subset O(n)$. Indeed, when G is reducible, by choosing suitable linear combinations ϕ'_c of the components of the fields $\phi_k(x)$, one can split the Hamiltonian into a sum of non-interacting but similar Hamiltonians, each one with a field of $n^{(\alpha)}$ components, (with $\sum_\alpha n^{(\alpha)} = n$) and with isotropy subgroups G_α, irreducible subgroups of $O(n^{(\alpha)})$. So one has only to study the Hamiltonians with an irreducible symmetry group.

As explained in Section 1, the deduction of the renormalization equations (1.5) requires only the weaker hypothesis Eq. (1.4), equivalent to $g \in \mathcal{T}_4^{(0)} \otimes \mathcal{T}_4^{(4)}$. However, the renormalization equations (1.1) are equivariant for the whole action of $O(n)$ on $\mathcal{T}_4$.

$$\forall \, u \in O(n) \quad , \quad U(u)_{\alpha\beta}\beta_\beta(g_{\alpha'}) = \beta_\alpha \, (U_{\alpha'\beta'}g_{\beta'}) \, . \tag{3.1}$$

The quartic term $g(\phi)$ is the value of $g(\lambda)$ for a fixed value λ_0 of λ (e. g. $\lambda_0 = 1$). If G is its isotropy group, Eq. (3.1) implies that for every value of λ, $g(\lambda)$ is invariant under G. We can also say that the trajectory of $g(\lambda)$ stays in the space $\mathcal{T}_4^G$. More precisely, the isotropy group of $g(\lambda)$ has to be independent of λ in a neighborhood of λ_0 when $\frac{dg}{d\lambda} \neq 0$ and it may become larger at the fixed points g^*. The physical requirement of stability is the positivity of the restriction of the matrix $\frac{\partial \beta_\alpha}{\partial g_\beta}$ to the subspace $\mathcal{T}_4^G$.

This is expressed by the equation for g^* , the stable fixed point of g of isotropy group G:

$$\beta_\alpha(g^*) = 0 \, , \tag{3.2.a}$$

$$\frac{\partial \beta_\alpha}{\partial g_\beta}\Big|_{\mathcal{T}_4^G} \geq 0 \, . \tag{3.2.b}$$

Since G is the isotropy group of $g(\lambda_0) \in \mathcal{T}_4^G$, G is also the centralizer $C_{O(n)}(\mathcal{T}_4^G)$ (equality in Eq. (2.6)); so from Eq. (2.8), the normalizer of $\mathcal{T}_4^G$ is the normalizer of G in $O(n)$

$$N_{O(n)}(\mathcal{T}_4^G) = N_{O(n)}(G) \, . \tag{3.3}$$

So $N_{O(n)}(G)$ acts on $\mathcal{T}_4^G$ through the quotient group

$$Q(G) = N_{O(n)}(G)/G \ , \tag{3.4}$$

which acts effectively (i.e. no element of $Q(G)$ different from the identity leaves fixed every point of $\mathcal{T}_4{}^G$). When $Q(G)$ is not trivial, from a solution g^* of Eq. (3.2.a), by the action of $Q(G)$ on $\mathcal{T}_4{}^G$ one builds in general an orbit of solutions. This was already noted in [8] and [10]. If $Q(G)$ is a Lie group of positive dimension n' , that is the dimension of the orbit $Q(G)(g^*)$ of g^* and the tangent plane at g^* to this orbit is in the kernel of $\left.\frac{\partial \beta_\alpha}{\partial g_\beta}\right|_{\mathcal{T}_4{}^G}$. We will show later that the stable fixed point, when it exists, is unique.

We have now to take into account the fact that $\beta(g)$ is known only through an ϵ expansion $\beta(\epsilon,g)$, so the solutions of $\beta(\epsilon,g^*) = 0$ define g^* as a function of ϵ. Only the solutions $g^*(\epsilon) \rightarrow 0$ when $\epsilon \rightarrow 0$ are physically relevant. It is difficult to study the convergence of the ϵ expansion for $\epsilon = 4 - d = 1$ (generally it is not convergent; it is an asymptotic expansion). We completely ignore this problem here.

Assume the expansion

$$g^*(\epsilon) = \epsilon \sum_{k=0}^{\infty} \epsilon^k \bar{g}_k \ .$$

The first term $\bar{g}_0$ is defined by a non-linear equation.

$$0 = \bar{\beta}(\epsilon,\bar{g}_0) \longleftrightarrow - \bar{g}_0 + \tfrac{3}{2} \bar{g}_0 {}_\wedge \bar{g}_0 = 0 \ . \tag{3.5}$$

The other terms are defined by a system of linear equations. For instance, that for $\bar{g}_1$ reads:

$$\epsilon^{-2} \frac{d\beta}{dg}(\epsilon,\bar{g}_0) \ \bar{g}_1 = \tfrac{1}{2} \bar{g}_0 \ . \tag{3.6}$$

This solution is unique if

$$\left.\frac{d\bar{\beta}}{dg}(\epsilon,\bar{g}_0)\right|_{\mathcal{T}_4{}^G}$$

is invertible. When this is not the case, one says that there is a bifurcation: indeed, in general, new solutions appear. We shall see that this is the case for $n = 4$.

In this paper we consider the cases without bifurcation. Then the solutions found to the first order of ϵ can be computed to the next orders. For some range of $\epsilon \geq 0$, $g^*(\epsilon)$ keeps the same isotropy group and does not change its stability character. We are only interested in these properties of the solutions and not in their precise location. We need only to study the extrema of $\Phi^{(1)}(g)$ defined in Eq. (1.14).

To simplify notations we will drop the index $\bar{g}_0$ and use the shorthand notation

$$H(g^*) = \frac{d^2\Phi^{(1)}}{dg^2}(g^*) \tag{3.7}$$

for the Hessian of a fixed point. We can now reformulate the simplified mathematical problem we have to solve for finding stable points (from first order in the ϵ expansion) in the renormalization of Landau theory of second order phase transitions.

One is given a positive quartic polynomial on $\mathcal{V}_n$:

$$0 \neq \phi \in \mathcal{V}_n \ , \quad g(\phi) > 0 \ , \quad g(\lambda\phi) = \lambda^4 \, g(\phi) \tag{3.8}$$

with an irreducible isotropy group $G = O(n)_g$. Find the extrema g^* of $\Phi^{(1)}(g)$ (defined in Eq. (1.11)) on $\mathcal{T}_4^G$. They are defined by:

$$g^* = \epsilon \, \tilde{g} \ , \quad \tilde{g}_\vee \tilde{g} = \tfrac{2}{3}\tilde{g} \ , \quad \tilde{g} \in \mathcal{T}_4^G \ . \tag{3.9}$$

Such an extremum is a stable fixed point if and only if the Hessian at g^* is strictly positive.

$$\epsilon > 0 \qquad H(g^*)\big|_{\mathcal{T}_4^G} > 0 \ . \tag{3.10}$$

Moreover this stable fixed point is physically acceptable for the given polynomial g if it is in the attraction basin of g^*, i.e., $\phi^{(1)}(g)$ never increases from g to g^* on the integral line of the gradient field. Finally one has also to verify that $g^*(\phi) > 0$ for $\epsilon \neq 0$.

4. THE $O(N)$ COVARIANT SYMMETRIC ALGEBRA $\mathcal{T}_4$ AND ITS IDEMPOTENTS.

Equation (3.9) means that the fixed points g^* are idempotents of the algebra defined by the symbol $\vee$. In this section we study some properties of this algebra similar to those studied in [12]. For the linear representation $O(n)$, $u \mapsto U(u)$ on $\mathcal{T}_4$, the expression:

$$\Theta(g) = g_{ijkl} \, g_{klpq} \, g_{pqij} = \Theta(U(u)g) \tag{4.1}$$

is a third degree polynomial invariant; by polarization we define a trilinear form

$$\tilde{\Theta}(u,v,w) = \tfrac{1}{6}\big[\Theta(u+v+w) - \Theta(u+v) - \Theta(v+w) - \Theta(w+u) + \Theta(u) + \Theta(v) + \Theta(w)\big] \tag{4.2.a}$$

$$= \big(u_{ijkl} \, v_{klpq} \, w_{pqij} + u_{ijkl} \, v_{klpq} \, w_{pqij}\big) \ , \tag{4.2.b}$$

which is invariant under any permutation of its three arguments and which is $O(n)$ invariant. If we fix u and v, $\tilde{\Theta}(u,v,w)$ is a linear form in w; so with the $O(n)$ invariant scalar product (g,g) defined in Eq. (1.12), it can be written as the scalar product of w by a fixed vector that we denote by $u_{\vee}v$:

$$\tilde{\Theta}(u,v,w) = (u_{\vee}v, w) \ . \tag{4.3}$$

The correspondence from the pair u, v to $u_{\vee}v$ is a linear map:

$$\mathcal{T}_4 \otimes \mathcal{T}_4 \xrightarrow{\vee} \mathcal{T}_4 \ ,$$

which defines an algebra on $\mathcal{T}_4$:

$$(u_{\vee}v)_{ijkl} = \frac{1}{6} \left\{ u_{ijpq} v_{pqkl} + v_{ijpq} u_{pqkl} + u_{ikpq} v_{pqjl} + v_{ikpq} u_{pqjl} \right.$$
$$\left. + u_{ilpq} v_{pqjk} + v_{ilpq} u_{pqjk} \right\} \ . \tag{4.4}$$

The symmetry of $\tilde{\Theta}$ in its argument implies:

$$(u_{\vee}v, w) = (v, u_{\vee}w) = (u, v_{\vee}w) = etc \ldots \tag{4.5}$$

For each $g \in \mathcal{T}_4$ we can define a linear operator $D_g \in \mathcal{L}(\mathcal{T}_4)$, on $\mathcal{T}_4$ by:

$$D_g w = g_{\vee}w \ .$$

The first equality of Eq. (4.5) shows that D_g is a symmetric operator

$$D_g^T = D_g \ , \tag{4.6}$$

and from the $O(n)$ invarinace of $\Theta(g)$, one proves that the linear map

$$\mathcal{T}_4 \xrightarrow{D} \mathcal{L}(\mathcal{T}_4) \tag{4.7}$$

is $O(n)$ covariant:

$$D_{U(u)g} = U(u) D_g U(u)^{-1} \ . \tag{4.8}$$

The Hessian $H(g)$ is simply:

$$\frac{d^2\Phi(g)}{dg^2} = H(g) = -\epsilon I + 3 D_g \ . \tag{4.9}$$

To compute easily with this algebra on fourth degree homogeneous polynomials $g(\phi)$ one can introduce the $n \times n$ matrix

$$T_g(\phi)_{ij} = \frac{1}{12} \frac{\partial^2 g}{\partial \phi_i \partial \phi_j} ,$$ (4.10)

which is quadratic in ϕ. Note that:

$$g(\phi) = T_g(\phi)_{ij} \, \phi_i \, \phi_j ,$$ (4.11)

$$T_g(\phi)_{ii} = \operatorname{Tr} T_g(\phi) = \frac{1}{12} \triangle g(\phi) .$$ (4.12)

The algebra product of g and h is simply:

$$g \vee h = \operatorname{Tr} T_g T_h .$$ (4.13)

As an example we easily compute for $s(\phi) = (\phi, \phi)^2$ (see Eq. (2.22.a))

$$T_s(\phi)_{ij} = \frac{1}{3} \left(\delta_{ij}(\phi, \phi) + 2 \, \phi_i \, \phi_j \right)$$ (4.14)

so

$$s \vee s = \frac{n + 8}{9} s$$ (4.15)

and more generally with

$$g = g^{(0)} + g^{(2)} + g^{(4)} ,$$ (4.16)

(see Eq. (2.18) for this $O(n)$ covariant decomposition)

$$D_s g = s \vee g = \frac{n + 8}{9} g^{(0)} + \frac{n + 16}{18} g^{(2)} + \frac{2}{3} g^{(4)} .$$ (4.17)

Eq. (4.15) implies that $\Phi^{(1)}(g)$ has an extremum for

$$s^* = \epsilon \, \tilde{s} \qquad , \qquad \tilde{s} = \frac{6}{n + 8} s ,$$ (4.18)

and from Eq. (4.17) we obtain its Hessian

$$H(s^*) = \epsilon \left(P^{(0)} + \frac{6}{n + 8} P^{(2)} + \frac{4 - n}{n + 8} P^{(4)} \right)$$ (4.19)

where $P^{(x)}$ is the orthogonal projector on $\mathcal{T}_4^{(x)}$, $x = 0, 2, 4$. Since $\mathcal{T}_4^{O(n)} = \mathcal{T}_4^{(0)}$, the fixed point s^* *is a stable fixed point for any n*, a well known result. For $n < 4$, s^*

is even a minimum on the whole space $\mathcal{T}_4$. We can prove the (purely mathematical) result, independent of the value of n:

Theorem 4.1: $\Phi^{(1)}(g)$ has no minimum $\neq s^*$ on $\mathcal{T}_4$ or on $\mathcal{T}_4^{(4)} + \mathcal{T}_4^{(0)}$. We first establish some general relations. From Eq. (4.5) and Eq. (4.17):

$$\gamma(g \vee g)n = (s, g \vee g) = (s \vee g, g) =$$
$$= \frac{n+8}{9}\left(g^{(0)}, g^{(0)}\right) + \frac{n+16}{18}\left(g^{(2)}, g^{(2)}\right) + \frac{2}{3}\left(g^{(4)}, g^{(4)}\right), \qquad (4.20.a)$$

$$\gamma(g \vee g)n = \frac{2}{3}(g, g) + \frac{n+2}{9}\left(g^{(0)}, g^{(0)}\right) + \frac{n+4}{18}\left(g^{(2)}, g^{(2)}\right). \qquad (4.20.b)$$

For an extremum $g^* = \epsilon \bar{g}$ not collinear to s, (i.e. $\left(g^{(2)}, g^{(2)}\right) + \left(g^{(4)}, g^{(4)}\right)$ > 0) with the use of Eq. (3.5) and Eq. (2.25.b) we obtain $\gamma(\bar{g})\{\gamma(\bar{s}) - \gamma(\bar{g})\}$ > 0, i.e.,

$$0 < \gamma(\bar{g}) < \gamma(\bar{s}) = 2\frac{n+2}{n+8}. \qquad (4.21)$$

The Hessian at g^* is, from Eq. (4.9):

$$H(g^*) = \epsilon\left\{3\,D_{\bar{g}}(0) - I + 3\,D_{\bar{g}}(2) + 3\,D_{\bar{g}}(4)\right\}$$
$$= \epsilon\frac{\gamma(\bar{g})}{\gamma(\bar{s})}H(\bar{s}) - \epsilon\left\{\left(1 - \frac{\gamma(\bar{g})}{\gamma(\bar{s})}\right)I - 3\,D_{\bar{g}}(2) - 3\,D_{\bar{g}}(4)\right\}. \qquad (4.22)$$

The trace of the product of 2 positive operators is > 0; for $n \geq 4$ we will show that $\mathrm{Tr}\,H(\epsilon\bar{g})P^{(4)} < 0$ so that $H(\epsilon\bar{g})P^{(4)}$ is not a positive matrix and this will prove the theorem for $n \geq 4$.

We just remark that $\mathrm{Tr}\,D_g$ and $\mathrm{Tr}\,\{D_g P^{(4)}\}$ are linear forms on $\mathcal{T}_4$; moreover Eq. (4.8) and the fact that $\mathcal{T}_4^{(4)}$ is an invariant space for $O(n)$ imply that these linear forms are $O(n)$ invariant. So they must be proportional to (s, g). This requires $\mathrm{Tr}\,\{D_{g^{(2)}}P^{(4)}\}$ $= 0$ and $\mathrm{Tr}\,\{D_{g^{(4)}}P^{(4)}\} = 0$; so Eq. (4.22) with Eq. (4.19) and Eq. (4.21) imply that $\mathrm{Tr}\,\{H(g^*)P^{(4)}\} < 0$ for $n \geq 4$. We did not need to compute the proportionality factors K' and K'' in $\mathrm{Tr}\,D_g = K'(s,g)$, $\mathrm{Tr}\,\{D_g P^{(4)}\} = K''(s,g)$. They can be computed from $g = s$ and the use of Eq. (4.19) and Eq. (2.17). One finds

$$\mathrm{Tr}\,D_g = \frac{(n+2)(n+3)}{12}(s, g), \qquad (4.23.a)$$

$$\mathrm{Tr}\,D_g P^{(4)} = -\frac{(n+6)(n+1)(n-1)(n-4)}{8(n+2)(n+4)}(s, g), \qquad (4.23.b)$$
and similarly

$$\frac{(s, g \vee s)}{(s, s)} = \mathrm{Tr}\,D_g P^{(0)} = \frac{n+8}{3\,n(n+2)}. \qquad (4.23.c)$$

The proof of the theorem when $n < 4$ will be a simple consequence of Theorem 5.2 in the next section.

5. EXTREMA OF $\Phi^{(1)}$

Let $g^* = \epsilon \tilde{g}$ be an extremum of $\Phi^{(1)} = -\frac{\epsilon}{2}(g, g) + \frac{1}{2}(g \vee g, g)$. It satisfies $\tilde{g} \vee \tilde{g} = \frac{2}{3}\tilde{g}$ so:

$$\Phi^{(1)}(g^*) = -\frac{\epsilon^3}{6}(\tilde{g}, \tilde{g}) . \tag{5.1}$$

From Eq. (4.20.b) we deduce:

$$(\tilde{g}, \tilde{g}) = \frac{n}{2}\gamma(\tilde{g})(2 - \gamma(\tilde{g})) - \frac{n+4}{12}(g^{(2)}, g^{(2)}) . \tag{5.2}$$

Assume we have a second extremum $h^* = \epsilon \tilde{h}$ and consider the restriction of $\Phi^{(1)}$ to the straight line containing g^* and h^*. It is a third degree polynomial in λ:

$$\psi(\lambda) = \Phi^{(1)}((1 - \lambda) g^* + \lambda h) = \frac{\epsilon^3}{6}\left\{ \left[(\tilde{h}, \tilde{h}) - (\tilde{g}, \tilde{g})\right]\lambda^2(2\lambda - 3) - (\tilde{g}, \tilde{g})\right\} . \tag{5.3}$$

Since $\lambda = 0$ and $\lambda = 1$ correspond respectively to the extrema g^* and h^* of $\Phi^{(1)}$ these values must be extrema of $\psi(h)$. Indeed:

$$\frac{d\psi}{d\lambda} = \epsilon^3\left[(\tilde{h}, \tilde{h}) - (\tilde{g}, \tilde{g})\right]\lambda(1 - \lambda) . \tag{5.4}$$

We know that a third degree polynomial has no other extrema. So when $(\tilde{h}, \tilde{h}) = (\tilde{g}, \tilde{g})$ we verify from Eq. (5.3) that ψ is constant. Then we can prove that the direction $\tilde{g} - \tilde{h}$ does not correspond to a zero eigenvalue of the Hessian $H(g^*)$ or $H(h^*)$. Indeed, assume that:

$$0 = H(g^*)(\tilde{g} - \tilde{h}) = \epsilon(3 D_{\tilde{g}} - I)(\tilde{g} - \tilde{h}) = \epsilon(\tilde{g} + \tilde{h} - 3\tilde{g} \vee \tilde{h}) = 0 . \tag{5.5}$$

By scalar multiplication with $\tilde{g}$ and $\tilde{h}$ one obtains $(\tilde{g}, \tilde{g}) = (\tilde{h}, \tilde{h}) = (\tilde{g}, \tilde{h})$, i.e., $\tilde{g} = \tilde{h}$ which is absurd. So $\tilde{g} - \tilde{h}$ is not an eigenvector with zero eigenvalue of $H(g^*)$ or $H(h^*)$ although the expectation value of these operators on $\tilde{g} - \tilde{h}$ vanishes when $(\tilde{g}, \tilde{g}) = (\tilde{h}, \tilde{h})$:

$$(\tilde{g} - \tilde{h}, H(g^*)(\tilde{g} - \tilde{h})) = \epsilon\left[(\tilde{g}, \tilde{g}) - (\tilde{h}, \tilde{h})\right] = -(\tilde{g} - \tilde{h}, H(h^*)(\tilde{g} - \tilde{h})) . \tag{5.6}$$

We can therefore conclude, if $G = O(n)_{\tilde{g}} \cap O(n)_{\tilde{h}}$

Lemma 5.1: If two extrema of $\Phi^{(1)}$ on $\mathcal{T}_4{}^G$ have the same length, they are not minima of $\Phi^{(1)}\big|_{\mathcal{T}_4{}^G}$.

Of course neither are they minima of the whole polynomial $\Phi^{(1)}$. When $(\tilde{g}, \tilde{g}) \neq (\tilde{h}, \tilde{h})$, since $\psi(\lambda)$ has no other extrema than 0 and 1, the extremum with the shortest length is in the attraction basin of that with the biggest length (see Eq. (5.1)). This is true for any pair of extrema. This discussion establishes the following theorem:

> **Theorem 5.2:** For any subspace $\mathscr{E} \in \mathscr{T}_4$, if $\Phi^{(1)}$ has a minimum on $\mathscr{E}$, this minimum is unique and any other extremum of $\Phi^{(1)}$ on $\mathscr{E}$ is on the boundary of the attractor basin of this minimum.

This completes the proof of Theorem 4.1 for $n < 4$ since in that case $\tilde{s}$ is a minimum. We also remark that for $n \leq 4$, $\gamma(\tilde{s}) \leq 1$ (see Eq. (4.21)) and from $\gamma(\tilde{g}) < \gamma(\tilde{s})$ and Eq. (4.21) we deduce $(\tilde{g}, \tilde{g}) < (\tilde{s}, \tilde{s})$ for any extremum $\epsilon \, \tilde{g}$ when $n \leq 4$.

The discussion in Section 2 on the action of the normalizer $N_{O(n)}(\mathscr{T}_4^G)$ on $\mathscr{T}_4^{(G)}$ and the equalities (2.6) and (2.8) gives the following addition to Theorem 2.

> **Corollary 5.3:** If g is an isotropy group on $\mathscr{T}_4^G$ and if $\Phi^{(1)}\big|_{\mathscr{T}_4^G}$ has a minimum, this minimum is unique and its invariant group is $\supseteq N_{O(n)}^4(G)$, the normalizer of G in $O(n)$.

The interesting question would be to decide for which conjugate classes $[G]$ of subgroups of $O(n)$, and more specifically for which conjugate classes of irreducible strict subgroups of $O(n)$, $\Phi^{(1)}\big|_{\mathscr{T}_4^G}$ has a minimum. Indeed this minimum yields a stable fixed point $\tilde{g}(\epsilon)$ of the renormalization problem when its Hessian is not degenerate; (when $detH(g^*) = 0$, a study of the bifurcation has to be done). For $n = 2$ or 3 we have seen that the class of irreducible strict subgroups is unique, it is $[B_n]$ and , as is well known, the isotropic fixed point is the only stable one. When $n = 4$, the list of the classes of irreducible subgroups which are isotropy groups on $\mathscr{T}_4$ is known [19]; we have shown here that $\tilde{s}$ is still the only minimum of $\Phi^{(1)}$, but it is degenerate. We are studying the case $n = 4$ in collaboration with J. C. and P. Toledano. When $n > 4$, we have seen that $\tilde{s}$ is never a minimum of $\Phi^{(1)}\big|_{\mathscr{T}_4^G}$ for G irreducible. In the next section we will construct a family of irreducible G such that $\Phi^{(1)}\big|_{\mathscr{T}_4^G}$ has a minimum when $n > 4$ and is not prime. For all $n > 4$ and $G = B_n$, there is a well known cubic invariant minimum.

To end this section we consider the case of a pair of extrema $\epsilon \, \tilde{u}$, $\epsilon \, \tilde{v}$ such that $\tilde{u} \vee \tilde{v}$ is a linear combination of $\tilde{u}$ and $\tilde{v}$:

$$\tilde{u} \vee \tilde{u} = \tfrac{2}{3} \, \tilde{u} \, , \qquad\qquad \tilde{v} \vee \tilde{v} = \tfrac{2}{3} \, \tilde{v} \, , \qquad\qquad \tilde{u} \vee \tilde{v} = \tfrac{1}{3} \left(\alpha \, \tilde{u} + \beta \, \tilde{v} \right) . \quad (5.7)$$

The scalar product with $\tilde{u}$ and $\tilde{v}$ yields

$$\alpha(\tilde{u}, \tilde{u}) = (2 - \beta) (\tilde{u}, \tilde{v}) , \tag{5.8.a}$$

$$\beta(\tilde{v}, \tilde{v}) = (2 - \alpha) (\tilde{u}, \tilde{v}) . \tag{5.8.b}$$

We note $\tilde{u} = \tilde{v} \nrightarrow 2 - \alpha - \beta \doteq 0$. The exact converse is not true, but one easily verifies that:

$$\alpha \beta \neq 0 \quad and \quad 2 - \alpha - \beta = 0 \quad \rightarrow \tilde{u} = \tilde{v} . \tag{5.9}$$

The Schwartz inequality yields from Eq. (5.9):

$$\alpha \beta \neq 0 \quad and \quad \tilde{u} \neq \tilde{v} \nrightarrow \alpha \beta (2 - \alpha - \beta) = 0 . \tag{5.10}$$

Moreover, when $\alpha \beta \neq 1$ there is a third extremum $\tilde{w}^* = \epsilon \tilde{w}$ in the 2-plane $\tilde{u}$, $\tilde{v}$:

$$\tilde{w} = \zeta \tilde{u} + \eta \tilde{v} , \qquad \zeta = \frac{1 - \alpha}{1 - \alpha \beta} , \qquad \eta = \frac{1 - \beta}{1 - \alpha \beta} . \tag{5.11}$$

This equation is also valid for $\alpha = \beta = 0$. Indeed, the three cases $\alpha = \beta = 0$; $\alpha = 0$, $\beta = 2$; and $\alpha = 2$, $\beta = 0$ correspond to the same 2-plane:

$$\tilde{u}_{\vee}\tilde{v} = 0 = (\tilde{u}, \tilde{v}) \quad , \quad \tilde{w} = \tilde{u} + \tilde{v} . \tag{5.12}$$

For any extremum g^* of $\Phi^{(1)}(g)$ we have

$$H(g^*)g^* = \epsilon g^* . \tag{5.13}$$

It is easy to compute the other eigenvalue and eigenvector in the 2-plane spanned by u^*, v^*, and w^* of the respective Hessians:

$$H(u^*) (\alpha \tilde{u} + (\beta - 2) \tilde{v}) = \epsilon (\beta - 1) (\alpha \tilde{u} + (\beta - 2) \tilde{v}) , \tag{5.14.a}$$

$$H(v^*) ((\alpha - 2) \tilde{u} + \beta \tilde{v}) = \epsilon (\alpha - 1) ((\alpha - 2) \tilde{u} + \beta \tilde{v}) , \tag{5.14.b}$$

$$H(w^*) (\alpha \tilde{u} - \beta \tilde{v}) = \epsilon \frac{(1 - \alpha) (1 - \beta)}{1 - \alpha \beta} (\alpha \tilde{u} - \beta \tilde{v}) . \tag{5.14.c}$$

We will see several examples of such 2-planes in the next section. Here we consider a 2-plane which contains $\tilde{s}$ and we use the notation x^*, y^*, s^* instead of u^*, v^* and w^* in order to distinguish this particular case. We denote by $\mathcal{E}_x$ this 2-plane and we require moreover

$$\mathscr{E}_x \subset \mathscr{E} = \mathscr{T}_4^{(0)} + \mathscr{T}_4^{(4)} \tag{5.15}$$

since we are interested in extrema invariant under an irreducible subgroup $G \subset O(n)$. We will also use the direct sum of orthogonal subspaces

$$\mathscr{E} = \mathscr{E}_x \oplus \mathscr{E}_x^\perp \ . \tag{5.16}$$

Then we get from Eq. (4.17) and Eq. (4.18)

$$g \in \mathscr{E} \quad \bar{s} \underset{\vee}{} g = \tfrac{1}{3} \left(\gamma(g)\bar{s} + \tfrac{12}{n+8} g \right) \ . \tag{5.17}$$

If we apply $D_{\bar{x}}$ and $D_{\bar{y}}$ to $\bar{s} = \zeta \bar{x} + \eta \bar{y}$ and use $\bar{x} \underset{\vee}{} \bar{y} = \tfrac{1}{3} \left(\alpha \bar{x} + \beta \bar{y} \right)$ we obtain $\alpha = \gamma(\bar{y})$, $\beta = \gamma(\bar{x})$, i.e.,

$$\bar{x} \underset{\vee}{} \bar{y} = \tfrac{1}{3} \left(\gamma(\bar{y}) \, \bar{x} + \gamma(\bar{x}) \, \bar{y} \right) \tag{5.18}$$

and

$$\zeta + \eta = \frac{12}{n+8} = \frac{2 - \gamma(\bar{x}) - \gamma(\bar{y})}{1 - \gamma(\bar{x}) \, \gamma(\bar{y})} \ . \tag{5.19}$$

From the inequalities (4.21) and (5.10), and from the preceding equation, we obtain the inequalities:

$$0 < \gamma(\bar{x}) + \gamma(\bar{y}) < \mathrm{Inf} \left(2, \frac{4 \, (n+2)}{n+8} \right) , \tag{5.20.a}$$

$$0 < \gamma(\bar{x}) \, \gamma(\bar{y}) < 1 \ . \tag{5.20.b}$$

Eq. (5.19) can also be written equivalently:

$$(\gamma(\bar{x}) - 1) \, (\gamma(\bar{y}) - 1) = - \frac{n-4}{n+8} \left(1 - \gamma(\bar{x}) \, \gamma(\bar{y}) \right) \ . \tag{5.21}$$

When $n = 4$ either $\gamma(\bar{x}) = 1$ or $\gamma(\bar{y}) = 1$ which implies that $\eta = 0$ or $\zeta = 0$, hence

Lemma 5.4: For $n = 4$, any 2-plane containing $\bar{s}$ can contain at most one other extremum.

When $n \neq 4$, any extremum $\epsilon \bar{x}$ defines a 2-plane $\mathscr{E}_x$ containing two other extrema $\epsilon \bar{s}$ and $\epsilon \bar{y}$. When $n < 4$ we verify again that $\epsilon \bar{s}$ is the minimum of $\Phi^{(1)}\big|_{\mathscr{E}_x}$. When $n > 4$, Eq. (5.20.b) implies that the right hand side of Eq. (5.21) is negative so either $\gamma(\bar{x}) - 1$ or $\gamma(\bar{y}) - 1$ is positive and equations (5.14.a), (5.14.b) and (5.14.c) show that

Lemma 5.5: When $n > 4$, either $\epsilon \tilde{x}$ or $\epsilon \tilde{y}$ is the minimum of $\Phi^{(1)}\big|_{\mathscr{C}_x}$.

We can have more knowledge of the Hessians of $\epsilon \tilde{x}$ and $\epsilon \tilde{y}$. Indeed, from $\tilde{s} = \zeta \, \tilde{x} + \eta \, \tilde{y}$ we have a linear relation among the Hessians

$$\zeta \, H(\epsilon \tilde{x}) + \eta \, H(\epsilon \tilde{y}) = H(\epsilon \tilde{s}) + \epsilon I(1 - \zeta - \eta) . \tag{5.22}$$

From Eq. (4.19) and Eq. (5.19), and the values of ζ and η given by Eq. (5.11) and Eq. (5.18) we obtain by projection on $\mathscr{C}_x^{\perp}$:

$$(\gamma(\tilde{y}) - 1) \, H(\epsilon \tilde{x})\big|_{\mathscr{C}_x^{\perp}} + (\gamma(\tilde{x}) - 1) \, H(\epsilon \tilde{y})\big|_{\mathscr{C}_x^{\perp}} = 0 \tag{5.23}$$

The two Hessians are proportional on $\mathscr{C}_x^{\perp}$ and the proportionality factor is positive when $n > 4$.

Let us apply these results to the 2-plane $\mathscr{T}_4{}^{B_n}$ spanned by $c = \sum_i \phi_i^4$ and $s = (\phi, \phi)^2$. Note that:

$$(c, c) = (s, c) = n \quad , \quad \gamma(c) = 1 \quad , \quad c_{\vee} c = c \quad , \quad s_{\vee} c = \tfrac{1}{3}(s + 2c) . \tag{5.24}$$

Hence $\bar{c} = \dfrac{2}{3} \, c$

$$\bar{c}' = \frac{2}{n} \, s + \frac{n-4}{3} \, c \quad \longleftrightarrow \quad \bar{s} = \frac{3}{n+8} \left\{ (4 - n) \, \bar{c} + n \, \bar{c}' \right\} . \tag{5.25}$$

When $n > 4$, $\epsilon c'$ is the minimum of $\Phi^{(1)}$ in $\mathscr{C}_c = \mathscr{T}_4{}^{B_n}$; indeed

$$H(c'^*) \, (s - 2c) = \epsilon \, \frac{n-4}{3n} \, (s - 2c) . \tag{5.26}$$

We can also write the whole spectrum of $H(c^*)$ and $H(c'^*)$ on $\mathscr{T}_4$. Indeed, let us denote by $\mathscr{C}_{(4)}$, $\mathscr{C}_{(3,1)}$, $\mathscr{C}_{(2,2)}$, $\mathscr{C}_{(2,1,1)}$, $\mathscr{C}_{(1,1,1,1)}$ the subspaces of $\mathscr{T}_4$ defined by the following properties of the values of indices of g_{ijkl} corresponding to non-vanishing values of this tensor: the four indices are equal, only three are equal, two different pairs of equal indices, only two equal indices, the four indices are different. We give below the dimension of these spaces and the eigenvalues of $H(c^*)$:

space:	$\mathscr{C}_{(4)}$	$\mathscr{C}_{(3,1)}$	$\mathscr{C}_{(2,2)}$	$\mathscr{C}_{(2,1,1)}$	$\mathscr{C}_{(1,1,1,1)}$
dimension:	n	$n(n-1)$	$\dfrac{n(n-1)}{2}$	$\dfrac{n(n-1)}{2}$	$\dbinom{n}{4}$
eigenvalues of $H(c^*)$:	ϵ	0	$-\dfrac{\epsilon}{3}$	$-\dfrac{2\epsilon}{3}$	$-\epsilon$

$H(c'^*)$ has the same eigenspaces in $\mathcal{E}_c$ and the eigenvalues are multiplied by $(n-4)/n$. Since B_n is finite, the dimension of the orbit of these extrema is $n(n-1)/2$. This is half the dimension of Ker $H(c^*)$ = Ker $H(c'^*)$; so these extrema also have an accidental degeneracy of dimension $n(n-1)/2$.

6. THE ISOTROPY GROUPS $G_{r_1, r_2, \ldots, r_k}$, $N = \prod_{i=1}^{k} r_i$, AND THEIR INVARIANT POLYNOMIALS.

Consider the polynomials $x_{pq} \in \mathcal{E} = \mathcal{T}_4^{(0)} + \mathcal{T}_4^{(4)}$ with

$$p \, q = n \qquad x_{p,q} = \sum_{j=1}^{q} \left\{ \sum_{i=1}^{p} \phi_{ij}^2 \right\}^2 . \tag{6.1}$$

For each value of j, $\left\{ \sum_{i=1}^{p} \phi_{ij}^2 \right\}^2$ has $O(p)$ as isotropy group; with the summation over j the isotropy group is the semidirect product

$$\Gamma_{p,q} = O(p)^q \quad \Pi_q \tag{6.2}$$

where $O(p)^q$ is the direct product of q factors isomorphic to $O(p)$ and Π_q is the permutation group of q objects. Particular cases are:

$$\Gamma_{n,1} = O(n) \quad , \quad \Gamma_{1,n} = B_n . \tag{6.3}$$

The group $\Gamma_{p,q}$ is realized as a subgroup of $O(n)$. The $n \times n$ orthogonal matrices are made of q^2 blocks of $p \times p$ submatrices; $O(p)^q$ has all its blocks zero except the diagonal ones and each of these diagonal blocks is a $p \times p$ orthogonal matrix $\in O(p)$. The permutation group Π_q is represented by matrices which have only q blocks different from zero, one per row and per column and each of these non-zero blocks is equal to the matrix I_p, the $p \times p$ unit matrix. We will prove later that the groups $\Gamma_{p,q}$ are irreducible subgroups of $O(n)$. We prove now that

$$\dim \mathcal{T}_4^{\Gamma_{p,q}} = 2 . \tag{6.4}$$

To be invariant under $O(p)^q$, the quartic polynomial $g(\phi)$ has to be a quadratic polynomial in the quadratic forms

$$Q_j = \sum_{i=1}^{p} \phi_{ij}^2 .$$

The quadratic invariants of the group Π_q of permutations of the Q_j are known, they are

$$s = \left\{ \sum_{j=1}^{q} \varrho_j \right\}^2 \quad \text{and} \quad x_{p,q} = \sum_{j=1}^{q} \varrho_j^2 \; .$$

It is convenient for our purposes to consider the n dimensional space $\mathcal{V}_n$ as a tensor product, $\mathcal{V}_n = \mathcal{V}_p \otimes \mathcal{V}_q$ so the coordinates ϕ_{ij} can be written as $\phi_{ij} = \rho_i \otimes \sigma_j$. Then:

$$x_{p,q} = \left\{ \sum_i \rho_i^2 \right\}^2 \otimes \sum_j \sigma_j^4 = s_p \otimes c_q \quad , \quad x_{n,1} = s \quad , \quad x_{1,n} = c \; ; \tag{6.5}$$

similarly one finds

$$s = s_p \otimes s_q \quad , \quad c = c_p \otimes c_q \; . \tag{6.6}$$

Then it is easy to compute:

$$(x_{p,q}, x_{p,q}) = \frac{p\,q(p+2)}{3} = (s, x_{p,q}) \quad , \quad\quad (c, x_{p,q}) = p\,q \; , \tag{6.7}$$

$$x_{p,q} \vee x_{p,q} = \frac{p+8}{9} \, x_{p,q} \; , \tag{6.8.a}$$

$$s \vee x_{p,q} = \frac{p+2}{9} \, s + \frac{2}{3} \, x_{p,q} \quad , \quad c \vee x_{p,q} = \frac{1}{3} \, x_{p,q} + \frac{2}{3} \, c \; . \tag{6.8.b}$$

In the particular cases $p = 1$ or $q = 1$ we find Eq. (5.24). We can also generalize these equations for different pairs p,q; e.g.

$$(x_{p,qr}, x_{pq,r}) = p\,q\,r\,\frac{p+2}{3} \quad , \quad\quad\quad x_{p,qr} \vee x_{pq,r} = \frac{2}{3} \, x_{p,qr} + \frac{p+2}{9} \, x_{pq,r} \; . \tag{6.9}$$

Now we can consider a family of groups $\Gamma_{p,q}$, $pq = n$ and their intersections. Let

$$r_1 \, r_2 \cdots r_k = n = \prod_{i=1}^{k} r_i \tag{6.10}$$

be a decomposition of n in k factors, not necessarily prime, listed in a fixed order. With $r_0 = 1$, define

$$p_l = \prod_{i=1}^{l} r_i \quad , \quad q_l = n/p_l \quad , \quad \text{so that} \quad p_0 = q_k = 1 \; . \tag{6.11}$$

The group $G_{r_1, \ldots r_k}$ is defined by:

$$G_{r_1, \ldots r_k} = \bigcap_{0 \le i \le k} \Gamma_{p_i, q_i} \; . \tag{6.12}$$

We could have defined it by recursion:

$$G_{r_1} = B_{r_1} \quad , \quad G_{r_1, r_2, \ldots, r_k} = \left(G_{r_1, r_2, \ldots, r_{k-1}} \right)^{r_k} \; \Pi_{r_k} \; . \tag{6.13}$$

The order of this subgroup of B_n is:

$$. \ |G_{r_1, r_2, \ldots, r_k}| = 2^n \prod_{i=1}^{k} (r_i !)^{q_i} . \tag{6.14}$$

It is interesting to note that for $r_1 = r_2 = \ldots = r_k = 2$,

$$G_{2,2, \ldots, 2} = \mathrm{Syl}_2 (B_2 k) , \tag{6.15}$$

i.e., $G_{2,2, \ldots, 2}$ is isomorphic to the Sylow-2 groups[1] of the Coxeter group $B_2 k$.

We verify that $G_{r_1, r_2, \ldots, r_k}$ is an irreducible subgroup of $O(n)$. Indeed it contains the Abelian subgroup $\triangle_n$ of diagonal matrices with elements ± 1. The matrices which commute with $\triangle_n$ are the diagonal matrices. We see that the permutation matrices of $G_{r_1, r_2, \ldots, r_k}$ act transitively on the basis $\{\phi_i\}$ of the n dimensional space $\mathcal{V}_n$; indeed the ordered set of n basis vectors can be decomposed into nested sets of $p_1, p_2, \ldots, p_{k-1}$ elements for the different levels of nesting. The permutations of Π_{r_k} permute the p_{k-1}-element sets[2]; inside one such set $\Pi_{r_{k-1}}$ is the group of permutations of the p_{k-2} element sets and so on. So ϕ_1 can be sent to the place of any ϕ_a, $1 \leq a \leq$ n. By conjugation the permutation matrices permute the diagonal elements of diagonal matrices; so the only matrices which commute with all elements of $G_{r_1}, \ldots, r_k$ are multiples of the identity, i.e., $G_{r_1}, \ldots, r_k$ is irreducible. This is also true of all $O(n)$ subgroups which contain it. So the subgroups $\Gamma_{p,q}$ are also irreducible.

7. THE STABLE FIXED POINTS OF THE POLYNOMIAL $\displaystyle\sum_{i=0}^{k} \lambda_i \, x_i(\phi)$.

We use a still more condensed notation

$$0 \leq i \leq k , \quad x_i = x_{p_i q_i}(\phi) , \quad x_0 = c , \quad x_k = s . \tag{7.1}$$

where p_i and q_i are defined in Eq. (6.11) and $x_{p,q}(\phi)$ in Eq. (6.1). The polynomials x_i span the space $\mathcal{T}_4^{G_{r_1, \ldots, r_k}}$ whose dimension is

$$\dim \mathcal{T}_4^{G_{r_1 \ldots r_k}} = k + 1 . \tag{7.2}$$

Eq. (6.9) can be written in the present notation:

---

[1]Given a finite group whose order (that is, the number of elements) is $|G|$, let $|G| = 2^{l_2} 3^{l_3} \ldots p^{l_p}$ be the decomposition of $|G|$ into prime factors. Then all subgroups of G of order p^{l_p} are conjugated and are called Sylow p subgroups of G.

[2]That is they transform any basis vector into any other.

$$(x_i , x_j) = n \frac{p_i + 2}{3} \ , \quad i \le j \ , \tag{7.3.a}$$

$$x_i \wedge x_j = \frac{2}{3} x_i + \frac{p_i + 2}{9} x_j \ , \quad i \le j \ . \tag{7.3.b}$$

For $i = j$ this equation shows that the x_i are idempotents; they yield the fixed points:

$$x_i^* = \epsilon \tilde{x}_i = \epsilon \frac{6}{p_i + 8} x_i \ , \qquad \tilde{x}_i \wedge \tilde{x}_i = \frac{2}{3} \tilde{x}_i \ . \tag{7.4}$$

From equations (7.3.a) and (7.3.b) we see that any pair of these fixed points satisfies an equation of the form of Eq. (5.7):

$$\tilde{x}_i \wedge \tilde{x}_j = \frac{1}{3} \left(\frac{12}{p_j + 8} \tilde{x}_i + 2 \frac{p_i + 2}{p_i + 8} \tilde{x}_j \right) \ , \quad i \le j \ . \tag{7.5}$$

From Eq. (7.3.a) we obtain the orthogonality relations:

$$(x_i , x_j - x_k) = 0 \ , \quad i \le j \le k \tag{7.6}$$

and similarly:

$$\tilde{x}_i \wedge (x_j - x_k) = \frac{2}{3} \frac{p_i + 2}{p_i + 8} (x_j - x_k) \ , \quad i \le j \le k \ . \tag{7.7}$$

We deduce immediately for the Hessian of x_i^* (see Eq. (4.9))

$$H(x_i^*)(x_j - x_k) = \epsilon \frac{p_i - 4}{p_i + 8} (x_j - x_k) \ , i \le j \le k \ . \tag{7.8}$$

For $i = 0$, $x_i^* = c^*$, we simply find that, in the hyperplane of $\mathcal{T}_4^{G_{r_1 \dots r_k}}$ orthogonal to c, $H(c^*)$ is $- (\epsilon/3) I$ (proportional to the identity); indeed this hyperplane is in $\mathscr{C}_{(2,2)}$. It is more interesting to consider the case $i = 1$. Then for $p_1 = r_1 > 4$ the Hessian $H(x_1^*)$ is a positive multiple of the unit operator on the $k-1$ dimensional subspace of $\mathcal{T}_4^{G_{r_1 \dots r_k}}$ orthogonal to $c = x_0$ and x_1. The restriction to the 2-plane spanned by c, x of the Hessians $H(c^*)$, $H(x_i^*)$ are not positive; however we know, from Eq. (5.11), how to form a third fixed point, that we call here z^* whose restriction of the Hessian is positive in the 2-plane; in Eq. (7.5) with

$$i = 0 \ , \quad j = 1 \ , \quad \alpha = \frac{12}{p_1 + 8} \ , \quad \beta = \frac{2}{3}$$

and

$$\tilde{z} = \frac{r_1 - 4}{r_1} \tilde{c} + \frac{r_1 + 8}{3 r_1} \tilde{x}_1 \tag{7.9}$$

(we have used $p_1 = r_1$). The eigenvectors and eigenvalues of $H(z^*)$ in the 2-plane $\tilde{c}$, $\tilde{x}_1$ are (we use Eq. (5.14.c))

$$H(z^*) \, \tilde{z} = \epsilon \, \tilde{z} \quad , \tag{7.10}$$

$$H(z^*)\left\{ \frac{12}{r_1 + 8} \tilde{c} - \frac{2}{3} \tilde{x}_1 \right\} = \frac{r_1 - 4}{3r_1} \left\{ \frac{12}{r_1 + 8} \tilde{c} - \frac{2}{3} \tilde{x}_1 \right\}.$$

The linear relation (7.9) implies for the Hessian:

$$H(z^*) = \frac{r_1 - 4}{r_1} H(c^*) + \frac{r_1 + 8}{3 \, r_1} H(x_1^*) + \epsilon \frac{r_1 - 4}{3r_1} I \tag{7.11}$$

and from Eq. (7.8)

$$H(z^*) \, (x_1 - x_j) = \epsilon \, \frac{r_1 - 4}{3r_1} I \quad , \quad 1 \le j \le k \, .$$

To summarize: when n is divisible by $r_1 > 4$ and n/r_1 can be written as the product of at least $k - 1$ factors, one can consider the Hamiltonian Eq.(1) with a quartic polynomial

$$\sum_{i=0}^{k} \alpha_i x_i$$

depending on $k + 1$ parameters α_i; it has a stable fixed point $z^*(\phi)$, given by Eq. (7.9), which is physically relevant since it is a positive polynomial (the coefficients in Eq. (7.9) are positive and the x_i are positive). Mathematically, $k + 1$ can be arbitrarily large. This example simply destroys the "conviction" in reference [11] that as "a general consequence of some kind of topological properties of the renormalization group" no stable fixed points can exist for $k > 2$. We will end this section by giving another counter example when $r_1 = 2$ or 3 (of course the case $r_1 = 4$ is excluded). If $r_1 = 2$, we require $r_2 > 2$ so that $p_2 = r_1 r_2 > 4$. Using [Eqs. (7.3.a) to (7.8)], we leave it to the reader to check that $b^* = \epsilon \, \tilde{b}$ where :

$$\tilde{b} = \frac{1}{r_1(r_1 r_2 + 8(r_2 - 2) + 16)} \left\{ (r_1 r_2 - 4)(r_1 + 8) \, \tilde{x}_1 - (r_1 - 4)(r_1 r_2 + 8) \, \tilde{x}_2 \right\} \tag{7.12}$$

is a stable fixed point when $r_1 < 4 < r_1 r_2$. Indeed:

$$\text{Spectrum } H(b^*) \Big|_{\mathcal{T}_4 G} = \left\{ \epsilon, \; k \text{ times } \; \epsilon \, \frac{(4 - r_1)(r_1 r_2 - 4)}{r_1 (r_1 r_2 + 8(r_2 - 2) + 16)} \right\} . \tag{7.13}$$

The corresponding minimum value of n is:

Table 7-1: Fixed points g^* in the 3-plane $\mathcal{C}_{hij}$ spanned by $\tilde{x}_h$, $\tilde{x}_i$ and $\tilde{x}_j$, $0 \leq h < i < j \leq k$.

g^*	Component on			Spectrum of $\epsilon^{-1}H(g^*)|_{g_{hij}}$			
	$6\epsilon x_h$	$6\epsilon x_i$	$6\epsilon x_j$				
0	0	0	0	-1	-1	-1	-1
$\tilde{x}_h$	$\dfrac{1}{p_h+8}$	0	0	1	1	$\dfrac{p_h-4}{p_h+8}$	$\dfrac{p_h-4}{p_h+8}$
$\tilde{x}_i$	0	$\dfrac{1}{p_i+8}$	0	1	1	$-\dfrac{p_i-4}{p_i+8}$	$\dfrac{p_i-4}{p_i+8}$
$\tilde{x}_j$	0	0	$\dfrac{1}{p_j+8}$	1	1	$-\dfrac{p_j-4}{p_j+8}$	$-\dfrac{p_j-4}{p_j+8}$
$\tilde{y}_{ij}$	0	$(p_j-4)\zeta_{ij}^{-1}$	$-(p_i-4)\zeta_{ij}^{-1}$	1	1	$-\zeta_{ij}(p_i-4)(p_j-4)$	$-\zeta_{ij}(p_i-4)(p_j-4)$
$\tilde{y}_{hj}$	$(p_j-4)\zeta_{hj}^{-1}$	0	$-(p_h-4)\zeta_{hj}^{-1}$	1	1	$-\zeta_{hj}(p_h-4)(p_j-4)$	$\zeta_{hj}(p_h-4)(p_j-4)$
$\tilde{y}_{hi}$	$(p_i-4)\zeta_{hi}^{-1}$	$-(p_h-4)\zeta_{hi}^{-1}$	0	1	1	$-\zeta_{hi}(p_h-4)(p_i-4)$	$-\zeta_{hi}(p_h-4)(p_i-4)$
$\tilde{z}_{hij}$	$(p_i-4)(p_j-4)\zeta^{-1}$	$-(p_h-4)(p_j-4)\zeta^{-1}$	$(p_h-4)(p_i-4)\zeta^{-1}$	1	1	$-\dfrac{1}{6}(p_h-4)(p_i-4)(p_j-4)\zeta^{-1}$	$\dfrac{1}{6}(p_h-4)(p_i-4)(p_j-4)\zeta^{-1}$

Notation: $\zeta_{ij}=p_ip_j-16p_i+8p_j+16>0$; similar expressions for ζ_{hi}, ζ_{hj}.

$$\zeta=\frac{1}{6}(p_hp_ip_j+8p_ip_j-16p_hp_j+8p_hp_i+16p_h-80p_i+16p_j+128)$$

Note that $\tilde{z}_{hij}$ is at the intersection of the three 2-planes spanned respectively by $\tilde{x}_h$, $\tilde{y}_{ij}$; $\tilde{x}_i$, $\tilde{y}_{hj}$; $\tilde{x}_j$, $\tilde{y}_{hi}$. This table is valid only if $(p_h-4)(p_i-4)(p_j-4)\neq0$. The minimum is $\tilde{y}_{ij}$ when $p_h<p_i<4$, $\tilde{y}_{hj}$ when $p_h<4<p_i$, and $\tilde{x}_h$ when $4<p_h$.

$$n \geq 2^{k-1} \times 3 . \tag{7.14}$$

Why both $H(z^*)$ and $H(b^*)$ have the eigenvalues $\neq \epsilon$ of multiplicity k is not clear to me. It is to be noted that the isotropy group of the fixed points is much larger than the normalizer of the isotropy group $G_{r_1, \ldots, r_k}$. Indeed

$$O(n)_{z^*} = \Gamma_{r_1, q_1} \quad , \quad O(n)_{b^*} = \Gamma_{p_1, q_1} \cap \Gamma_{p_2, q_2} . \tag{7.15}$$

The proof is easy in the second case since $\dim O(n)_{b^*} = n(r_1 - 1)/2$ while the normalizer of $G_{r_1, \ldots, r_k}$ is finite since $G_{r_1, \ldots, r_k}$ is finite and irreducible and $\subset O(n)$. The polynomials x_i appear often in actual studies of Landau transitions.

Finally, I hope that the powerful method developed in this paper for computing fixed points will also be useful for practical applications.

ACKNOWLEDGEMENTS

I am grateful to M. J. Jarić; J. C. and P. Toledano, and J. Tits; I benefited from discussions with them about this work.

APPENDIX

In Section 5 we studied the fixed points of some 2-planes. We give here similar results for 3-planes spanned by x_i, x_j and x_h with $0 \leq h < i < j \leq k$; this implies that $1 \leq p_h < p_i < p_j \leq n$; we add the condition that every p is different from 4. An interesting particular case is $h = 0$, $j = n$, so $p_0 = 1$, $p_k = n$, $\bar{x}_h = c$ and $\bar{x}_j = s$. The component of the fixed points and the spectrum of the restriction of their Hessians to the 3-plane are given in Table 7.1.

We find that such a 3-plane has 8 fixed points. They can be considered as the vertices of a polyhedron which a has the same faces and edges as a cube. Similarly, we conjecture that in the $k + 1$ dimensional space $\mathcal{T}_4^{G_{r_1, \ldots, r_k}}$ there are 2^{k+1} fixed points forming a cube-like figure.

REFERENCES

[1]. E. Brézin, Les Houches, session 28 p. 330, North Holland (1976).

[2]. E. Brézin, J. C. Le Guillou, J. Zinn-Justin a) Phys. Rev. B10 (1974) 892;
     b) in Phase transitions and critical phenomena 6 edit. C. Domb, M. S. Green.

[3]. L. Landau, E. M. Lifshitz, Statistical Physics, tr. Pergamon Press N. Y. 1958.

[4]. A. Aharony, Phys. Rev. B8 (1973) 4270.

[5]. S. A. Brazovski, I. E. Dzyaloshinskii, JETP Lett. 21 (1975) 164 (ZhETF Pis. Red. 21 (1975) 360).

[6]. D. Mukamel, Phys. Rev. Lett. 34 (1975) 481.

[7]. V. A. Alessandrini, A. P. Cracknell, J. A. Przystawa, Commun. Phys. 1 (1976:I).

[8]. R. P. K. Zia, D. J. Wallace, J. Physics A8 (1975) 1089.

[9]. A. L. Korzhenevskii, Sov. Phys. JETP 44 (1976) 751 (ZhEksp. Teor. Fiz. 71 (1976) 1434).

[10]. M. Jarić, Phys. Rev. B18 (1978) 2237 and 2391.

[11]. I. E. Dyzaloshinskii, Sov. Phys. JETP 45 (1977) 1014.

[12]. Gell-Mann, Phys. Rev. 125 (1962) 1097.

[13]. L. Michel, L. A. Radicati a) in *Evolution of Particle Physics*, ed. M. Conversi (1970) 191, Academic Press, N.Y.; b) Ann. Phys. (N.Y.) 66 (1971) 758; c) Ann. Inst. H. Poincaré 18 (1973) 185.

[14]. D. H. Sattinger, Bull. Am. Math. Soc. 3 (1980) 779.

[15]. D. J. Wallace, R. P. K. Zia, Phys. Lett. 48A (1974) 325.

[16]. D. Montgomery, C. T. Yang, Ann. Math. 65 (1957) 108.

[17]. H. S. M. Coxeter, W. O. J. Moser, *Generators and Relations for Discrete Groups* (Springer Verlag, 1957).

[18]. L. Michel, in Group Theoretical Methods in Physics, ed. Sharp and Kolman (Academic Press, N. Y. 1977) p. 75.

[19]. L. Michel, J. C. Toledano, P. Toledano, in *Symmetries and Broken Symmetries in Condensed Matter Physics*, edit. Boccara, IDSET (Paris) 1981, p. 253 .

RELATIVISTIC HEAVY ION COLLISIONS AND FUTURE PHYSICS

T. D. Lee

Department of Physics
Columbia University,
New York, NY 10027

1. TWO PUZZLES

At present, most physicists feel that we have finally arrived at a closed system of physical laws, with *QCD* for the strong interaction and a unifying gauge theory for the weak and electromagnetic forces, plus of course Einstein's theory of general relativity. However, there are a few things that are not completely satisfactory. Of the participating fundamental particles, only the leptons and the photon have been observed directly. The intermediate bosons and the graviton, we hope, can be detected in the near future. All the rest, the quarks and the gluon, we believe can never come out in the open and therefore direct observation will always be impossible.

In view of this, our strong belief in this grand scheme must not be due entirely to direct experimental evidence, but rather based on the esthetic simplicity of the theoretical foundation and the compelling conclusion of our mathematical deduction.

1.1. Missing Symmetry

The basis of all these theories rests entirely on the symmetry [1] under local transformations with respect to either the internal gauge variables or the space time variables. Yet, in reality, almost all the symmetry quantum numbers are found to be, or believed to be, not conserved. Even the best-established conservation law, that of the baryon number, is now also believed to be violated. Surely, this is somewhat puzzling.

T. D. Lee

1.2. Color Confinement

, Another puzzle is the problem of quark or color confinement, which makes half of
the elementary particles, quarks and gluons, non-direct observables. The explanation
of both puzzles is to invoke the properties of the vacuum.

In the first case of the missing symmetry we require the vacuum, though Lorentz
invariant, to be a coherent mixture of states of different quantum numbers. Therefore
the vacuum expectation value of any quantum-number-carrying spin-0 field ϕ can be
different from zero:

$$<\phi>_{VAC} \neq 0 . \tag{1.1}$$

In the second case, color confinement, we assume the *QCD* vacuum to be a condensed
state of gluon pairs so that it is a perfect color dia-electric (i.e., color dielectric constant $\kappa = 0$). This is in complete analogy to the description of a superconductor as a
condensed state of electron pairs in *BCS* theory, which results in making the superconductor a perfect dia-magnet (with magnetic susceptibility $\mu = 0$). When we switch
from *QED* to *QCD* we replace the magnetic field $\vec{H}$ by the color electric field $\vec{E}$, the
superconductor by the *QCD* vacuum, and the *QED* vacuum by the interior of the
hadron. As shown in Figure 1-1, the roles of the inside and the outside are inter-
changed. Just as the magnetic field is expelled outward from the superconductor, the
color electric field is pushed into the hadron by the *QCD* vacuum, and that leads to
color confinement. This situation is summarized in Table 1-1 .

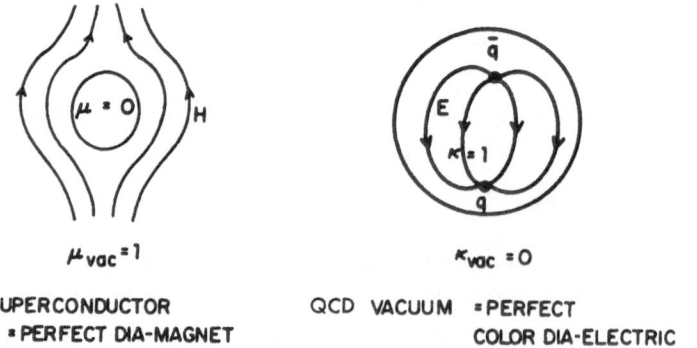

Figure 1-1:   Superconductivity in QED vs. quark confinement in QCD.

Table 1-1: Analogies between Superconductivity and the QCD Vacuum

QED superconductivity as a perfect diamagnet		QCD vacuum as a perfect color dielectric
$\vec{H}$	$\longleftrightarrow$	$\vec{E}$
$\mu_{inside} = 0$	$\longleftrightarrow$	$\kappa_{vacuum} = 0$
$\mu_{vacuum} = 1$	$\longleftrightarrow$	$\kappa_{inside} = 1$
inside	$\longleftrightarrow$	outside
outside	$\longleftrightarrow$	inside

2. MICROSCOPIC VS. MACROSCOPIC WORLD

In both of the above cases the system of elementary particles no longer forms a self-contained unit. The microscopic particle physics depends on the coherent properties of the macroscopic world, represented by these operator averages in the physical vacuum state.

If we pause and think about it, this represents a rather startling conclusion, contrary to the traditional view of particle physics which holds that the microscopic world can be regarded as an isolated system. To a very good approximation it is separate and uninfluenced by the macroscopic world at large. Now, however, we need these vacuum averages; they are due to some long-range ordering in the state vector. At present our theoretical technique for handling such coherent effects is far from being developed. Each of these vacuum averages appears as an independent parameter, and that accounts for the twenty-some constants in the present weak and electromagnetic gauge theory. A comparable number of parameters is also needed in any of the grand unified theories. Consequently, we must view our present theoretical framework as at least partly phenomenological. After all, who has ever heard of a fundamental theory that requires a grand total of twenty-some parameters?

On the experimental side, there has hardly been any direct investigation of these coherent phenomena. This is because hitherto in any high energy experiment, the higher the energy the smaller has been the spatial region we are able to examine. Likewise, in nuclear physics we have so far concentrated mostly on nuclear matter at a constant density. In order to explore physics in this fundamental area, relativistic heavy ion collisions offer an important new direction.

3. RELATIVISTIC HEAVY ION COLLISIONS

3.1. General Discussions

A normal nucleus of baryon number A has an average radius $r_A \approx 1.2\ A^{1/3}\ fm$ and an average energy density

$$\mathscr{E}_A \approx \frac{m_A}{(4\pi/3)\ r_A^{\ 3}} \approx 130\ MeV/fm^3\ . \tag{3.1}$$

Each of the A nucleons inside the nucleus can be viewed as a smaller bag which contains three relativistic quarks inside; the nucleon radius is $r_N \approx 0.8\ fm$ and its average energy density is

$$\mathscr{E}_N \approx \frac{m_N}{(4\pi/3)\ r_N^{\ 3}} \approx 440\ MeV/fm^3\ . \tag{3.2}$$

Our purpose is to study the physics of objects which, in their rest frames, can extend over several *fm* and have energy densities much greater than that inside a single nucleon. If this is possible, then inside each such an object the usual concept of nuclear physics (regarding nucleons as units) must break down; new physics may thereby emerge.

When two heavy nuclei collide, to begin with there is the problem of Coulomb repulsion which must be overcome in order to reach the nuclear surface; that requires a c. m. energy/nucleon $\gtrsim 20\ MeV$. Next, in order to penetrate the nucleonic surface, it is necessary to overcome the strong repulsive force between nucleons due to ω and ϕ exchange. Thus, *we must have a c. m. energy of about several GeV/nucleon or higher*. This then brings us to accelerators with an energy of several hundreds of *GeV* in the center of mass.

3.2. Coherence vs. Incoherence

The total high-energy cross section of a proton-nucleus collision is known to be given by the geometrical area of the nucleus; that of a high-energy nucleus-nucleus collision is therefore also expected to be similarly determined. We have

$$\sigma_{A'\ A}(total) \propto \left\{\ A'^{1/3} + A^{1/3}\ \right\}^2$$

where A and A' are the baryon numbers of these two nuclei. More specific information concerning coherence or incoherence can be obtained by examining some inclusive reactions such as

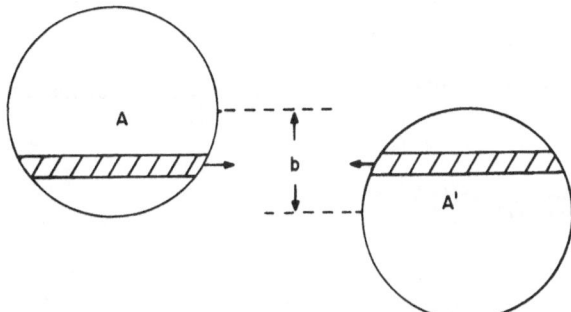

Figure 3-1:   Collision between two heavy nuclei A and A'.

$$A' + A \rightarrow \pi + \ldots \tag{3.3}$$

at high energy.

For orientation purposes, let us first assume complete incoherence during the reaction; i. e., each collision is like a single nucleon on a single nucleon and there is no shadowing, no absorption and no re-scattering. The process at a given impact parameter b may be illustrated by Figure 3-1. The shaded regions in A and A' denote two circular cylinders of nucleons facing each other. The number of nucleons contained in these two cylinders is respectively proportional to $A^{1/3}$ and $A'^{1/3}$. Thus, for incoherent collisions their rate for reaction Eq. (3.3) is proportional to $A^{1/3} A'^{1/3}$. Adding the contributions due to all such cylinders and averaging over the impact parameter, we see that

$$d\sigma_{A'A}(incoh) \propto A^{1/3} A'^{1/3} (A^{1/3} + A'^{1/3})^2 , \tag{3.4}$$

so that for $A' = A$

$$d\sigma_{A'A}(incoh) \propto A^{4/3} \tag{3.5}$$

and for a single proton ($A' = 1$) on A

$$d\sigma_{pA}(incoh) \propto A . \tag{3.6}$$

We now define any deviation from these A and A' dependences as due to *coherent* processes. The coherence may either serve as a suppression (such as the shadowing effect), or act as an enhancement (e.g., due to multiple scattering). In any case, its operational meaning is clear.

A case of interest is to examine the A-dependence of

$$p + A \to \pi^{\pm} + \ldots$$

for $\pi^{\pm}$ with a high perpendicular momentum $k_{\perp}$. The experimental result [2] expressed in the form

$$\left(\frac{d\sigma}{dk_{\perp}}\right)_{pA} = \left(\frac{d\sigma}{dk_{\perp}}\right)_{pp} A^{a}(k_{\perp}) \tag{3.7}$$

gives for $\pi^{\pm}$

$$a(k_{\perp}) \sim 0.9 \qquad when \ k_{\perp} \lesssim 1 \ GeV$$

and

$$a(k_{\perp}) \sim 1.1 \qquad when \ k_{\perp} \sim 4{-}6 \ GeV \ .$$

The former is a suppression, the latter an enhancement. The corresponding values for other particles in the high $k_{\perp} \sim 4{-}6 \ GeV$ region are

$a(k_{\perp})$		
	~ 1.15	*for* K^{-}
	$\sim 1.2 - 1.3$	*for* K^{+}
	$\sim 1.3 - 1.4$	*for* p *and* $\bar{p}$

$$\tag{3.8}$$

The slight suppression in the low $k_{\perp}$ region can be readily understood by taking into account the shadowing effect. The relatively large enhancement in the high $k_{\perp}$ region, especially for p and $\bar{p}$, is more complicated.

Since $d\sigma$ decreases very rapidly with $k_{\perp}$, at high $k_{\perp}$ one is more sensitive to rare but more interesting events. The rapid rise of $a(k_{\perp})$ is due partly to multiple scattering. But it may also be due in part to a completely different mechanism (one that is similar to the well-known formation of antideuteron and anti-He^3 in a high energy $p{-}p$ collision.) In Figure 3-1, when the two Lorentz-contracted (shaded) columns of particles hit each other, the quark and the antiquark that form the final pion may come from two separate collisions. Since the probability of having an additional collision is proportional to $A^{1/3} A'^{1/3}$, one should therefore multiply Eq. (3.4) by such a factor. In a $p + A$ reaction, such a mechanism gives an enhancement factor of $A^{1/3}$, making $d\sigma_{pA} \propto A^{4/3}$ for π and K. For p and $\bar{p}$, there is the possibility of a triple collision, which carries an enhancement factor of $A^{2/3}$, giving $d\sigma_{pA} \propto A^{5/3}$. Similarly, in an $A + A$ collision, the corresponding enhancement factor would be $A^{2/3}$ for π or K (making $d\sigma_{A\,A} \propto A^2$), and $A^{4/3}$ for p or $\bar{p}$ (making $d\sigma_{pA} \propto A^{8/3}$).

3.3. The Central Problem

The most crucial questions in an ultra-relativistic heavy ion collision are:

1. Can extended objects of very high energy densities ($\geqslant$ energy density inside a proton) be produced?

2. If produced, how can they be detected and how can we analyze their properties?

These two problems have recently been studied by several people [3-5]. The answers to both questions are affirmative, especially with colliders that have energies of several hundreds of *GeV*.

Consider the collision between two relativistic heavy ions. In the center-of-mass system, most of the final energy goes to fragments with very large rapidity; these particles hadronize outside nuclear matter and are therefore not relevant for our purposes. However, at least $\sim$ 10-15% of the energy will be trapped in the two projectile nuclei and in the central rapidity region. But if the energy is many hundreds of *GeV*, 10% of it per nucleon is much greater than the nucleon mass. Therefore it is not surprising that we can create extended objects of energy density an order larger than that inside the proton.

We now turn to the second question. If these objects are produced, how can we detect them and how can we analyze their properties?

4. A POSSIBLE EXPERIMENTAL PROGRAM

4.1. Quark-Antiquark and Gluon Plasma

Consider now such an extended object whose energy density is $\mathscr{E}$, radius R and lifetime τ (all in its rest frame). Its interior is assumed to be in a new plasma state, but its exterior consists essentially of normal hadrons. Ideally, it looks like the object shown in Figure 4-1. In its interior, because

$$\mathscr{E} \gg \mathscr{E}_N$$

the usual hadron bags disappear. Instead there should be a large bag of quark, antiquark and gluon plasma. Normal small hadron bags can appear only on the surface. These different properties are summarized in the following table [Table 4-1] . However, because the collision is a violent one, there will be all kinds of instability (due to *QCD* vacuum dynamics, phase transitions, non-central collisions, etc.). In most cases, the inside will simply erupt like volcanoes. Hence, as shown in Figure 4-2, a fair

Table 4-1:   Interior and Exterior Properties of Metastable Extended Object

Interior	Exterior
energy density $\mathcal{E} \gg \mathcal{E}_N$	energy density $\mathcal{E} \approx \mathcal{E}_N \approx \frac{1}{2}$ GeV/fm^3
almost exact flavor $SU(3)$ symmetry	almost exact flavor $SU(2)$ symmetry
approximate $SU(4)$ symmetry	approximate $SU(3)$ symmetry
average momentum	average momentum
$<k> \sim \left(\frac{\mathcal{E}}{\mathcal{E}_N}\right)^{1/4} 300$ MeV	$<k> \sim 300$ MeV

fraction of the plasma can come out directly without thermalization with the exterior. The lifetime τ of such an unstable object is $\sim$ its radius R.

4.2. Detection of Volcanoes

To detect such an eruption we take advantage of the fact that the total mesons produced from such an object would be on the order of $10^4 - 10^5$. Hence in a heavy-ion collision we may divide the central and the near-central regions of rapidity y into, say, ten sections. In each Δy sector we may again divide the azimuthal angle Φ around the

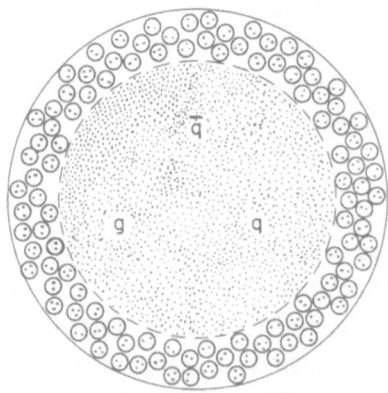

Figure 4-1:   An object whose interior is a quark-antiquark and gluon plasma, but whose exterior consists essentially of normal hadrons.

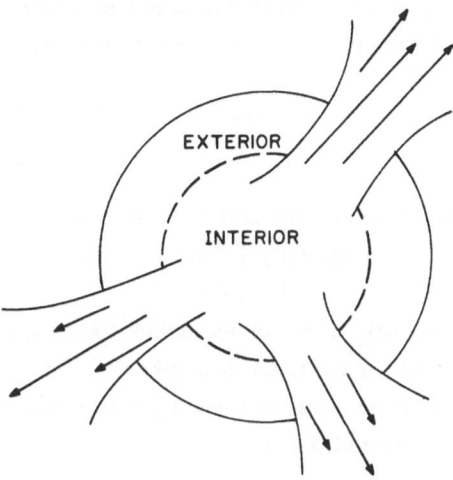

Figure 4-2: A schematic view of the decay of the extended object when $\tau \approx R$.

line of collision in the center-of-mass frame into, say, ten angular intervals. Per collision, in each Δy-$\Delta\Phi$ section we would still expect a statistically significant number of mesons, $\sim 10^2 - 10^3$. We can then plot the average K/π ratio in each section. If this ratio has violent variations, from ~ 1 in some sectors to $\sim 1/10$ in others, then this is a clear sign of "volcanoes" with $K/\pi \sim 1$ indicating a direct hadronization of the interior. For corroboration, we may plot $\Sigma |k_\perp|$ in each of the Δy-$\Delta\Phi$ sectors, where $k_\perp$ is the meson momentum component perpendicular to the initial line of collision and the sum extends to all mesons emitted in that sector. Again, we should expect that in a volcano-like event the average $|k_\perp|$ could be much bigger than 300 MeV in the Δy-$\Delta\Phi$ sector that has a high K/π ratio, but ~ 300 MeV if the K/π ratio is about 1/10.

4.3. Detection of New Metastable Objects

Occasionally, we may find that the K/π ratio and $\Sigma |k_\perp|$ are uniform in all Δy-$\Delta\Phi$ sectors. This is then a good indication that there is an absence of eruption, and that a metastable object of a large bag of quarks-antiquarks and gluons is formed, as shown in Figure 4-1. Its lifetime τ can be quite a bit longer. [If there is no bag pressure, no surface tension, and quarks and gluons behave like free particles, then τ should be approximately $3^{1/2}R$. Since none of these assumptions is correct, we expect τ to be $\geqslant 3^{1/2}R$.]

We shall now examine the question of how to determine τ for such a metastable object. The important decay mode of such an object is through the radiation of mesons from its exterior, plus emission of photons and lepton pairs from its interior. The rate of soft-pion radiation is proportional to its surface, and therefore the total number of soft pions emitted is

$$N_\pi = 4 \pi \kappa R_\pi^2 \tau \tag{4.1}$$

where κ can be calculated theoretically and R_π, the mean radius of the π-emitting exterior, can be measured experimentally [6] through the Hanbury-Brown and Twiss method, as we shall see.

Consider the emission of, say, a π^- of momentum $\vec{k}$ from a volume element $d^3 r_a$ near the surface together with the emission of another π^- with momentum $\vec{k} + \vec{q}$ from $d^3 r_b$. Because these are identical particles obeying Bose statistics, the total probability of seeing these two pions is proportional to

$$\frac{1}{2} \left| e^{\{i\vec{k} \cdot \vec{r}_a + i(\vec{k} + \vec{q}) \cdot \vec{r}_b\}} + e^{\{i\vec{k} \cdot \vec{r}_b + i(\vec{k} + \vec{q}) \cdot \vec{r}_a\}} \right|^2 .$$

The interference term between these two terms in the sum gives the deviation of the correlation function from unity. Hence, the measurement of such correlation functions can give a direct measurement of the geometry of the surface region. The radius R_π can be determined. By using Eq. (4.1), together with the experimental values of N_π and R_π, we can infer the lifetime τ. Similar correlation functions can also be observed by using $\pi^+ \pi^+$, $K^+ K^+$, $K^- K^-$, etc. Each of these yields an independent measurement of the same lifetime τ.

In addition, the γ and lepton pairs are emitted directly from the interior. Their emission rates are proportional to the volume of the metastable object. Similar studies of $\gamma \gamma$ correlation and lepton pair-lepton pair correlations can determine the geometry of the interior. Just as in Eq. (4.1), the total number of γ emitted is given by

$$N_\gamma = \frac{4\pi}{3} \lambda R_\gamma^3 \tau \tag{4.2}$$

where R_γ is determined experimentally through the $\gamma \gamma$ correlation function and λ can be calculated according QCD. Thus τ can again be independently determined. In an entirely similar way we can also use lepton pair-lepton pair to measure the lifetime. [See Figure 4-3.]

In this idealized experimental program, we take full advantage of the large multiplicity of particles produced. The spirit is to regard each collision as an experiment, similar to the observation of a stellar object. Our strategy is to replace photons radiated from a star by mesons, and neutrinos by photons and lepton pairs. The large

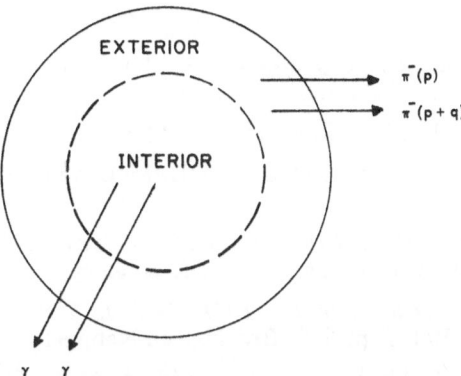

Figure 4-3: The radius of the surface can be determined by the $\pi\,\pi$ (or other hadron-hadron) correlation function, and the radius of the inside volume by the $\gamma\,\gamma$ (or lepton pair-lepton pair) correlation. From these radii and Eqs. (9) and (10), one can derive the lifetime τ.

radius of these objects and the enormous number of particles involved makes it possible for us to measure their geometrical size through correlation functions. The total multiplicity gives us the equivalent of the integrated luminosity of a star. Thereby their lifetimes can be determined. In this way we can discover an entirely new class of metastable states of large baryon number and high energy density. Their equations of state would gives us insight into the physics of the early universe. Remembering that the total heavy ion cross section is $\sim 10^{-24}\ cm^2$, even an extremely rare event of $\sim 10^{-10}$ probability has a rate that can be compared favorably with a typical neutrino event.

So far we have refrained from discussing any speculative phenomena in order to show that even at the very minimum we can expect, through relativistic heavy ion collisions, the creation of a new state of matter. Its energy density should be greater than that in the interior of neutron stars. Our usual concepts of nuclear physics are not applicable, but new principles are yet to be uncovered.

It is not inconceivable that in a decade or two, a fair percentage of the high energy physicists and nuclear physicists will be engaged in this exciting area of research.

ACKNOWLEDGMENTS

I wish to thank A. H. Mueller, L. Radicati and W. Willis for discussions.

REFERENCES

[1]. F. Gürsey, "Symmetries of Quarks and Leptons", 1059 in *The Whys of Subnuclear Physics*, Erice Proceedings, 1977.

[2]. J. W. Cronin, et. al. Phys. Rev. D11 (1975) 3105.

[3]. See e.g. R. Anishetty, P. Koehler and L. McLerran, Phys. Rev. D22 (1980) 2793.

[4]. A. H. Mueller, p. 636 in *Proceedings of the 1981 ISABELLE Summer Workshop*, Vol. II, Brookhaven National Laboratory.

[5]. W. Willis, in *Proceedings of the 1981 ISABELLE Summer Workshop*, Vol. I, p. 84 and Vol. II p. 652, Brookhaven National Laboratory.

[6]. G. Goldhaber, S. Goldhaber, W. Lee and A. Pais, Phys. Rev. 120 (1960) 300. M. Gyulassy, S. K. Kauffmann and L. W. Wilson, Phys. Rev. C20 (1979) 2267.

THE FERMION DETERMINANT
IN
MASSLESS TWO-DIMENSIONAL QCD

Ralph Roskies

Department of Physics and Astronomy
University of Pittsburgh
Pittsburgh, PA 15260

INTRODUCTION

I was very lucky to have been a colleague of Feza Gürsey's from 1965 to 1971 at Yale. As a young Ph. D. I learned a great deal from him--about field theory, group theory, and mathematics in general--and I came to admire his originality and the elegance of his ideas. Most people seem to associate Feza's physics with aspects of group theory. But what struck me most about him in the late 60's was his conviction that field theory would be the source of our new insights, even into strong interaction physics. This was by no means the party line at the time.

The developments in non-Abelian gauge theories have proven him right. This area of physics combines field theory and group theory in a wonderfully elegant and economical way. I am delighted to have the chance of honoring Feza with an article in a field which combines his two major interests. My only hesitation is that my results are still incomplete and inelegant. That's not appropriate for Feza, but he will have to forgive me.

1. GENERAL FRAMEWORK

Two dimensional QCD (QCD_2) with massless quarks is tantalizing because it lies between the physically interesting four-dimensional theory and the soluble two-dimensional Schwinger model. We hope that its properties will resemble those of the realistic four-dimensional case, but that it is simple enough to be understood thoroughly.

The Schwinger model [1] is soluble because of two independent properties. First, in two space-time dimensions, given any field A_μ, one can always find Φ (a linear combination of 1 and γ_5) satisfying [1]

$$\gamma^\mu A_\mu = \gamma^\mu \partial_\mu \Phi. \tag{1.1}$$

This enables us to evaluate the fermion determinant arising from the functional integration over the quark fields. Second, the effective boson Lagrangian thus obtained is quadratic in the boson fields, and so is easily solved. The expectation is that in the non-Abelian case, the fermion determinant can be evaluated, but the effective Lagrangian in terms of gluons is not quadratic. Indeed, it is probably not a polynomial in the gluon fields. But one might be able to make some non-perturbative sense of it.

Several authors have studied the fermion determinant [2] in QCD_2. While they have given implicit expressions for the determinant in terms of the gluon fields, explicit expressions eluded them because the non-Abelian nature of the theory forced them to consider path ordered integrals, which are hard to evaluate. This can be rephrased by saying that their expressions for the fermion determinant in the presence of an external gluon field was not the exponential of a space-time integral of a quantity local in the gluon fields. This is already seen in the Schwinger model where the fermion determinant takes the value

$$\exp\left\{ \frac{-e^2}{2\pi} \int d^2x\, A_\mu(x)\, \left(g_{\mu\nu} - \frac{\partial_\mu \partial_\nu}{\partial^2}\right) A_\nu(x) \right\} \tag{1.2}$$

and $\frac{1}{\partial^2}$ is not a local operator. However, in the Schwinger model, there is a gauge ($\partial_\nu A_\nu = 0$) which makes the expression in Eq. (1.2) local. Similarly I will show that for $SU(2)$, at least, there exists a gauge in the non-Abelian case which yields a fermion determinant which is the exponential of the integral of a quantity local in the gluon field. While I can characterize that gauge abstractly, I do not yet have an elegant differential characterization for it.

My approach is a generalization of the techniques that Schaposnik and I [3] used to study the Schwinger model. The essential observation was that because of Eq. (1.1), the photon field in its coupling to fermions in two dimensions looks like a pure gauge field. Naively one might imagine performing a gauge transformation on the fermion fields in the functional integral

$$\psi \to e^{-i\Phi(x)} \psi, \tag{1.3}$$

which would decouple the fermion and the photon. However, because Φ also has γ_5

pieces, this is really a local axial transformation. Fujikawa [4] has shown that such transformations induce extra terms in the Lagrangian because of the axial vector anomaly. In the Schwinger model, these extra terms are just the mass terms for the photons. I will extend these ideas to QCD_2 below.

In the Euclidean formulation, the generating functional for QCD_2 with massless quarks is

$$Z(J, \eta, \bar{\eta}) = \int [DA_\mu] [D\psi] [D\bar{\psi}] \exp\left\{- \mathscr{S}[A, \psi, \bar{\psi}]\right.$$

$$\left. + \int d^2x \left(A_\mu{}^a(x) J_\mu^a(x) + \bar{\eta}(x) \psi(x) + \bar{\psi}(x) \eta(x)\right) \right\} \tag{1.4}$$

with

$$\mathscr{S}[A, \psi, \bar{\psi}] = \int d^2x \left(- \bar{\psi} \not{D} \psi + \frac{1}{4} F_{\mu\nu}^a F_{\mu\nu}^a + \text{gauge fixing terms}\right) \tag{1.5}$$

and

$$\not{D} = \gamma^\mu (i\partial_\mu + \vec{\tau} \cdot \vec{A}_\mu). \tag{1.6}$$

The conventions are that $\{\gamma^\mu, \gamma^\nu\} = 2 \delta_{\mu\nu}$ and $\gamma_5 = i \gamma^1 \gamma^2$.

I will restrict my attention to an $SU(2)$ internal symmetry group, and use vector notation for $SU(2)$ throughout. The fermion part of the functional integral

$$\int D\psi \, D\bar{\psi} \, \exp\{\int d^2x \, \bar{\psi} \not{D} \psi\}$$

can be trivially integrated to give

$$\text{Det } \not{D} = \text{Det}\left\{ \gamma^\mu (i\partial_\mu + \vec{\tau} \cdot \vec{A}_\mu)\right\} = \left(\text{Det } \gamma^\mu \, i \, \partial_\mu\right) \left\{\text{Det } \frac{1}{i \, \gamma^\nu \, \partial_\nu} \left(\gamma^\mu (i\partial_\mu + \vec{\tau} \cdot \vec{A}_\mu)\right)\right\}. \tag{1.7}$$

I will refer to the second factor in Eq. (1.7) as the "fermion determinant" and I will evaluate it in a special gauge to be defined in the next section. In the following section I will evaluate the fermion determinant in this gauge. In Section 4, I discuss a differential characterization of the gauge, and show that it involves not more than two derivatives of the gauge field.

2. CHOICE OF GAUGE

The non-Abelian generalization of Eq. (1.1) is

$$i\gamma^\mu \, \vec{A}_\mu \cdot \vec{\tau} = \gamma^\mu \, (\partial_\mu U) U^{-1}, \tag{2.1}$$

where U is an element of the Lie group whose Lie algebra is spanned by $\vec{\tau}$ and $i\gamma_5 \, \vec{\tau}$. Realizing that

$$j \equiv - i \, \gamma_5 \tag{2.2}$$

behaves as a complex unit $(j^2 = -1)$, we can define

$$z = x_1 + j \, x_2 \, , \qquad \bar{z} = x_1 - j \, x_2$$
$$\vec{A} = \vec{A}_1 + j \, \vec{A}_2, \tag{2.3}$$

and rewrite Eq. (2.1) as

$$\vec{A} \cdot \vec{\tau} = -2i(\partial_{\bar{z}} U) U^{-1}, \tag{2.4}$$

where U is an element of $SL(2,C)$ (the j complexification of $SU(2)$). This equation is just the equation for a holomorphic principal bundle over the complex plane. All such bundles are trivial [5], which means that a global solution U of this equation exists.

Any element U in $SL(2,C)$ can be written uniquely as the product of a unitary matrix and a positive definite Hermitian matrix. Multiplication by a unitary matrix just corresponds to an ordinary $SU(2)$ gauge transformation. So we can pick a gauge such that U is Hermitian. Thus we have shown that we can choose a gauge for $\vec{A}_\mu$ where

$$i\gamma^\mu \, \vec{A}_\mu \cdot \vec{\tau} = \gamma^\mu \, (\partial_\mu U \,) U^{-1} \tag{2.5}$$

with U being a positive definite Hermitian matrix (with respect to the complex unit j) of determinant 1, i.e.,

$$U = \exp\{-ij \, \vec{b} \cdot \vec{\tau}\} = \exp\{- \, \gamma_5 \, \vec{b} \cdot \vec{\tau}\} \tag{2.6}$$

with $\vec{b}$ real. This defines our choice of gauge.

The relation between $\vec{A}_\mu$ and $\vec{b}$ is

$$\vec{A}_\mu = \frac{\vec{b} \times \partial_\mu \vec{b} \, \sinh^2 b}{b^2} + \epsilon_{\mu\nu} \left(\partial_\nu \vec{b} - \frac{\vec{b} \, \partial_\nu b}{b} \right) \frac{\sinh b \, \cosh b}{b} .$$
$$+ \, \epsilon_{\mu\nu} \frac{\vec{b} \, \partial_\nu b}{b} \tag{2.7}$$

The conditions on $\vec{A}_\mu$ which allow it to be written in the form Eq. (2.7) will be studied further in Section 4.

3. FERMION DETERMINANT

The idea now is to assume that $\vec{A}_\mu$ has been chosen in a gauge such that Eqs. 2.5 and 2.6 are valid. We would then like to perform the change of variables

$$\psi'(x) = \exp\{\gamma_5 \, \vec{b} \cdot \vec{\tau}\} \, \psi(x)$$

$$\bar{\psi}'(x) = \bar{\psi}(x) \, \exp\{\gamma_5 \, \vec{b} \cdot \vec{\tau}\} \tag{3.1}$$

in Eq. (1.4). This will result in a decoupling of the fermion from the gluon. However, since this is a local axial transformation there are extra terms induced in the Lagrangian because of the axial vector anomaly. Fujikawa [4] has shown how to compute these terms for infinitesimal transformations. So we must write Eq. (3.1) as a sequence of infinitesimal transformations and integrate the final result.

Define [6]

$$\psi(x,\alpha) = \exp\{\alpha \, \gamma_5 \, \vec{b} \cdot \vec{\tau}\} \, \psi(x)$$

$$\bar{\psi}(x,\alpha) = \bar{\psi}(x) \, \exp\{\alpha \, \gamma_5 \, \vec{b} \cdot \vec{\tau}\}. \tag{3.2}$$

One can easily show that

$$\bar{\psi}(x,\, 0) \, \gamma^\mu \, (i\partial_\mu + \vec{\tau} \cdot \vec{A}_\mu) \, \psi(x,\, 0) = \bar{\psi}(x,\, \alpha) \, \gamma^\mu \, (i\partial_\mu + \vec{\tau} \cdot \vec{A}_\mu(\alpha)) \, \psi(x,\alpha) \tag{3.3}$$

where

$$\vec{A}_\mu(\alpha) = \vec{A}_\mu + \sinh^2 \alpha b \left[2 \, \vec{A}_\mu + \frac{\vec{b} \times \partial_\mu \vec{b}}{b^2} - \frac{2 \, \vec{b} \, (\vec{b} \cdot \vec{A}_\mu)}{b^2} \right] + \epsilon_{\nu\mu} \frac{\vec{b} \, \partial_\nu \, b}{b}$$

$$+ \epsilon_{\nu\mu} \sinh \alpha b \cosh \alpha b \left[-\frac{2\vec{b} \times \vec{A}_\nu}{b} - \frac{\vec{b} \, \partial_\nu \, b}{b^2} + \frac{\vec{b} \, \partial_\nu \vec{b}}{b} \right]. \tag{3.4}$$

Note that

$$\vec{A}_\mu(0) = \vec{A}_\mu, \qquad\qquad\qquad \vec{A}_\mu(1) = 0 \tag{3.5}$$

by virtue of Eq. (2.7). So the sequence of axial transformations indexed by α has effectively decoupled the gluons and fermions as promised.

Following Fujikawa it is straightforward to demonstrate that

$$[D\bar{\psi}(x,\, \alpha+\delta\alpha) \, D\psi(x,\, \alpha+\delta\alpha)] = [D\bar{\psi}(x,\, \alpha) \, D\psi(x,\, \alpha)] \left\{ 1 + \frac{\epsilon_{\mu\nu}}{\pi} \frac{\delta\alpha}{\pi} \int d^2x \, \vec{b} \cdot \vec{F}_{\mu\nu}(\alpha) \right\} \tag{3.6}$$

with

$$\vec{F}_{\mu\nu}(\alpha) = \partial_\mu \vec{A}_\nu(\alpha) - \partial_\nu \vec{A}_\mu(\alpha) + 2 \vec{A}_\mu(\alpha) \times \vec{A}_\nu(\alpha) \tag{3.7}$$

so that

$$[D\bar{\psi}(x, \alpha) \, D\psi(x, \alpha)] = [D\bar{\psi}(x, 0) \, D\psi(x, 0)] \exp\left\{\frac{-\epsilon_{\mu\nu}}{\pi} \int d^2x \int_0^\alpha d\alpha \, \vec{b} \cdot \vec{F}_{\mu\nu}(\alpha)\right\}. \tag{3.8}$$

Applying this for $\alpha = 1$ finally yields the desired result

$$\int [D\bar{\psi} \, D\psi] \exp\left\{\int d^2x \, \bar{\psi}(x) \, \gamma^\mu \, (i\partial_\mu + \vec{\tau} \cdot \vec{A}_\mu) \, \psi(x)\right\}$$

$$= \left(\int [D\bar{\psi}' \, D\psi'] \exp\left\{\int d^2x \, \bar{\psi}'(x) \, i\gamma^\mu\partial_\mu \, \psi'(x)\right\}\right)$$

$$\times \exp\left\{\frac{-\epsilon_{\mu\nu}}{\pi} \int d^2x \int_0^\alpha d\alpha \, \vec{b} \cdot \vec{F}_{\mu\nu}(\alpha)\right\} \tag{3.9}$$

with the fermions totally decoupled from the gluons and the last factor in Eq. (3.9) representing the fermion determinant.

Given the explicit α dependence of $\vec{A}_\mu(\alpha)$ in Eq. (3.4), it is elementary but tedious to evaluate the α integral in Eq. (3.9). The final result is

$$Fermion\ Determinant = \exp\left\{\frac{-\epsilon_{\mu\nu}}{\pi} \int d^2x \int_0^1 d\alpha \, \vec{b} \cdot \vec{F}_{\mu\nu}(\alpha)\right\}$$

$$= \exp\left\{\frac{-\epsilon_{\mu\nu}}{\pi} \int d^2x \, \vec{b} \, \{\partial_\mu \vec{A}_\nu - \partial_\nu \vec{A}_\mu + \vec{A}_\mu \times \vec{A}_\nu \, (\frac{1}{b \tanh b} - \frac{1}{\sinh^2 b})\}\right\}. \tag{3.10}$$

If we knew how to determine $\vec{b}$ from $\vec{A}_\mu$, this would be a complete evaluation of the fermion determinant. That is addressed in the next section.

4. GAUGE CONDITIONS ON A_μ

Equation (2.7) expresses six functions (the components of $\vec{A}_\mu$) in terms of three functions $\vec{b}$. There must then be three subsidiary conditions on $\vec{A}_\mu$ which are the gauge conditions. To find these conditions, we can rewrite Eq. (2.7) as six first order differential equations for $\partial_\mu \vec{b}$, whose coefficients are functions of $\vec{A}_\mu$. The integrability conditions will be the gauge conditions.

If we define

$$\vec{z} = \frac{\vec{b} \tanh b}{b}, \tag{4.1}$$

then Eq. (2.7) can be rewritten as

$$\partial_\mu \vec{z} = - \vec{A}_\mu \times \vec{z} + \epsilon_{\mu\nu} \{ \vec{z} (\vec{z} \cdot \vec{A}_\nu) - \vec{A}_\nu \}. \tag{4.2}$$

The first integrability condition is

$$\partial_1 (\partial_2 \vec{z}) = \partial_2 (\partial_1 \vec{z}) \tag{4.3}$$

which leads to

$$\partial_\mu \vec{A}_\mu + (\epsilon_{\mu\nu} \partial_\nu \vec{A}_\mu) \times \vec{z} = 0. \tag{4.4}$$

Further gauge conditions are derived by taking further derivatives of Eq. (4.4), and using Eq. (4.2) to eliminate $\partial_\mu \vec{z}$. Unfortunately, these conditions all involve $\vec{z}$, and are not yet conditions on $\vec{A}_\mu$.

Before studying the gauge conditions any further, it is worth noting that in the Abelian case (all $SU(2)$ vectors are parallel), the gauge condition Eq. (4.4) is

$$\partial_\mu A_\mu = 0, \tag{4.5}$$

and since this is z independent, no other conditions are necessary. Looking at Eq. (2.7), we find

$$A_\mu = \epsilon_{\mu\nu} \partial_\nu b, \tag{4.6}$$

and the Fermion determinant in Eq. (3.10) is just the expression we derived [3] for the Schwinger model with two flavors.

Returning to the full non-Abelian theory, define

$$\vec{F} = \partial_\mu \vec{A}_\mu , \qquad\qquad \vec{G} = \epsilon_{\mu\nu} \partial_\nu \vec{A}_\mu \tag{4.7}$$

so that Eq. (4.4) reads

$$\vec{F} + \vec{G} \times \vec{z} = 0 \tag{4.8}$$

so that

$$\vec{F} \cdot \vec{G} = 0 \tag{4.9}$$

is necessary. This is one of the gauge conditions. It also follows that if $\vec{G} \neq 0$,

$$\vec{z} = \frac{\vec{G} \times \vec{F}}{G^2} + \beta \vec{G} \tag{4.10}$$

Ralph Roskies

is the general solution of Eq. (4.8), with β an arbitrary function. If we substitute Eq. (4.10) back into Eq. (4.2) we get

$$\partial_\mu \left(\frac{\vec{G} \times \vec{F}}{G^2} \right) + \beta \, (\partial_\mu \vec{G}) + (\partial_\mu \beta) \, \vec{G}$$

$$= - \vec{A}_\mu \times \left(\frac{\vec{G} \times \vec{F}}{G^2} + \beta \, \vec{G} \right) + \epsilon_{\mu\nu} \left[\left(\frac{\vec{G} \times \vec{F}}{G^2} + \beta \vec{G} \right) \left(\frac{\vec{G} \times \vec{F}}{G^2} + \beta \vec{G} \right) \cdot \vec{A}_\nu - \vec{A}_\nu \right]. \qquad (4.11)$$

If we cross these equations with $\vec{G}$, the terms in $\partial_\mu \beta$ on the left hand side vanish, as do those quadratic in β on the right hand side. We then have a set of linear equations for β in terms of $\vec{A}_\mu$. Solving any one of those for β and reinserting the solution into Eq. (4.11), gives the complete set of consistency conditions that $\vec{A}_\mu$ must satisfy. This is a complete differential specification of the gauge conditions on $\vec{A}_\mu$. From Eq. (4.10) we also have the explicit dependence of $\vec{z}$ on $\vec{A}_\mu$, and so of $\vec{b}$ on $\vec{A}_\mu$ from Eq. (4.1). If we insert that into Eq. (3.10), the fermion determinant is expressed as a function of $\vec{A}_\mu$ in the appropriate gauge. I have not been able to find an elegant differential characterization of the gauge. However, my argument shows that these conditions are local, not involving higher than second derivatives in the fields $\vec{A}_\mu$.

CONCLUSION

I have evaluated the fermion determinant in massless *QCD* in two dimensions in a special gauge. I have not found an elegant differential characterization of the gauge. It can be characterized by conditions on the potential not involving more than two derivatives.

ACKNOWLEDGEMENTS

I wish to thank F. Schaposnik for sending me a copy of his work (reference 6) which was the immediate stimulus for this work. I also wish to acknowledge fruitful conversations with R. Horsley, E. T. Newman, G. Sparling and U. Wolff. I wish to thank the particle theory group at the Weizmann Institute for their hospitality. This work was supported in part by the National Science Foundation under grant PHY-80-24638.

REFERENCES

[1]. J. Schwinger, Phys. Rev. 128 (1962) 2425.

[2]. N. K. Nielsen, K. D. Rothe and B. Schroer, Nucl. Phys. B160 (1979) 330; H. Arodz, Acta Phys. Pol. B10 (1979) 983; C. Sorensen and G. H. Thomas, Phys. Rev. D21, (1980) 1625.

[3]. R. Roskies and F. Schaposnik, Phys. Rev. D23 (1981) 558.

[4]. K. Fujikawa, Phys. Rev. Lett. 42 (1979) 1195; Phys. Rev. D21 (1980) 2848.

[5]. See e.g. H. Grauert, Math. Annalen 135 (1958) 266. I am indebted to G. Sparling for explaining this point to me.

[6]. The approach to this problem was suggested by the preprint *Path Integral Formulation of Two Dimensional Gauge Theories with Massless Fermions* by R. E. Gamboa Saravi, F. A. Schaposnik and J. E. Solomin, subsequently published in Nucl. Phys. B185 (1981) 239.

DYNAMIC MASS GENERATION
FOR FERMIONS

Abdus Salam and *J. Strathdee*

International Center for Theoretical Physics, Trieste, Italy

Dedicated to Feza Gürsey on his sixtieth birthday
with warmest good wishes.

Fermionic mass is commonly represented in Lagrangian field theories either by direct mass terms or by Yukawa couplings to scalar fields which have the potential to develop non-zero expectation values. Both types involve *a priori* undetermined parameters. In gauge theories there arises the further problem of explaining why some of the fermions are relatively light (the gauge hierarchy problem [1]). The purpose of this note is to suggest a mechanism, appropriate to a class of gauge theories, which might go some way to solving these problems. Our considerations, though couched for a particular group, are more generally applicable.

Consider a gauge theory with a chiral structure such as $SU(3)_L \times SU(3)_R$ (color) or $SU(2)_L \times SU(2)_R$ (flavor). It is possible to envision a spontaneous breakdown, for example, of the local color symmetry to $SU(3)_{L+R}$ such that the axial vector gauge fields acquire mass. In bringing this about a number of Higgs fields are of course involved. But suppose that there is no mass term for the fermions. This could be for various reasons: no suitable Higgs field, a discrete symmetry, or simply the absence unrelated to symmetry of a bare mass or Yukawa coupling in the Lagrangian. In all such cases the vanishing fermionic mass will persist in finite orders of perturbation theory. However, it would be unrealistic to trust this perturbative behavior. Our suggestion is that one must sum an infinite selection of graphs to get a more realistic view. The simplest way to do this is by approximating to the Dyson-Schwinger equations.

Now, the crucial point of this note is that the graphs which contribute to fermion mass must contain a line representing the mixed propagator[1] $<T\ W_{L\mu} W_{R\nu}>$. The

---

[1]The method described in this note was pioneered by F. Englert, J. M. Frere and P. Nicoletopolelos, Phys. Lett. 52B, (1974) 433. The authors learned of their work after their own investigation had been completed, in International Center Preprints, IC/80/67 and IC/80/95.

115

components of this propagator decrease like $\mathcal{O}(k^{-4})$ at large momentum, and in consequence, the integrals are ultraviolet convergent. *No subtraction is needed and the
fermion mass is computable in principle.*[2] Of course, one meets the usual difficulties
of approximation schemes which are not strictly perturbative. In particular, gauge invariance tends to be lost.[3]

To illustrate the main point we use the example of a pair of chiral triplets ψ_L and ψ_R
interacting with the color $SU(3)_L \times SU(3)_R$ gauge fields W_L and W_R. A matrix of
Higgs fields, ϕ, in the representation $(3, \bar{3})$ serves to break the symmetry to
$SU(3)_{L+R}$. The Lagrangian is given by

$$\mathcal{L} = \bar{\psi}_L \gamma_\mu (i\partial_\mu - gW_{L\mu})\psi_L + \bar{\psi}_R \gamma_\mu (i\partial_\mu - gW_{R\mu})\psi_R$$

$$+ 2\,\mathrm{Tr}\,[\partial_\mu\phi + ig\,W_{L\mu}\phi - ig\,\phi W_{R\mu}]^2 - V(\phi, \phi^\dagger)$$

$$+ 2\,\mathrm{Tr}\,[-\tfrac{1}{4}\,W_{L\mu\nu}^2 - \tfrac{1}{2\beta_L}\,(\partial_\mu W_{L\mu})^2 - \tfrac{1}{4}\,W_{R\mu\nu}^2 - \tfrac{1}{2\beta_R}\,(\partial_\mu W_{R\mu})^2\,]\,,$$

where $W_L = W_L^\alpha\,\lambda^\alpha/2$ with the standard $SU(3)$ matrices. The parameters β_L and β_R
define the gauge. We assume that the potential V develops a stable minimum at

$$\phi = \frac{M}{g\sqrt{2}} \times \textit{unit matrix}$$

so that the axial vector combination $\frac{1}{\sqrt{2}}\{W_L - W_R\}$ acquires the mass M while the
vectors $\frac{1}{\sqrt{2}}\{W_L + W_R\}$ remain massless. Note that the Yukawa coupling, $\bar{\psi}_L\,\phi\,\psi_R$
+ *H.C.*, has been omitted. In every finite order of perturbation theory the fermions
will remain without a mass.

It is straightforward to obtain the gauge field propagators from the above Lagrangian. They are

$$-\frac{i}{\hbar} <T\,W_{L\mu}\,W_{L\nu}> = \frac{k^2 - M^2/2}{k^2(k^2 - M^2)}\,(-\eta_{\mu\nu} + \frac{k_\mu k_\nu}{k^2}) - \beta_L\,\frac{k_\mu k_\nu}{k^4}\,,$$

$$-\frac{i}{\hbar} <T\,W_{L\mu}\,W_{R\nu}> = \frac{-M^2/2}{k^2(k^2 - M^2)}\,(-\eta_{\mu\nu} + \frac{k_\mu k_\nu}{k^2})\,,$$

---

[2]The convergence of the self-mass calculation can also be seen as the result of a partial cancellation
between vector and axial vector contributions to the fermion's self energy. This fact was noted by Budini
[2] who brought it to our attention after the completion of this work.

[3]Lack of gauge invariance is a serious weakness of the approximation scheme and must be overcome
before applying it to realistic models. The gauge problem presents itself for situations where Green's
functions are involved; e.g. for many of the renormalization group calculations. A "gauge technique" to
overcome this problem was elaborated in [3]. However, in this note this technique has not been applied.

Figure 1: Approximate Dyson-Schwinger equation for the fermion propagator. The fermion self-energy Σ is here expressed in terms of the unmodified vertices $g\gamma_\mu(1\pm i\gamma_5)/2$ and vector propagator $D_{\mu\nu}$, but with the full fermion propagator $S=(\not p-\Sigma)^{-1}$.

$$- \frac{i}{\hbar} <T\, W_{R\mu}\, W_{R\nu}> = \frac{k^2 - M^2/2}{k^2(k^2 - M^2)}\left(-\eta_{\mu\nu} + \frac{k_\mu k_\nu}{k^2}\right) - \beta_R\, \frac{k_\mu k_\nu}{k^4}\,.$$

It is amusing that the mixed components $<T\, W_L W_R>$ of Feynman-like gauges do not depend on the parameters β_L and β_R and to the order we are working (see below) the gauge problem is circumvented. However this circumstance is not repeated in other types of gauge.

The fermion self-energy $\Sigma(p) = A(p^2) + \not p\, B(p^2)$ appears in the propagator

$$S = (\not p - \Sigma(p))^{-1}\,.$$

It is to be obtained by solving an integral equation. The approximate Dyson-Schwinger equation is presented graphically in Fig. 1. Write, for the mixed components,

$$D_{\mu\nu}(k) = \frac{-M^2/2}{k^2(k^2 - M^2)}\left(-\eta_{\mu\nu} + \frac{k_\mu k_\nu}{k^2}\right)\,.$$

Then the amplitude $A(p^2)$ is given by the convergent projection of the full equation:

$$-\frac{1 + i\gamma_5}{2}\Sigma(p)\frac{1 - i\gamma_5}{2} = \frac{\hbar}{i}\int \frac{d^4k}{(2\pi)^4}\, g\gamma_\mu \frac{1 - i\gamma_5}{2}\frac{\lambda^\alpha}{2}(\not p - \not k - \Sigma)^{-1} g\gamma_\nu \frac{1 + i\gamma_5}{2}\frac{\lambda^\alpha}{2}D_{\mu\nu}(k)$$

in the form:[4]

$$-A(p^2) = \frac{g^2\hbar}{i}\int \frac{d^4k}{(2\pi)^4}\, \frac{A(\{p-k\}^2)}{(p-k)^2\,(1-B)^2 - A^2}\, \frac{4}{3}\, D_{\mu\mu}(k)\,.$$

To obtain an estimate of the fermion mass, m, we reduce this to an algebraic problem

---

[4]The group theoretical factor $\sum_\alpha \frac{\lambda^\alpha}{2}\frac{\lambda^\alpha}{2} = \frac{4}{3}$ occurs with the standard normalizations of the λ^α, the SU(3) matrices.

by making the simplifying assumptions

$$A(p^2) = m \quad ; \quad B(p^2) = 0$$

and taking $p = 0$.

$$- m = \frac{g^2 \hbar}{i} \int \frac{d^4 k}{(2\pi)^4} \frac{m}{k^2 - m^2} \frac{3M^2/2}{k^2(k^2 - M^2)} \frac{4}{3} .$$

The result is

$$1 = \frac{\alpha_s}{\pi} \frac{M^2}{M^2 - m^2} \ln \frac{M^2}{m^2} ,$$

where $\alpha_s = g^2 \hbar/(4\pi)$. With the approximations of this note the fermion mass is therefore given by

$$m \approx M \exp\{-\pi/(2\alpha_s)\} .$$

Thus the axial gluon mass M is of the order of 10^7 Gev if $\alpha_s \approx 1/10$ and the fermionic mass m is ≈ 1 Gev.

In view of the approximations which went into the derivation of this result, and in view of the fact that the starting model itself is not fully realistic[5] the numbers above are only a rough guide to the orders of magnitude involved. Also, as noted earlier, the ratio m/M is sensitive to gauge through the factor of $D_{\mu\mu}(k)$ in the Dyson-Schwinger equation. (Presumably one is not allowed to replace, as we have done, a full vertex Γ_ν by a point approximation $g\gamma_\nu \frac{1 + i\gamma_5}{2}$ while at the same time keeping the full propagator $(\not{p} - \Sigma)^{-1}$ for the fermion.) However, in spite of this serious technical difficulty we believe that *the basic computability of fermion mass in this type of gauge theory is a fact of some importance.* Further, the particular form of our result lends credence to the idea that light fermions can coexist with heavy gauge mesons in gauge theories. Similar ideas can be used to compute, in principle, Cabibbo and other mixing angles finitely and self-consistently.

To show this, we shall consider as an example, a system of two families, say the e and the u families of fermions. Each is gauged separately, e.g., $[SU(3) \times SU(2) \times U(1)]_e \times [SU(3) \times SU(2) \times U(1)]_u$. At the present low energies only the

---

[5]For example, we have left out of consideration extra fermions needed to cancel anomalies. This will be done elsewhere.

diagonal sum of the two types of gauges manifests itself. We wish to compute the general fermion mass matrix dynamically, the only input being the chiral symmetry for the gauge mesons; e.g., our color group is not just $SU(3)_e \times SU(3)_u$ but $[SU(3)_L \times SU(3)_R]_e \times [SU(3)_L \times SU(3)_R]_u$. To simplify the discussion even more we consider just two fermions, and the color group $U(1)$.

The two Dirac spinors ψ_1 and ψ_2 interact with gauge fields W_{1L}, W_{1R}, W_{2L} and W_{2R}. In terms of two-component spinors

$$\psi_L = \begin{pmatrix} \psi_1 \\ \psi_2 \end{pmatrix}_L ; \quad \psi_R = \begin{pmatrix} \psi_1 \\ \psi_2 \end{pmatrix}_R$$

and gauge fields

$$W_L = \begin{pmatrix} W_{1L} & 0 \\ 0 & W_{2L} \end{pmatrix} ; \quad W_R = \begin{pmatrix} W_{1R} & 0 \\ 0 & W_{2R} \end{pmatrix}$$

the Lagrangian takes the form

$$\begin{aligned}
\mathcal{L} = &\; \bar{\psi}_L(i\slashed{\partial} - gW_L)\psi_L + \bar{\psi}_R(i\slashed{\partial} - gW_R)\psi_R \\
&- \tfrac{1}{4} \, \text{Tr}(W_{L\mu\nu}^2 - W_{R\mu\nu}^2) \; + \textit{gauge fixing terms} \\
&+ \frac{M_1^2}{4}(W_{1L} - W_{1R})^2 + \frac{M_2^2}{4}(W_{2L} - W_{2R})^2 + \frac{M_3^2}{8}(W_{1L} + W_{1R} - W_{2L} - W_{2R})^2.
\end{aligned} \tag{1}$$

Here the gauge fixing terms have not been specified. It is assumed that the gauge symmetry breaks spontaneously to $U(1)$ so that three of the four gauge fields acquire mass by the Higgs mechanism. We are not interested in the details of the Higgs system and have suppressed all scalar field-containing terms. (In general there can be scalar vector mixing terms which affect the structure of the vector propagators and should therefore be taken into account. Here, for simplicity, we shall adopt the Landau gauge where such couplings do not occur.)

The gauge propagators are easily deduced from those of the orthogonal combinations

$$\begin{aligned}
A_1 &= \tfrac{1}{\sqrt{2}}(W_{1L} - W_{1R}) \;, \quad V_3 = \tfrac{1}{2}(W_{1L} + W_{1R} - W_{2L} - W_{2R}) \,, \\
A_2 &= \tfrac{1}{\sqrt{2}}(W_{2L} - W_{2R}) \;, \quad V_0 = \tfrac{1}{2}(W_{1L} + W_{1R} + W_{2L} + W_{2R}) \,,
\end{aligned} \tag{2}$$

which invert to give

$$\begin{aligned}
W_{1L} &= \tfrac{1}{\sqrt{2}}A_1 + \tfrac{1}{2}V_3 + \tfrac{1}{2}V_0 \;, \quad W_{1R} = -\tfrac{1}{\sqrt{2}}A_1 + \tfrac{1}{2}V_3 + \tfrac{1}{2}V_0 \,, \\
W_{2L} &= \tfrac{1}{\sqrt{2}}A_2 - \tfrac{1}{2}V_3 + \tfrac{1}{2}V_0 \;, \quad W_{2R} = -\tfrac{1}{\sqrt{2}}A_2 - \tfrac{1}{2}V_3 + \tfrac{1}{2}V_0 \,,
\end{aligned} \tag{3}$$

In the Landau gauge, then

$$
\begin{aligned}
<T\, W_{1L\mu} W_{1R\nu}> &= -\tfrac{1}{2} <T\, A_{1\mu} A_{1\nu}> + \tfrac{1}{4} <T\, V_{3\mu} V_{3\nu}> + \tfrac{1}{4} <T\, V_{0\mu} V_{0\nu}> \qquad (4)\\
&= \tfrac{\hbar}{i} \Big(\eta_{\mu\nu} - \tfrac{k_\mu k_\nu}{k^2} \Big) \Big[-\tfrac{1}{2} \tfrac{1}{k^2 + M_1^{\,2}} + \tfrac{1}{4} \tfrac{1}{k^2 + M_3^{\,2}} + \tfrac{1}{4} \tfrac{1}{k^2} \Big]\\
&= \tfrac{\hbar}{i} \Big(\eta_{\mu\nu} - \tfrac{k_\mu k_\nu}{k^2} \Big) \tfrac{1}{k^2} \Big[-\tfrac{1}{2} \tfrac{M_1^{\,2}}{k^2 + M_1^{\,2}} + \tfrac{1}{4} \tfrac{M_3^{\,2}}{k^2 + M_3^{\,2}} \Big]
\end{aligned}
$$

$$
\begin{aligned}
<T\, W_{1L\mu} W_{2R\nu}> &= <T\, W_{2L\mu} W_{1R\nu}>\\
&= -\tfrac{1}{4} <T\, V_{3\mu} V_{3\nu}> + \tfrac{1}{4} <T\, V_{0\mu} V_{0\nu}>\\
&= -\tfrac{\hbar}{i} \Big(\eta_{\mu\nu} - \tfrac{k_\mu k_\nu}{k^2} \Big) \tfrac{1}{4} \tfrac{1}{k^2} \tfrac{M_3^{\,2}}{k^2 + M_3^{\,2}}
\end{aligned}
$$

$$
\begin{aligned}
<T\, W_{2L\mu} W_{2R\nu}> &= -\tfrac{1}{2} <T\, A_{2\mu} A_{2\nu}> + \tfrac{1}{4} <T\, V_{3\mu} V_{3\nu}> + \tfrac{1}{4} <T\, V_{0\mu} V_{0\nu}>\\
&= \tfrac{\hbar}{i} \Big(\eta_{\mu\nu} - \tfrac{k_\mu k_\nu}{k^2} \Big) \Big[-\tfrac{1}{2} \tfrac{1}{k^2 + M_2^{\,2}} + \tfrac{1}{4} \tfrac{1}{k^2 + M_3^{\,2}} + \tfrac{1}{4} \tfrac{1}{k^2} \Big]\\
&= \tfrac{\hbar}{i} \Big(\eta_{\mu\nu} - \tfrac{k_\mu k_\nu}{k^2} \Big) \tfrac{1}{k^2} \Big[-\tfrac{1}{2} \tfrac{M_2^{\,2}}{k^2 + M_2^{\,2}} + \tfrac{1}{4} \tfrac{M_3^{\,2}}{k^2 + M_3^{\,2}} \Big] .
\end{aligned}
$$

Since the object of the calculation is to compute the fermion mass matrix, μ, we must use this matrix in the parametrization of the fermion propagator. To obtain the appropriate expressions for the propagator components consider the effective Lagrangian

$$
\mathcal{L}_{Fermi} = \bar\psi_L \not{\partial} \psi_L + \bar\psi_R \not{\partial} \psi_R - \bar\psi_L \mu \psi_R - \bar\psi_R \mu^\dagger \psi_L \ .
$$

The mass matrix can be represented in the canonical form

$$
\mu = \mathcal{U} \, m \, \mathcal{V}^{-1} , \qquad (5)
$$

where $\mathcal{U}$ and $\mathcal{V}$ are unitary and m is real and diagonal. It is straightforward to derive the expressions

$$
< T\, \psi_L \bar\psi_L > = -\tfrac{\hbar}{i} \tfrac{1 + i\gamma_5}{2} \not{p} \, \mathcal{U} \, (p^2 - m^2)^{-1} \, \mathcal{U}^{-1} = -\tfrac{\hbar}{i} \tfrac{1 + i\gamma_5}{2} \not{p} \, (p^2 - \mu\mu^\dagger)^{-1} \quad (6)
$$

$$< T \, \psi_L \, \bar{\psi}_R > \; = \; - \frac{\hbar}{i} \frac{1 + i\gamma_5}{2} \, \mathcal{U} \, m \, (p^2 - m^2)^{-1} \, \mathcal{V}^{-1} = \; - \frac{\hbar}{i} \frac{1 + i\gamma_5}{2} \, \mu (p^2 - \mu\mu^\dagger)^{-1}$$

$$= \; - \frac{\hbar}{i} \frac{1 + i\gamma_5}{2} \, (p^2 - \mu\mu^\dagger)^{-1} \mu \; ,$$

etc. In the approximation where corrections to the vector propagators and the vertices are discarded, the Dyson equation for the fermion propagator takes the form

$$\Sigma(p) = -\frac{g^2 \hbar}{i} \int \frac{d^4 k}{(2\pi)^4} \, \gamma_\mu \, t^a \, (\not{p} - \not{k} - \Sigma)^{-1} \, \gamma_\nu \, t^b \, D^{ab}_{\mu\nu}(k) \; . \tag{7}$$

In this equation the vertices derive from the interaction Lagrangian

$$g \, \bar{\psi} \, \gamma_\mu \, t^a \, \psi \, W^a_\mu$$

in which the matrices t^a generally include the chiral projector $\dfrac{1 \pm i\gamma_5}{2}$. The free vector propagators, Eq. (4), are included in the general expressions

$$<T \, W^a_\mu \, W^b_\nu > \; = \; - \frac{\hbar}{i} \, D^{ab}_{\mu\nu}(k) \; . \tag{8}$$

The complete fermion propagator is written

$$<T \, \psi \, \bar{\psi} > \; = \; - \frac{\hbar}{i} \, (\not{p} - \Sigma)^{-1} \; . \tag{9}$$

The diagonal terms ($a=b$) in the vector propagator $D^{ab}_{\mu\nu}(k)$ must be $\mathcal{O}(k^{-2})$ for $k \to \infty$ and they will be associated with ultraviolet divergent terms in the Dyson equation. Off-diagonal terms D^{ab} will be $\mathcal{O}(k^{-4})$ if the corresponding element of the mass matrix $(M^2)^{ab}$ is non-vanishing, otherwise they will decrease even faster. All such terms contribute finite pieces to the right-hand side of eqn.(7). If the equation is decomposed into chiral pieces then the mass term Σ_{LR} will be given by a convergent integral if there is no component among the vectors which couples to both left and right spinors, i.e., if the gauging is fully chiral. (The kinetic terms Σ_{LL} and Σ_{RR} must inevitably diverge and be subtracted by wave function renormalizations.)

If the mass matrix is determined by a convergent integral equation then, at least in principle, one can compute it. That is one can express the fermion masses in terms of a set of given vector boson masses. This is a more modest program than the fully self-consistent generation of dynamical symmetry breakdown in which the vector masses are also to be computed by exploiting the appropriate Dyson equations.

To obtain an estimate for the mass matrix μ we set $p = 0$ in Eq. (7) and in the integration let Σ be represented just by the mass matrix itself,

$$\Sigma(k) \approx \frac{1 - i\gamma_5}{2}\, \mu + \frac{1 + i\gamma_5}{2}\, \mu^\dagger\,.$$

It is a simple matter to eliminate the Dirac matrices and extract the equation

$$\mu = -\frac{g^2 \hbar}{i} \int \frac{d^4k}{(2\pi)^4}\, t^a\, \mu\, (k^2 - \mu^\dagger \mu\,)^{-1}\, t^b\, D^{ab}_{\mu\mu}(k)\,, \tag{10}$$

where it is understood that t^a $(a = 1, 2)$ refers to W_L and t^b $(b = 1, 2)$ to W_R. The propagator components D^{ab} needed for eqn.(10) are just those listed in Eq. (4). On making the substitutions,

$$\mu = 3\frac{g^2\hbar}{i} \int \frac{d^4k}{(2\pi)^4} \left[\{t^1\,\mu\,(k^2 - \mu^\dagger\mu)^{-1}\,t^1\,\frac{1}{k^2}\,(-\frac{1}{2}\,\frac{M_1^2}{k^2 + M_1^2} + \frac{1}{4}\,\frac{M_3^2}{k^2 + M_3^2})\} \right.$$

$$+ \{(t^1\,\mu\,(k^2 - \mu^\dagger\mu)^{-1}\,t^2 + t^2\,\mu\,(k^2 - \mu^\dagger\mu)^{-1}\,t^1)\,\frac{1}{k^2}\,(-\frac{1}{4}\,\frac{M_3^2}{k^2 + M_3^2})\} \tag{11}$$

$$\left. + \{t^2\,\mu\,(k^2 - \mu^\dagger\mu)^{-1}\,t^2\,\frac{1}{k^2}\,(-\frac{1}{2}\,\frac{M_2^2}{k^2 + M_2^2} + \frac{1}{4}\,\frac{M_3^2}{k^2 + M_3^2})\} \right]\,.$$

Make a Wick rotation $k_o \rightarrow ik_4$ and integrate out the angles, i.e., write $d^4k = 2\pi^2\,k^2dk^2$ to obtain

$$\mu = \frac{3g^2\hbar}{8\pi^2} \int_0^\infty dk^2 \left[\{t^1\,\mu\,(k^2 + \mu^\dagger\mu)^{-1}\,t^1\,(\frac{1}{2}\,\frac{M_1^2}{k^2 + M_1^2} - \frac{1}{4}\,\frac{M_3^2}{k^2 + M_3^2})\} \right.$$

$$+ \{(t^1\,\mu\,(k^2 + \mu^\dagger\mu)^{-1}\,t^2 + t^2\,\mu\,(k^2 + \mu^\dagger\mu)^{-1}\,t^1)\,(\frac{1}{4}\,\frac{M_3^2}{k^2 + M_3^2})\}$$

$$\left. + \{t^2\,\mu\,(k^2 - \mu^\dagger\mu)^{-1}\,t^2\,(\frac{1}{2}\,\frac{M_2^2}{k^2 + M_2^2} - \frac{1}{4}\,\frac{M_3^2}{k^2 + M_3^2})\} \right]$$

$$= \frac{3\alpha}{8\pi} \left[t^1\mu\,\frac{2M_1^2}{M_1^2 - \mu^\dagger\mu}\,\ln\frac{M_1^2}{\mu^\dagger\mu}\,t^1 + t^2\mu\,\frac{2M_2^2}{M_2^2 - \mu^\dagger\mu}\,\ln\frac{M_2^2}{\mu^\dagger\mu}\,t^2 \right.$$

$$\left. - (t^1 - t^2)\mu\,\frac{M_3^2}{M_3^2 - \mu^\dagger\mu}\,\ln\frac{M_3^2}{\mu^\dagger\mu}\,(t^1 - t^2) \right]\,, \tag{12}$$

where $\alpha = g^2\,\hbar/4$. Alternatively, this equation can be given in terms of eigenvalues and mixing angles. Quite generally, the mass matrix can be expressed in the form

$$\mu = \mathcal{U}\,m\,\mathcal{V}^{-1} = \begin{pmatrix} \cos\Theta_L & -\sin\Theta_L \\ \sin\Theta_L & \cos\Theta_L \end{pmatrix} \begin{pmatrix} m_1 & 0 \\ 0 & m_2 \end{pmatrix} \begin{pmatrix} \cos\Theta_R & \sin\Theta_R \\ -\sin\Theta_R & \cos\Theta_R \end{pmatrix} \tag{13}$$

(any complex phases in μ can be absorbed by appropriately re-defining ψ_L and ψ_R). It is convenient to introduce the rotated coupling matrices

$$t^1(\Theta) = \begin{pmatrix} \cos\Theta & \sin\Theta \\ -\sin\Theta & \cos\Theta \end{pmatrix} t^1 \begin{pmatrix} \cos\Theta & -\sin\Theta \\ \sin\Theta & \cos\Theta \end{pmatrix} = \begin{pmatrix} \cos^2\Theta & -\sin\Theta\cos\Theta \\ -\sin\Theta\cos\Theta & \sin^2\Theta \end{pmatrix}$$

and

$$t^2(\Theta) = \begin{pmatrix} \sin^2\Theta & \sin\Theta\cos\Theta \\ \sin\Theta\cos\Theta & \cos^2\Theta \end{pmatrix} \tag{14}$$

eqn.(12) now takes the form

$$m \doteq \frac{3\alpha}{8\pi} \left[t^1(\Theta_L) \frac{2M_1^2 m}{m_1^2 - m^2} \ln \frac{M_1^2}{m^2} t^1(\Theta_R) + t^2(\Theta_L) \frac{2M_2^2 m}{m_2^2 - m^2} \ln \frac{M_2^2}{m^2} t^2(\Theta_R) \right.$$

$$\left. + (t^1(\Theta_L) - t^2(\Theta_L)) \frac{M_3^2 m}{m_3^2 - m^2} \ln \frac{M_3^2}{m^2} t^2 (t^1(\Theta_R) - t^2(\Theta_R)) \right], \tag{15}$$

where m is a diagonal matrix. There are four distinct solutions here for the four unknowns, m_1, m_2, Θ_L and Θ_R. To solve them we shall make some approximations. Thus, we shall assume that the fermion masses are small relative to the vector masses

$$m/M \ll 1 . \tag{16}$$

Then eqn.(15) reduces to

$$m = \frac{3\alpha}{8\pi} [t_L^1 2m \ln \frac{M_1^2}{m^2} t_R^1 + t_L^2 2m \ln \frac{M_2^2}{m^2} t_R^2 - \{t_L^1 - t_L^2\}/2m \ln \frac{M_3^2}{m^2} \{t_R^1 - t_R^2\}]$$

or, in detail

$$\begin{pmatrix} m_1 & 0 \\ 0 & m_2 \end{pmatrix} = \frac{3\alpha}{8\pi} \begin{pmatrix} a_{11} & a_{12} \\ a_{21} & a_{22} \end{pmatrix} \tag{17}$$

where,

$$a_{11} = m_1 \ln \frac{M_1 M_2}{m_1^2} + (c_L + c_R) m_1 \ln \frac{M_1}{M_2} + (c_L c_R m_1 + s_L s_R m_2) \ln \frac{M_1 M_2}{M_3^2} ,$$

$$a_{12} = (-s_L c_R m_1 + c_L s_R m_2) \ln \frac{M_1 M_2}{M_3^2} - (s_L m_1 + s_R m_2) \ln \frac{M_1}{M_2} ,$$

$$a_{21} = (-c_L s_R m_1 + s_L c_R m_2) \ln \frac{M_1 M_2}{M_3^2} - (s_R m_1 + s_L m_2) \ln \frac{M_1}{M_2} ,$$

$$a_{22} = m_2 \ln \frac{M_1 M_2}{m_2^2} - (c_L + c_R) m_2 \ln \frac{M_1}{M_2} + (s_L s_R m_1 + c_L c_R m_2) \ln \frac{M_1 M_2}{M_3^2} ,$$

where $c_L = \cos2\Theta_L$, $c_R = \cos2\Theta_R$, etc. Consider first the two off-diagonal components

$$0 = \left(-s_L c_R \ln \frac{M_1 M_2}{M_3^2} - s_L \ln \frac{M_1}{M_2} \right) m_1 + \left(c_L s_R \ln \frac{M_1 M_2}{M_3^2} - s_R \ln \frac{M_1}{M_2} \right) m_2$$

$$0 = \left(-c_L s_R \ln \frac{M_1 M_2}{M_3^2} - s_R \ln \frac{M_1}{M_2} \right) m_1 + \left(s_L c_R \ln \frac{M_1 M_2}{M_3^2} - s_L \ln \frac{M_1}{M_2} \right) m_2 .$$

If m_1 and/or m_2 are non-vanishing, compatibility requires

$$0 = \left\{ \left(-s_L c_R \ln \frac{M_1 M_2}{M_3^2} - s_L \ln \frac{M_1}{M_2} \right) \left(s_L c_R \ln \frac{M_1 M_2}{M_3^2} - s_L \ln \frac{M_1}{M_2} \right) \right.$$

$$\left. - \left(-c_L s_R \ln \frac{M_1 M_2}{M_3^2} \; s_R \ln \frac{M_1}{M_2} \right) \left(c_L s_R \ln \frac{M_1 M_2}{M_3^2} - s_R \ln \frac{M_1}{M_2} \right) \right\} \tag{18}$$

$$0 = \left\{ \left(c_L^2 s_R^2 - s_L^2 c_R^2 \right) \ln^2 \frac{M_1 M_2}{M_3^2} + \left(s_L^2 - s_R^2 \right) \ln^2 \frac{M_1}{M_2} \right\}$$

$$= \left\{ \left(s_L^2 - s_R^2 \right) \left(-\ln^2 \frac{M_1 M_2}{M_3^2} + \ln^2 \frac{M_1}{M_2} \right) \right\} .$$

Since the vector masses are presumably independent we must take

$$s_L^2 = s_R^2 . \tag{19}$$

For convenience, in the remaining equations we shall adopt a particular solution

$$\Theta_L = -\Theta_R = \Theta , \tag{20}$$

where Θ is given by

$$0 = -\sin 2\Theta \cos 2\Theta \, (m_1 + m_2) \ln \frac{M_1 M_2}{M_3^2} - \sin 2\Theta \, (m_1 - m_2) \ln \frac{M_1}{M_2} ,$$

i. e.

$$\cos 2\Theta \, \frac{m_1 + m_2}{m_1 - m_2} = -\frac{\ln (M_1/M_2)}{\ln\{(M_1 M_2)/M_3^2\}} . \tag{21}$$

The remaining two equations are

$$m_1 = \frac{3\alpha}{8\pi} \left[m_1 \ln \frac{M_1 M_2}{m_1^2} + 2\cos 2\Theta \, m_1 \ln \frac{M_1}{M_2} + (\cos^2 2\Theta \, m_1 - \sin^2 2\Theta \, m_2) \ln \frac{M_1 M_2}{M_3^2} \right]$$

$$\tag{22}$$

$$m_2 = \frac{3\alpha}{8\pi} \left[m_2 \ln \frac{M_1 M_2}{m_2^2} - 2\cos 2\Theta \, m_2 \ln \frac{M_1}{M_2} + (\cos^2 2\Theta \, m_2 - \sin^2 2\Theta \, m_1) \ln \frac{M_1 M_2}{M_3^2} \right] .$$

To simplify these we make an assumption about the relative magnitudes of the vector masses

$$\left| \ln \frac{M_1 M_2}{M_3^2} \right| \ll \left| \ln \frac{M_1}{M_2} \right| , \tag{23}$$

since this serves to decouple the equations (22). They give

$$m_1^2 = M_1^{1 + 2c} M_2^{1 - 2c} \exp\left\{-\frac{8\pi}{3\alpha}\right\}$$

$$m_2^2 = M_1^{1 - 2c} M_2^{1 + 2c} \exp\left\{-\frac{8\pi}{3\alpha}\right\} . \tag{24}$$

Here c is the solution of the transcendental equation obtained which results by substituting Eq. (24) into Eq. (21)

$$c \frac{(M_1/M_2)^c + (M_2/M_1)^c}{(M_1/M_2)^c - (M_2/M_1)^c} = - \frac{\ln (M_1/M_2)}{\ln\{(M_1 M_2)/M_3\}} . \tag{25}$$

An amusing feature of this calculation is the emergence of the relation Eq. (19) $\sin^2\Theta_L = \sin^2\Theta_R$.

For the two family system treated here, we started with three spin-one masses and recovered three parameters for the fermion mass matrix. For three families, the number of spin-one masses will be five, while the Fermi mass matrix which we recover contains seven parameters. Thus the results of the calculation will give testable constraints, if the individual fermions are replaced by the known e, μ and τ families.

REFERENCES

[1]. S. Weinberg, Phys. Lett. 82B (1979) 387.

[2]. P. Budini, Acta Physica Austriaca, Suppl. XV (1976) 499.

[3]. Abdus Salam, Phys. Rev. 130 (1963) 1287. Abdus Salam and R. Delbourgo, Phys. Rev. 135B (1964) 1398. J. Strathdee, Phys. Rev. 135B (1964) 1428. R. Delbourgo, J. Phys. A 10 (1977) 1369.

TOMOGRAPHIC REPRESENTATION OF QUANTIZED FIELDS

Charles M. Sommerfield

Department of Physics
Yale University
P. O. Box 6666
New Haven, CT 06511

1. INTRODUCTION

One of the disconcerting things about the breadth of Feza Gürsey's knowledge of physics is his skill in recognizing some apparently "new" idea as a restatement or re-working of concepts that are well known, at least to him. And so it was about ten years ago when I showed him some work I was doing using the representation to be described in this paper. At that time the application was in connection with the behavior of the S matrix in the infinite momentum frame (a topic more in vogue then than now). He immediately spotted my procedure as the Radon transformation.

Not realizing that "Radon" was the proper name of an Austrian mathematician [1] my initial reaction was to think that the nomenclature was connected with the Greek version of the Latin word "radius", which would not, by coincidence, have been inappropriate. (I am not such a fool as to make a connection with the chemical element). Feza straightened me out by referring me to the beginning of a book that I had, in fact, read the second half of, namely, "Generalized Functions: Volume 5," by Gel'fand, Graef and Vilenkin [2].

More recently the transform has been applied with much well-deserved fanfare to the medical diagnostic procedure known as computer-assisted tomography (usually re‑ ferred to as the CAT scan) [3], and a whole book devoted to its mathematical properties has appeared [4].

Since the definition of the transform to be used here differs somewhat from the standard mathematical one, I will refer to it as the tomographic transform. The application to quantum field theory has its origin in the attempt of Alan Luther to construct Fermi fields in three spatial dimensions out of Bose fields [5]. Although not appearing explicitly in Luther's work, the tomographic transform is actually the fundamental tool used.

The usual Radon transform of a mass distribution, for example, is defined by indicating how much mass lies in a given plane cross section of the distribution. The plane is specified by its normal and a one-dimensional displacement in the direction of the normal. Analytically if $\rho(\vec{r})$ is the mass distribution then the Radon transform is

$$\tilde{\rho}(y,\hat{n}) = \int d^3r \; \delta(y-\hat{n}\cdot\vec{r})\rho(\vec{r}) \; .$$

The inverse transform is

$$\rho(\vec{r}) \; = \; -\frac{1}{8\,\pi^2} \int d^2n \; \; \partial_y^2 \; \tilde{\rho}(y,\,\hat{n})\Big|_{y=\hat{n}\cdot\vec{r}} \; . \tag{1.1}$$

The validity of Eq. (1.1) requires that $\tilde{\rho}$ and $\partial_y \tilde{\rho}$ vanish sufficiently rapidly as $|y| \to \infty$. This is proved formally in references [2] and [4]. A less rigorous demonstration will be given below in a field-theoretical context.

The organization of this paper is as follows. In Section 2 are presented the orthogonality and completeness relations that allow the tomographic transform to be viewed as a canonical transformation. This is done for the cases of scalar, spinor and vector fields. The tomographic transforms are seen to have many of the characteristics of fields in one space dimension. Derivatives of only one of the three coordinates appear. The other two coordinates behave as internal indices. In addition, the scale dimensionality and number of components are as in one dimension.

In Section 3 the properties of the transforms with respect to translations and rotations of the original rectangular coordinate system are considered. Behavior under translations is, as usual, quite simple. With rotations, however, the spin properties of the spinor and vector fields do not show up in terms of multi-dimensional irreducible representations. Instead, all of the fields transform as one-dimensional representations. This is possible because the tomographic transform has affected the topology of the angular coordinates in exactly the same way that a magnetic monopole does for the angular coordinates of a charged particle.

Also examined is the behavior of the transformed fields under Lorentz boosts. This is much more complicated than with the usual fields.

The reader should be aware of the following notational conventions. Unless otherwise indicated, all fields are evaluated at a given time and the time coordinate is suppressed. The summation convention is in effect for repeated indices. Spinor indices are sometimes suppressed and matrix notation is used. Integrals over the trans-form coordinate y are understood to go from $-\infty$ to $+\infty$.

2. THE TOMOGRAPHIC TRANSFORM AS A CANONICAL TRANS-FORMATION.

2.1. Scalar Field.

The completeness and orthonormality relations which govern the use of the tomogra-phic transform are

$$\delta(\vec{r}-\vec{r}') = \frac{1}{8\pi^2} \int dy \, d^2n \, \delta'(y-\hat{n}\cdot\vec{r}) \, \delta'(y-\hat{n}\cdot\vec{r}') \tag{2.1}$$

and

$$\frac{1}{4\pi^2} \int d^3r \, \delta'(y-\hat{n}\cdot\vec{r}) \, \delta'(y'-\hat{n}'\cdot\vec{r}) = \delta(y-y') \, \delta(\hat{n},\hat{n}') - \delta(y+y') \, \delta(\hat{n},-\hat{n}') \ . \tag{2.2}$$

The proof of Eq. (2.1) is by means of the following steps:

$$\delta(\vec{r}-\vec{r}') = \frac{1}{8\pi^3} \int d^3k \, e^{i\vec{k}\cdot(\vec{r}-\vec{r}')}$$

$$= \frac{1}{8\pi^3} \int_0^\infty dk \, k^2 \int d^2\hat{k} \, e^{ik\hat{k}\cdot(\vec{r}-\vec{r}')}$$

$$= \frac{1}{16\pi^3} \int_{-\infty}^\infty dk \, k^2 \int d^2\hat{k} \, e^{ik\hat{k}\cdot(\vec{r}-\vec{r}')}$$

$$= -\frac{1}{8\pi^2} \int d^2\hat{k} \, \delta''[\hat{k}\cdot(\vec{r}-\vec{r}')]$$

$$= \frac{1}{8\pi^2} \int dy \, d^2\hat{k} \, \delta'(y-\hat{k}\cdot\vec{r}) \, \delta'(y-\hat{k}\cdot\vec{r}') \ .$$

The $\vec{k}$ integral was first written in terms of polar coordinates and the identity of the points $(k,\hat{k})$ and $(-k,-\hat{k})$ was then used in the next line.

A derivation of Eq. (2.2) is

$$\frac{1}{4\,\pi^2} \int d^3r \; \delta'(y-\hat{n}\cdot\vec{r}) \; \delta'(y'-\hat{n}'\cdot\vec{r})$$

$$= \frac{1}{16\,\pi^4} \int d^3r \int_{-\infty}^{\infty} dk\, k \int_{-\infty}^{\infty} dk'\, k'\, e^{iky \,-\, ik'y'} \, e^{-i\vec{r}\cdot(k\hat{n}-k'\hat{n}')}$$

$$= \frac{1}{2\,\pi} \int_{-\infty}^{\infty} dk\, k \int_{-\infty}^{\infty} dk'\, k'\, e^{iky \,-\, ik'y'} \, |k|^{-2}$$

$$\times \left[\, \delta(k-k')\, \delta(\hat{n},\hat{n}') + \delta(k+k')\, \delta(\hat{n},-\hat{n}')\, \right]$$

$$= \frac{1}{2\,\pi} \int_{-\infty}^{\infty} dk \left[e^{ik(y-y')}\, \delta(\hat{n},\hat{n}') - e^{ik(y+y')}\, \delta(\hat{n},-\hat{n}')\, \right]$$

$$= \delta(y-y')\, \delta(\hat{n},\hat{n}') - \delta(y+y')\, \delta(\hat{n},-\hat{n}') \; .$$

The tomographic transform of a scalar field $\phi(\vec{r})$ is defined as

$$\tilde{\phi}(y,\hat{n}) = \frac{1}{2\,\pi} \int d^3r \; \delta'(y-\hat{n}\cdot\vec{r}) \; \phi(\vec{r})$$

so that Eq. (2.1) leads to the inverse formula

$$\phi(\vec{r}) = \frac{1}{4\,\pi} \int dy\, d^2n \; \delta'(y-\hat{n}\cdot\vec{r}) \; \tilde{\phi}(y,\hat{n}) \; . \tag{2.3}$$

Correspondingly, if a field $\tilde{\phi}(y,\hat{n})$ can be given which vanishes sufficiently rapidly as $|y| \to \infty$ then the tomographic transformation of the field $\phi(\vec{r})$ given by Eq. (2.3) will give back the original field $\tilde{\phi}(y,\hat{n})$. This is evidenced by Eq. (2.2).

There are several points worth noting. First, the identity of the points $(y,\hat{n})$ and $(-y,-\hat{n})$ implies

$$\tilde{\phi}(y,\hat{n}) = -\tilde{\phi}(-y,-\hat{n}) \; . \tag{2.4}$$

Second, since the dimension of a canonical scalar field is L^{-1} its tomographic transform has dimension L^0 which befits a scalar field in one space dimension.

If $\Pi(\vec{r})$ is the field canonically conjugate to $\phi(\vec{r})$ then the tomographic transforms $\tilde{\phi}(y,\hat{n})$ and $\tilde{\Pi}(y,\hat{n})$ obey the following commutation relations

$$[\tilde{\phi}(y,\hat{n}),\, \tilde{\Pi}(y',\hat{n}')] = [\delta(y-y')\delta(\hat{n},\hat{n}') - \delta(y+y')\delta(\hat{n},-\hat{n}')] \; , \tag{2.5}$$

$$[\tilde{\phi}(y,\hat{n}),\, \tilde{\phi}(y',\hat{n}')] = [\tilde{\Pi}(y,\hat{n}),\, \tilde{\Pi}(y',\hat{n}')] = 0 \; . \tag{2.6}$$

One feature of the tomographic transform that makes it particularly attractive is that the variable $\hat{n}$ behaves in all respects as an internal index rather than as a coordinate. Thus if the Hamiltonian for a scalar field theory is of the form

$$H = \frac{1}{2} \int d^3r \left\{ \vec{\nabla}\phi(\vec{r}) \cdot \vec{\nabla}\, \phi(\vec{r}) + [\Pi(\vec{r})]^2 \right\} + V[\phi]\,,$$

where V contains no derivatives, then the same Hamiltonian expressed in terms of the tomographic transforms will be

$$H = \frac{1}{4} \int dy\, d^2n \left\{ [\partial_y\, \tilde{\phi}(y,\hat{n})]^2 + [\tilde{\Pi}(y,\hat{n})]^2 \right\} + \tilde{V}[\tilde{\phi}]$$

where $\tilde{V}$ contains no derivatives (although it will, in general, be non-local in y and $\hat{n}$).

The equation of motion is then

$$(\partial_0^2 - \partial_y^2)\tilde{\phi}(y,\hat{n}) = - \frac{\delta}{\delta\tilde{\phi}(y,\hat{n})}\, \tilde{V}[\tilde{\phi}]$$

and it is seen that there are no derivatives with respect to the components of $\hat{n}$. This feature is present in the theories of spin 1/2 and spin 1 fields as well.

If ϕ is a free field of mass m the equation of motion for $\tilde{\phi}$ is the same as for a one-dimensional field $\phi^{(1)}$ whose coordinate is y, namely

$$(\partial_0^2 - \partial_y^2 + m^2)\, \tilde{\phi}(y,\hat{n}) = 0\,.$$

The canonical commutation relations Eq. (2.5) and Eq. (2.6) then guarantee that the two-point function is the same as it would be in one space dimension except insofar as it contains an extra factor $\delta(\hat{n},\hat{n}')$ and an extra term that incorporates the duplication symmetry of Eq. (2.4):

$$<0|\tilde{\phi}(y,\hat{n},t)\, \tilde{\phi}(y',\hat{n}',t')|0> = <0|\phi^{(1)}(y,t)\, \phi^{(1)}(y',t')|0> \delta(\hat{n},\hat{n}')$$

$$- <0|\phi^{(1)}(y,t)\, \phi^{(1)}(-y',t')|0> \delta(\hat{n},-\hat{n}')\quad.$$

Similar statements hold for other matrix elements. It is in this sense that the tomographic transform behaves as if it were describing a one-dimensional theory with the angular variable $\hat{n}$ playing the role of a continuous internal symmetry index.

2.2. Spinor Field.

The generalization of Eq. (2.1) needed for application to spinor fields is obtained by writing Eq. (2.1) multiplied by the Kronecker delta $\delta_{\alpha\beta}$ as

$$\delta(\vec{r}-\vec{r}') \, \delta_{\alpha\beta} = \frac{1}{4\,\pi^2} \int dy \, d^2n \, \delta'(y-\hat{n}\cdot\vec{r}) \, \delta'(y-\hat{n}\cdot\vec{r}') \, \frac{1}{2} \left(1 + \vec{\alpha}\cdot\hat{n}\right)_{\alpha\beta} \tag{2.7}$$

where $\vec{\alpha}$ are the three 4×4 Hermitian Dirac matrices that satisfy

$$\{\alpha_i \, , \, \alpha_j\} = 2 \, \delta_{ij} \, , \qquad i, j = 1,2,3.$$

Since the $\hat{n}$ dependence of the $\vec{\alpha}\cdot\hat{n}$ term is odd, it integrates to zero in the integral. Now $\frac{1}{2}\left(1 + \vec{\alpha}\cdot\hat{n}\right)$ is a projection operator onto the subspace of four component spinors which are eigenspinors of $\vec{\alpha}\cdot\hat{n}$ with eigenvalue $+1$. Since $[\vec{\sigma}\cdot\hat{n}, \, \vec{\alpha}\cdot\hat{n}] = 0$, with $\vec{\sigma} = -\frac{1}{2} \, \vec{\alpha} \times \vec{\alpha}$, normalized spinors $u^a(\hat{n})$ that are eigenspinors of $\vec{\sigma}\cdot\hat{n}$ with eigenvalue $+1$ for $a = 1$ and -1 for $a = 2$ may be taken as a basis for this subspace. Then there is orthogonality,

$$u^{\dagger a}(\hat{n}) \, u^b(\hat{n}) = \delta^{ab} \, , \tag{2.8}$$

and the projection operator takes the form

$$\frac{1}{2} \left(1 + \vec{\alpha}\cdot\hat{n}\right)_{\alpha\beta} = u_{\alpha}{}^a(\hat{n}) \, u_{\beta}{}^{\dagger a}(\hat{n}) \tag{2.9}$$

while Eq. (2.7) becomes

$$\delta(\vec{r}-\vec{r}') \, \delta_{\alpha\beta} = \frac{1}{4\,\pi^2} \int dy \, d^2n \, \delta'(y-\hat{n}\cdot\vec{r}) \, \delta'(y-\hat{n}\cdot\vec{r}') \, u_{\alpha}{}^a(\hat{n}) \, u_{\beta}{}^{\dagger a}(\hat{n}) \quad .$$

The tomographic transform of the four component spinor field is then defined to be

$$\tilde{\psi}^a(y,\hat{n}) = \frac{1}{2\,\pi} \int d^3r \, \delta'(y-\hat{n}\cdot\vec{r}) \, u_{\alpha}{}^{\dagger a}(\hat{n}) \, \psi_{\alpha}(\vec{r}) \tag{2.10}$$

so that $\psi_{\alpha}(\vec{r})$ is represented as

$$\psi_{\alpha}(\vec{r}) = \frac{1}{2\,\pi} \int dy \, d^2n \, \delta'(y-\hat{n}\cdot\vec{r}) \, u_{\alpha}{}^a(\hat{n}) \, \tilde{\psi}^a(y,\hat{n}) \quad .$$

The orthogonality of the transform is contained in the statement

$$\delta^{ab} \, \delta(y-y') \, \delta(\hat{n},\hat{n}') = \frac{1}{4\,\pi^2} \int d^3r \, \delta'(y-\hat{n}\cdot\vec{r}) \, \delta'(y'-\hat{n}'\cdot\vec{r}) \, u_{\alpha}{}^{\dagger a}(\hat{n}) \, u_{\alpha}{}^b(\hat{n}')$$

which is easily derived using Eq. (2.2), Eq. (2.8) and

$$u_{\alpha}{}^{\dagger a}(\hat{n}) \, u_{\alpha}{}^b(-\hat{n}) = 0 \quad .$$

This last statement follows from the fact that the two spinors $u^a(\hat{n})$ and $u^b(-\hat{n})$ are eigenspinors of the operator $\vec{\alpha}\cdot\hat{n}$ with different eigenvalues.

If the Dirac matrices are written in terms of 2×2 Pauli matrices as

$$\tilde{\alpha} = \begin{pmatrix} 0 & \vec{\sigma} \\ \vec{\sigma} & 0 \end{pmatrix}$$

then explicit expressions for u^1 and u^2 are

$$u^1(\hat{n}) = 2^{-1/2} \begin{pmatrix} u(\hat{n}) \\ -u(\hat{n}) \end{pmatrix} \qquad (2.11)$$

$$u^2(\hat{n}) = 2^{-1/2} \begin{pmatrix} u(-\hat{n}) \\ u(-\hat{n}) \end{pmatrix} \qquad (2.12)$$

where, in terms of the angles θ and ϕ which parametrize n,

$$u(\hat{n}) = \begin{pmatrix} \cos\theta/2 \\ \sin\theta/2 \, e^{i\phi} \end{pmatrix}. \qquad (2.13)$$

Note that $u(\hat{n})$ is not properly defined when $\theta = \pi$. However, since $u(\hat{n})$ may be multiplied by an arbitrary phase depending on $\hat{n}$, the singularity may be eliminated if $u(\hat{n})$ is treated as a section after the manner of Yang and Wu [6]. The approach taken here will be not to worry explicitly about such things.

The canonical anticommutation relations for the transforms

$$[\tilde{\psi}^a(y,\hat{n}), \, \tilde{\psi}^{\dagger b}(y',\hat{n}') \,] = \delta^{ab} \, \delta(y-y') \, \delta(\hat{n}, \, \hat{n}') \qquad (2.14)$$

$$[\tilde{\psi}^a(y,\hat{n}), \, \tilde{\psi}^b(y',\hat{n}')] = [\tilde{\psi}^{\dagger a}(y,\hat{n}), \, \tilde{\psi}^{\dagger b}(y', \, \hat{n}')] = 0 \qquad (2.15)$$

then follow from the usual anticommutators, and Eq. (2.9) and Eq. (2.10).

If the Hamiltonian for the spinor field has the form

$$H = \int d^3r \, [i\bar{\psi}(\vec{r}) \, \gamma^\mu \partial_\mu \, \psi(\vec{r})] + V[\psi,\bar{\psi}]$$

where V does not depend on derivatives (and where, for the sake of brevity, no attempt has been made to symmetrize the operator products explicitly and to make them Hermitian), then in terms of the tomographic transform it is

$$H = \int dy \, d^2n \, \{ i\tilde{\psi}^{\dagger a}(y,\hat{n}) \, (\partial_0 + \partial_y) \, \tilde{\psi}^a(y,\hat{n}) \} + \tilde{V}[\tilde{\psi},\tilde{\psi}^\dagger] . \qquad (2.16)$$

Therefore the Dirac operator $-i\gamma^\mu \partial_\mu$, which appears in the equation of motion operating on the four component spinor, is replaced by $-i(\partial_0 + \partial_y)$ acting on the two component transform, and the matrices have disappeared. As a tradeoff, however, there may be nonlocality due to $\tilde{V}$.

For a free particle of mass m the Dirac equation written in terms of the transform would be

$$- i \, (\partial_0 + \partial_y) \, \tilde{\psi}^1(y,\hat{n}) = m \, \tilde{\psi}^2(-y,-\hat{n})$$

$$- i \, (\partial_0 + \partial_y) \, \tilde{\psi}^2(y,\hat{n}) = m \, \tilde{\psi}^1(-y,-\hat{n}) \; .$$

To make a comparison with a one-dimensional theory one calculates the two point function, using Eq. (2.16) and the anticommutation relations Eq. (2.14) and Eq. (2.15), as

$$<0|\tilde{\psi}_c(y,\hat{n},t) \, \tilde{\psi}_d^\dagger(y',\hat{n}',t')|0> = <0|\psi_c^{(1)}(y,t) \, \psi_d^{\dagger(1)}(y',t')|0> \, \delta(\hat{n},\hat{n}') \quad c, d = \pm \; .$$

In this equation the $\pm$ components of the three-dimensional field are

$$\tilde{\psi}_+(y,\hat{n}) = \tilde{\psi}^1(y,\hat{n}) \quad ; \quad \tilde{\psi}_-(y,\hat{n}) = \tilde{\psi}^2(-y,-\hat{n})$$

while those of the related one-dimensional Fermi field are symbolized by $\psi_\pm^{(1)}$ in a representation of the 2×2 Dirac matrices such that $\gamma^5 = \gamma^0\gamma^1$ is diagonal.

2.3. Vector Field

The generalization of Eq. (2.1) to include a polarization vector is

$$\delta_{ij}\delta(\vec{r}-\vec{r}') = \frac{1}{8\,\pi^2} \int dy \, d^2n \, \delta'(y-\hat{n}\cdot\vec{r}) \, \delta'(y-\hat{n}\cdot\vec{r}') \left\{ n_i n_j + \epsilon_i^a(\hat{n}) \, \epsilon_j^a(\hat{n}) \right\}$$

where n_i and $\epsilon_i^a(\hat{n})$, ($a = 1,2$; $i = 1,2,3$) are a right-handed orthonormal set of basis vectors tied to $\hat{n}$. For definiteness one may take $\epsilon_i^a(-\hat{n}) = -(-1)^a \, \epsilon_i^a(\hat{n})$.

A three-vector field $\vec{A}(\vec{r})$ may thus be expressed as

$$\vec{A}(\vec{r}) = \frac{1}{4\,\pi} \int dy \, d^2n \, \delta'(y-\hat{n}\cdot\vec{r}) \left\{ \hat{n} \, {}^L\tilde{A}(y,\hat{n}) + \hat{\epsilon}^a(\hat{n}) \, {}^T\tilde{A}^a(y,\hat{n}) \right\} \; .$$

The pre-superscripts L and T denote the longitudinal and transverse parts of the tomographic transform. A fourth component $A^0(\vec{r})$ of a four-vector field, which behaves as a scalar under rotations, is written

$$A^0(\vec{r}) = \frac{1}{4\,\pi} \int dy \, d^2n \, \delta'(y-\hat{n}\cdot\vec{r}) \, {}^S\tilde{A}(y,\hat{n}) \; .$$

The inverse formulas for the vector parts are

$$^L\tilde{A}(y,\hat{n}) = \frac{1}{2\pi} \int d^3r \; \delta'(y-\hat{n}\cdot\vec{r}) \; \hat{n}\cdot \vec{A}(\vec{r}) = {}^L\tilde{A}(-y,-\hat{n})$$

$$^T\tilde{A}^a(y,\hat{n}) = \frac{1}{2\pi} \int d^3r \; \delta'(y-\hat{n}\cdot\vec{r}) \; \hat{e}^a(\hat{n})\cdot\vec{A}(\vec{r}) = (-1)^a \; {}^T\tilde{A}^a(-y,-\hat{n}) \; .$$

There is no sum over a in this last expression.

These transforms will now be applied to the equations of motion of massive and massless vector fields coupled to an external current $J^\mu(\vec{r})$. It is convenient to use three-dimensional notation for Maxwell's equations:

$$\vec{\nabla}\cdot\vec{B}(\vec{r}) = 0$$

$$\partial_0 \vec{B}(\vec{r}) = - \vec{\nabla}\times\vec{E}(\vec{r})$$

$$\vec{\nabla}\times\vec{B}(\vec{r}) - \partial_0 \vec{E}(\vec{r}) = \vec{J}(\vec{r}) - m^2 \vec{A}(\vec{r})$$

$$\vec{\nabla}\cdot\vec{E}(\vec{r}) = J^0(\vec{r}) - m^2 A^0(\vec{r})$$

with

$$\vec{B}(\vec{r}) = \vec{\nabla}\times\vec{A}(\vec{r})$$

$$\vec{E}(\vec{r}) = - \vec{\nabla}A^0(\vec{r}) - \partial_0 \vec{A}(\vec{r}) \; .$$

In terms of the appropriate tomographic transforms and with the introduction of the complex combination $^T\tilde{A}(y,\hat{n}) = {}^T\tilde{A}^1(y,\hat{n}) + i \; {}^T\tilde{A}^2(y,\hat{n})$ (and similar definitions for the transverse parts of the other vector fields), these become

$$^L\tilde{B}(y,\hat{n}) = 0$$

$$\partial_0 \, {}^T\tilde{B}(y,\hat{n}) = - i \, \partial_y \, {}^T\tilde{E}(y,\hat{n})$$

$$i \, \partial_y \, {}^T\tilde{B}(y,\hat{n}) - \partial_0 \, {}^T\tilde{E}(y,\hat{n}) = {}^T\tilde{J}(y,\hat{n}) - m^2 \, {}^T\tilde{A}(y,\hat{n})$$

$$- \partial_0 \, {}^L\tilde{E}(y,\hat{n}) = {}^L\tilde{J}(y,\hat{n}) - m^2 \, {}^L\tilde{A}(y,\hat{n})$$

$$\partial_y \, {}^L\tilde{E}(y,\hat{n}) = {}^S\tilde{J}(y,\hat{n}) - m^2 \, {}^S\tilde{A}(y,\hat{n})$$

with

$${}^{T}\tilde{B}(y,\hat{n}) = i\,\partial_{y}\,{}^{T}\tilde{A}(y,\hat{n})$$

$${}^{T}\tilde{E}(y,\hat{n}) = -\,\partial_{0}\,{}^{T}\tilde{A}(y,\hat{n})$$

$${}^{L}\tilde{E}(y,\hat{n}) = -\,\partial_{y}\,{}^{S}\tilde{A}(y,\hat{n}) - \partial_{0}\,{}^{L}\tilde{A}(y,\hat{n})$$

together with the appropriate complex conjugates, and the transverse parts are completely uncoupled from the scalar and longitudinal parts. Of course, a choice of gauge
for the case $m = 0$ involves only the scalar and longitudinal parts of the vector potential.

There are one-dimensional analogues in this case as above, but they will not be
treated here in any detail. They are immediately evident for the longitudinal and scalar parts, while a more complicated analog holds for the transverse parts.

3. BEHAVIOR OF THE TOMOGRAPHIC TRANSFORM UNDER THE POINCARE GROUP

3.1. Translations

Scalar, spinor and vector fields all transform the same way under spatial translations. If the translation generator is denoted by $\vec{P}$ and if $\chi(\vec{r})$ is a component of such a
field, then

$$[\chi(\vec{r})\,,\,\vec{P}\,] = -\,i\,\vec{\nabla}\,\chi(\vec{r})\,.$$

When this formula is used to determine the commutator of $\vec{P}$ with the various tomographic transforms, the following results:

$$[\tilde{\chi}(y,\hat{n})\,,\,\vec{P}\,] = -\,i\,\hat{n}\,\partial_{y}\,\tilde{\chi}(y,\hat{n})$$

where $\tilde{\chi}(y,\hat{n})$ is the appropriate tomographic transform. The basic ingredient in the
derivation is the identity

$$\vec{\nabla}\,\delta'(y-\hat{n}\cdot\vec{r}) = -\,\hat{n}\frac{\partial}{\partial y}\,\delta'(y-\hat{n}\cdot\vec{r})\,.$$

3.2. Rotations

The really interesting behavior appears when rotations are considered. If the rotation generator is denoted by $\vec{J}$, then for a scalar field

$$[\phi(\vec{r}) \, , \, \vec{J}\,] = - \, i \, \vec{r} \times \vec{\nabla} \, \phi(\vec{r})$$

and so

$$[\tilde{\phi}(y,\hat{n}) \, , \, \vec{J}\,] = - \frac{i}{2\,\pi} \int d^3r \; \delta'(y - \hat{n} \cdot \vec{r}) \; \vec{r} \times \vec{\nabla} \; \phi(\vec{r}) \; .$$

Integration by parts and subsequent use of the identity

$$\vec{r} \times \vec{\nabla} \, \delta'(y - \hat{n} \cdot \vec{r}) = - \, \vec{r} \times \hat{n} \, \delta''(y - \hat{n} \cdot \vec{r}) = - \, \hat{n} \times \vec{\nabla}_n \, \delta'(y - \hat{n} \cdot \vec{r})$$

gives rise to the expected result

$$[\tilde{\phi}(y,\hat{n}) \, , \, \vec{J}\,] = - \, i \, \hat{n} \times \vec{\nabla}_n \, \tilde{\phi}(y,\hat{n}) \; .$$

The situation is more complicated for a spinor field. Starting from the usual transformation property

$$[\psi(\vec{r}) \, , \, \vec{J}\,] = - \, i \, \vec{r} \times \vec{\nabla} \, \psi(\vec{r}) + \frac{1}{2} \, \vec{\sigma} \, \psi(\vec{r})$$

and proceeding as with the scalar field one comes to the stage

$$[\tilde{\psi}^a(y,\hat{n}) \, , \, \vec{J}\,] = \frac{1}{2\,\pi} \int d^3r \left[- \, i \, \hat{n} \times \vec{\nabla}_n \, \delta'(y - \hat{n} \cdot \vec{r}) \right] u_\alpha^{\dagger a}(\hat{n}) \, \psi_\alpha(\vec{r})$$

$$+ \frac{1}{4\,\pi} \int d^3r \; \delta'(y - \hat{n} \cdot \vec{r}) \, u_\alpha^{\dagger a}(\hat{n}) \, \vec{\sigma}_{\alpha\beta} \, \psi_\beta(\vec{r}) \; .$$

This can be written as

$$[\tilde{\psi}^a(y,\hat{n}) \, , \, \vec{J}\,] = - \, i \, \hat{n} \times \vec{\nabla}_n \, \tilde{\psi}^a(y,\hat{n}) \qquad (3.1)$$

$$+ \frac{1}{2\,\pi} \int d^3r \; \delta'(y - \hat{n} \cdot \vec{r}) \left[i \, \hat{n} \times \vec{\nabla}_n \, u_\alpha^{\dagger a}(\hat{n}) \, \delta_{\alpha\beta} + \frac{1}{2} \, u_\alpha^{\dagger a}(\hat{n}) \, \vec{\sigma}_{\alpha\beta} \right] \psi_\beta(\vec{r}) \; .$$

Now consider the spin-differential operator $- \, i \, \hat{n} \times \vec{\nabla}_n + \frac{1}{2} \, \vec{\sigma}$. It is easy to show that it commutes with both $\vec{\alpha} \cdot \hat{n}$ and $\vec{\sigma} \cdot \hat{n}$. Hence, acting on $u^a(\hat{n})$, which is a simultaneous eigenspinor of $\vec{\alpha} \cdot \hat{n}$ and $\vec{\sigma} \cdot \hat{n}$, it must produce a spinor proportional to $u^a(\hat{n})$. If the proportionality factor is called $\vec{S}^a(\hat{n})$ then

$$- \, i \, \hat{n} \times \vec{\nabla}_n \, u^a(\hat{n}) + \frac{1}{2} \, \vec{\sigma} u^a(\hat{n}) = \vec{S}^a(\hat{n}) \, u^a(\hat{n}) \; , \quad (no \; sum \; on \; a) \; . \qquad (3.2)$$

Since $\vec{S}^a(\hat{n})$ is the eigenvalue of a Hermitian operator, it must be real. Using the Hermitian conjugate of Eq. (3.2) in Eq. (3.1) one finally obtains the transformation property

$$[\tilde{\psi}^a(y,\hat{n}) , \vec{J}] = - i \, \hat{n} \times \vec{\nabla}_n \, \tilde{\psi}^a(y,\hat{n}) + \vec{S}^a(\hat{n})\tilde{\psi}^a(y,\hat{n}) \ , \quad (\textit{no sum on a}) . \qquad (3.3)$$

Note that $\tilde{\psi}$ transforms as a one-dimensional representation of the rotation group. This is possible because $\vec{S}^a(\hat{n})$ has a singularity. The explicit form of $\vec{S}^a(\hat{n})$, computed from Eq. (2.11) - Eq. (2.14) may be written suggestively as

$$\vec{S}^a(\hat{n}) = - \, \hat{n} \times \vec{A}^a(\hat{n}) - \vec{B}^a(\hat{n})$$

where

$$\vec{A}^a(\hat{n}) = \tfrac{1}{2} \, \hat{\phi} \, (-1)^a \, \frac{1 - \cos\theta}{\sin\theta}$$

and

$$\vec{B}^a(\hat{n}) = \tfrac{1}{2} \, (-1)^a \, \hat{n} .$$

Then $\vec{A}^a(\hat{n})$ will be recognized as embodying the angular dependence of the vector potential due to a magnetic monopole of strength $1/2 \, (-1)^a$ and at the same time $\vec{B}^a(\hat{n})$ is seen to be the coefficient of r^{-2} of the magnetic field produced by such a monopole. The singularity of $\vec{S}^a(\hat{n})$ is then precisely that of the Dirac string [7].

If Eq. (3.3) is therefore reexpressed as

$$[\tilde{\psi}^a(y,\hat{n}) , \vec{J}] = \big\{ - i \, \hat{n} \times \vec{\nabla}_n \, - \hat{n} \times \vec{A}^a (\hat{n}) - \vec{B}^a(\hat{n}) \ \big\} \tilde{\psi}^a(y,\hat{n}) \quad (\textit{no sum on a}) ,$$

one observes that the rotational properties of $\tilde{\psi}$ are just those of a field that represents a scalar particle of charge 1 moving under the influence of a magnetic monopole of charge $1/2 \, (-1)^a$. The properties of such a field have been discussed by Wu and Yang [8], Biedenharn and Louck [9], and Goldhaber [10].

It should be emphasized that there are no real magnetic fields present here. It is only the angular dependence induced in the various functions by the nature of the basis spinor $u(\hat{n})$ that exhibits the same topological and analytic properties as those in a magnetic monopole configuration.

The rotational transformation properties of the vector fields embody no new features. The scalar and longitudinal tomographic transforms behave as scalar fields under rotations, while the two transverse components can be analyzed into fields behaving as scalars of unit charge moving in a monopole field of strength ± 1.

3.3. Boosts

Matters are not as simple when behavior under a Lorentz boost is considered. For a scalar field that transforms under the boost symbolized by $\vec{K}$ as

$$[\phi(\vec{r}),\,\vec{K}] = -\,it\,\vec{\nabla}\,\phi(\vec{r}) - i\vec{r}\,\partial_0\,\phi(\vec{r})$$

the tomographic transform satisfies

$$[\tilde{\phi}(y,\hat{n}),\,\vec{K}] = -\,i\hat{n}\,t\,\partial_y\,\tilde{\phi}(y,\hat{n}) - i\hat{n}\,y\,\partial_0\,\tilde{\phi}(y,\hat{n}) - i\hat{n}\,\partial_y^{-1}\,\partial_0\,\tilde{\phi}(y,\hat{n})$$

$$-\,i\,\hat{n}\times(\hat{n}\times\vec{\nabla}_n)\,\partial_y^{-1}\,\partial_0\,\tilde{\phi}(y,\hat{n})$$

where the nonlocal operator ∂_y^{-1} stands for integration with respect to y.

For the spinor field that transforms as

$$[\psi(\vec{r}),\,\vec{K}] = -\,it\,\vec{\nabla}\,\psi(\vec{r}) - i\vec{r}\,\partial_0\,\psi(\vec{r}) + \tfrac{i}{2}\,\vec{\alpha}\,\psi(\vec{r})$$

the behavior of the tomographic transform is quite complicated and only the result will be given here. It is

$$[\tilde{\psi}(y,\hat{n}),\,\vec{K}] = -\,it\,\hat{n}\,\partial_y\,\tilde{\psi}(y,\hat{n}) - i\hat{n}\,y\,\partial_0\,\tilde{\psi}(y,\hat{n}) + \tfrac{i}{2}\,\hat{n}\,\tilde{\psi}(y,\hat{n}) - i\hat{n}\,\partial_y^{-1}\,\partial_0\,\tilde{\psi}(y,\hat{n})$$

$$+\,\left\{\,\hat{n}\times[\,-\,i\hat{n}\times\vec{\nabla}_n + \Gamma^5\,\vec{S}(y,\hat{n})\,]\right\}\partial_y^{-1}\,\partial_0\,\tilde{\psi}(y,\hat{n})$$

$$-\,\tfrac{i}{2}\,(\Gamma^5\,\theta + i\,\phi)\,\left\{\tilde{\psi}(-y,-\hat{n}) - \partial_y^{-1}\,\partial_0\,\tilde{\psi}(-y,-\hat{n})\right\}\,.$$

A matrix notation has been used for the index on the Fermi field. The matrix Γ^5 is diagonal with elements $\pm\,1$.

A similar complication is involved for the vector field, and as with rotations, the formulas will not be presented here.

4. CONCLUSION

The attractive features of the tomographic transform are:

1. There are derivatives only with respect to time and the coordinate y.

2. The kinetic part of the Dirac operator involves no matrices.

3. For vector fields there is a natural separation into transverse and longitudinal parts.

4. The spin properties involve a topology identical to that engendered by a magnetic monopole.

The unpleasant features are:

1. Terms in the Hamiltonian that are of degree greater than 2 in the fields will be nonlocal in both the $\hat{n}$ and y coordinates.

2. Transformation properties with respect to Lorentz boosts are nonlocal and quite complicated.

It is a pleasure to thank Feza Gürsey for his observations concerning the matters presented here. I also would like to acknowledge helpful conversations with Itzhak Bars and Alan Chodos.

This work was supported in part by the United States Department of Energy under contract No. DE-AC 02-76 ERO 3075.

REFERENCES

[1]. J. Radon, *Über die Bestimmung von Funktionen durch ihre Integralwerte längs gewisser Mannigfaltigkeiten*, Ber. Verh. Sachs Akad. 69, 262 (1917).

[2]. I. M. Gel'fand, M. I. Graef and N. Y. Vilenkin, *Generalized Functions: Volume 5. Integral Geometry and Representation Theory"*, Academic Press, New York and London, (1966.

[3]. G. DiChiro and R. A. Brooks, *The 1979 Nobel Prize in Physiology and Medicine*, Science 206, 1062 (1979).

[4]. S. Helgason, *The Radon Transform*, Birkhäuser Boston, Cambridge, Mass. (1980).

[5]. A. Luther, *Bosonized Fermions in Three Dimensions*, Physics Reports 49, 261 (1979).

[6]. C. N. Yang and T. T. Wu, *Concept of Nonintegrable Phase Factors and Global Formulation of Gauge Fields*, Phys. Rev. D12, 3845 (1975).

[7]. P. A. M. Dirac, *Quantized Singularities in the Electromagnetic Field*, Proc. Roy. Soc. A133, 60 (1931).

[8]. T. T. Wu and C. N. Yang, *Dirac Monopoles without Strings: Monopole Harmonics*, Nucl. Phys. B107, 365 (1976).

[9]. L. C. Biedenharn and J. D. Louck, *Encyclopedia of Mathematics and its Applications: Volume 9. The Racah-Wigner Algebra in Quantum* Theory, Addison-Wesley, Reading, Mass. 1981.

[10]. A. S. Goldhaber, *Connection of Spin and Statistics for Charge-Monopole Composites*,B Phys. Rev. Lett. 36, 1122 (1976).

ON SPONTANEOUSLY BROKEN SUPERSYMMETRY

G. Domokos and S. Kövesi-Domokos

Department of Physics
Johns Hopkins University
Baltimore, Maryland, 21218

Abstract

We develop an effective action formalism for theories with a spontaneously broken supersymmetry. The use of the formalism is illustrated by a derivation of the Goldstone theorem and of various decoupling theorems for the Goldstone mode.

PREFACE

The set of Feza Gürsey's students can be naturally divided into two subsets. The first, finite, subset consists of those who studied the physics of elementary particles under his supervision and guidance at one time or another. The complement of this set is denumerable but not necessarily finite: its elements are those physicists who came to appreciate the importance of symmetries in particle physics from his papers and lecture notes without, however, the benefit of a personal contact at first.

Not having had the good fortune of becoming elements of the above-mentioned finite subset, it is a particular pleasure for us to be able to contribute to this volume celebrating Feza's sixtieth birthday. Although the results we are able to report at this time are less complete than we would like them to be, we hope that, nevertheless, he will not be entirely displeased by the approach to a subject to which he himself made several important contributions [1].

1. INTRODUCTION

Invariance under the inhomogeneous Lorentz (Poincaré) group is the fundamental symmetry every quantum field theory in flat space has to obey. Various generalizations of this basic symmetry requirement have been considered. For instance, Poincaré invariance may be substituted by invariance under a De Sitter group in certain space-times of constant curvature [2], in some other types of field theories conformal invariance plays a key role [3], etc. The Wess-Zumino (WZ) supersymmetry [4], in a sense, is just another generalization of Poincaré invariance: the WZ group is the semi-direct product of a (graded, non-Abelian) group of "translations" with the homogeneous Lorentz group. The similarity does not end here: just as the Poincaré group can be obtained from a De Sitter group by Wigner-Inönü contraction, the WZ group is the contraction of a graded De Sitter group. Moreover, as Poincaré-covariant field quantities can be viewed as representations of the inhomogeneous Lorentz group induced on the stability group of the origin of space-time, superfields are induced representations of the WZ group on the stability group of the origin of flat superspace [5]. (Both stability groups are isomorphic to the homogeneous Lorentz group.) We know, however, that if supersymmetry has anything to do with Nature, it must be broken, probably spontaneously. In view of a renewed interest in the possible role played by (broken) supersymmetry in curing some diseases of grand unified theories [6], it is worth taking a fresh look at the problem of the spontaneous breakdown of supersymmetries as well. In particular, it is desirable to understand the general structure of theories with broken supersymmetry, independently of specific models. In this work, we want to take a step in that direction.

Let us recall a few well-known elementary facts about theories with a spontaneously broken supersymmetry. Let Q_A, $\bar{Q}_{\dot{A}}$ stand for the Fermi generators (generators of "supertranslations") of the WZ group, where $A = 1, 2$ is a Weyl spinor index. If the vacuum is not annihilated by Q_A and $\bar{Q}_{\dot{A}}$, the anticommutator

$$\{Q_A, \bar{Q}_{\dot{A}}\} = 2\, \sigma^\mu{}_{A\dot{A}}\, P_\mu \tag{1.1}$$

implies that the vacuum is not translation invariant either. The only subgroup of the WZ group which has a chance of remaining unbroken is the homogeneous Lorentz group. It indeed will remain unbroken [7] if, for instance, supersymmetry is broken by giving a non-vanishing vacuum expectation value (VEV) to a chiral superfield, $<\Phi> = \phi$, which is constant over (ordinary) space-time. In this case the VEV of the energy-momentum tensor is of the form

$$<T_{\mu\nu}> = C\,\eta_{\mu\nu} \tag{1.2}$$

where $\eta_{\mu\nu}$ is the usual Minkowski metric tensor and C is a positive constant. It is easy to verify (by a suitable limiting procedure) that

$$<M_{\mu\nu}> = \int d\sigma^\rho \left\{ x_\mu <T_{\nu\rho}> - x_\nu <T_{\mu\rho}> \right\} = 0 \tag{1.3}$$

for any space-like surface with surface element $d\sigma^\rho$. As a consequence, we can always choose a vacuum such that $<P_0> \neq 0$, $<P_k> = 0$, $(k = 1, 2, 3)$. However, with Eq. (1.2) $<P_0>$ is actually infinite! This is somewhat disturbing and one may even argue that the proper setting for an investigation of broken supersymmetry is in the framework of supergravity, where Eq. (1.2) would give rise to a cosmological constant, to be compensated, we hope, by a contribution of equal magnitude arising from a simultaneous breakdown of internal symmetries at the present epoch. Formally, in a flat superspace, we can effectively ignore this problem by subtracting the infinite vacuum energy. Even so, we have to make the vacuum with broken supersymmetry *stable*, by forbidding the field equations to have supersymmetric vacuum solutions; see, in particular, O'Raifeartaigh [8]). If, as we assume from now on, the *VEV* of a chiral superfield is constant in space-time, we can always choose a vacuum such that the Fermi component of ϕ vanishes. (This is necessary in order to preserve invariance under the homogeneous Lorentz group throughout and it was implicitly assumed before.) Indeed, let us decompose $<\Phi> = \phi(\theta)$ into its components:

$$\phi(\theta) = a + 2^{1/2}\,(\theta\psi) + (\theta\theta)\,f \tag{1.4}$$

where θ^A is the usual Grassmann coordinate and $(\phi\chi)$ stands for the scalar product of two anticommuting Weyl spinors, $(\phi\chi) = \epsilon_{AB}\,\phi^A\,\chi^B$. By performing on ϕ a "supertranslation" with Grassmann coordinates ξ^A the components transform as follows:

$$a' = a + (\xi\xi)f - 2^{1/2}\,(\xi\psi)$$

$$\psi' = \psi - 2^{1/2}\,\xi f \tag{1.5}$$

$$f' = f \quad .$$

Thus, by choosing $\xi = 2^{-1/2}\,\psi/f$, we have $\psi' = 0$ but, in general, $a' \neq 0$. Consequently, unless specific models are manufactured otherwise, a breakdown of supersymmetry occurs simultaneously with a breakdown of *some* internal symmetry. (Here and in what follows, we suppress internal symmetry labels carried by superfields).

Our main purpose in this work is the construction of a formalism appropriate for the investigation of some properties of theories with a spontaneously broken supersymmetry in a model-independent fashion. Such a formalism is provided by the construction of various effective actions and the exploitation of their invariance properties under supersymmetry transformations. This formalism has been proposed and used for the investigation of the spontaneous breakdown of internal symmetries long ago [9]; we find that it can be generalized to supersymmetric theories as well. For the sake of simplicity, in this work we consider chiral superfields only; however, as we go along, it will become obvious that a generalization of the formalism so as to include, for instance, vector superfields is quite straightforward. Manifest covariance under supersymmetry transformations is maintained until we actually break the *WZ* group: only then will it be necessary to decompose any superfield into its components. (From a technical point of view, such a "component-free" approach entails a considerable gain in economy.)

The plan of this paper is the following. In Section 2, we review the variational calculus for chiral superfields. This is a straightforward matter; the only non-trivial technical problem to be overcome is that the variations of a chiral superfield have to respect the chiral constraints. In Section 3 we define various types of effective actions for supersymmetric theories and derive the Ward identities they satisfy; the solution of the Ward identities is briefly discussed. The use of the effective action formalism is illustrated in Section 4: we consider a supersymmetric theory with a non-vanishing *VEV* of a chiral superfield and give an (essentially) algebraic derivation of the Goldstone theorem altogether with various decoupling theorems for the Goldstone mode. The results are discussed in Section 5.

Notation: Our notation and conventions are essentially identical with the ones used in the Wess lecture notes [4], with the difference that tensor indices are denoted by lower case Greek letters, while spinor indices are denoted by capital Roman letters. Points in flat superspace with coordinates $(X_i^\mu, \theta_i^A, \bar{\theta}_i^A)$ are often denoted by capital letters (P_i), or if no misunderstanding can arise, simply by the subscripts. Thus for instance, if $dV_i \equiv d^4x_i\, d^2\theta_i\, d^2\bar{\theta}_i$ stands for the volume element of the supertranslation group at point P_i, we may use the shorthand $d(1)\, d(2)\, d(3) \ldots$ instead of $dV_1\, dV_2\, dV_3 \ldots$, etc. The symbol $\partial_{A\dot{A}}$ stands for the spinor form of the derivative with respect to x^μ, $\partial_{A\dot{A}} = \sigma^\mu_{A\dot{A}} \partial_\mu$ whereas ∂_A and $\bar{\partial}_{\dot{A}}$ stand for *left derivatives* with respect to the Grassmann variables θ^A and $\bar{\theta}^A$ respectively.

2. VARIATIONAL CALCULUS ON CHIRAL SUPERFIELDS IN A NUTSHELL

A complex chiral superfield, $\Phi(P)$, obeys the constraints:

$$\bar{D}_{\dot{A}} \Phi(P) \equiv - \{ \bar{\partial}_{\dot{A}} + i\, \theta^A\, \partial_{A\dot{A}} \} \Phi(P) = 0$$

$$D_{\dot{A}} \Phi^{\dagger}(P) \equiv - \{ \partial_A + i\, \partial_{A\dot{A}}\, \theta^{\dot{A}} \} \Phi^{\dagger}(P) = 0 \; . \qquad (2.1)$$

Eq. (2.1) restricts the dependence of Φ on $\bar{\theta}$, *viz.* $\Phi(P) = \Phi(y,\theta)$, where $y^{\mu} = x^{\mu} + i\theta\sigma^{\mu}\bar{\theta}$, and likewise for $\Phi^{\dagger}$.

Consider now a functional of a chiral superfield $\Phi(P)$, denoted by $F[\Phi(P)]$: it is a map from the bundle of chiral superfields over superspace onto the complex numbers. The functional derivative of F is defined by the formula [10]:

$$\frac{\delta F}{\delta \Phi(P_1)} = \left(\frac{d}{d\lambda} F[\Phi(P) + \lambda\delta(P, P_1)] \right)_{\lambda=0} , \qquad (2.2)$$

where $\delta(P_1, P_2)$ is an invariant, *chiral* δ-function; it can be written in several equivalent forms, *viz.*

$$\delta(P, P_1) = \delta^{(4)}(x_1 - x_2 + i\theta_2\sigma\bar{\theta}_1 - i\theta_1\sigma\bar{\theta}_2) \, \delta(\theta_1 - \theta_2)$$

$$= \delta^{(4)}(y_1 - y_2) \, \delta(\theta_1 - \theta_2) \qquad (2.3)$$

$$and \;\; \delta(\theta) \equiv (\theta\theta) \;\; .$$

Higher functional derivatives and derivatives with respect to right-handed fields, $\Phi^{\dagger}$, are defined analogously; for instance, we have:

$$\frac{\delta^2 F}{\delta\Phi(1)\,\delta\Phi(2)} = \left(\frac{\partial^2}{\partial\lambda_1 \partial\lambda_2} F[\Phi(P) + \lambda_1\delta(P, P_1) + \lambda_2\delta(P, P_2)] \right)_{\lambda_1 = \lambda_2 = 0} . \qquad (2.4)$$

The usual properties of functional derivatives (chain rule, commutativity, etc.) are easily verified from the definition. Superficially, it appears that paying so much attention to the very definition of a functional derivative with respect to a superfield is just hair-splitting; however, "naive" or careless definitions may lead to nonsensical results. Consider, for instance, an F-term in some supersymmetric action which is just a monomial in Φ. Its manifestly invariant form is:

$$F_n = \int dV \, \delta(\bar{\theta}) \, (\Phi(P))^n \; . \qquad (2.5)$$

Naively, one might think that its functional derivative is obtained just by "omitting a factor $\Phi(P)$ in n possible ways", i.e.

$$\frac{\delta F_n}{\delta\Phi(P_1)} \overset{(?)}{=} n\delta(\bar{\theta})\,(\Phi(P_1))^{n-1}\,. \qquad (2.6)$$

This is clearly incorrect, as one readily verifies, e.g. by writing out everything in component fields. (All but the "dynamical" scalar components of Φ are absent from the right hand side of of Eq. (2.6)!). By contrast, using Eq. (2.2) we have:

$$\frac{\delta F_n}{\delta\Phi(P_1)} = n\int dV\,\delta(\bar{\theta})\,\delta(\theta-\theta_1)\,\delta^{(4)}(y-y_1)\,(\Phi(P))^{n-1}$$

$$= n\int d^4x\,\delta^{(4)}(x-x_1-i\theta_1\sigma\bar{\theta}_1)\,(\Phi(x,\theta_1))^{n-1}$$

$$= n\,(\Phi(y_1,\theta_1))^{\cdot-1} \qquad (2.7)$$

which is obviously the correct result.

A concept related to a functional derivative is that of the "infinitesimal variation" (more precisely, the functional differential) of a functional of chiral superfields. Again, it is easily verified that if $\delta\Phi(y,\theta)$ stands for the differential of a chiral superfield, i.e. in terms of component fields

$$\delta\,\Phi(P) = \delta A(y) + \sqrt{2}(\theta\,\delta\psi(y)) + (\theta\,\theta)\,\delta F(y)\,, \qquad (2.8)$$

the correct expression of the differential of $F[\Phi(P)]$ is

$$\delta F[\Phi] = \int dV_1\,\delta\Phi(P_1)\,\frac{\delta F[\phi]}{\delta\Phi(P_1)}\,, \qquad (2.9)$$

using Eq. (2.2) to compute the functional derivative. (Notice that in this approach the $(\theta\theta)$ component of a chiral superfield--the "auxiliary scalar field"--is treated on an equal footing with the "dynamical" fields. This is as it should be: a manifestly supersymmetric formalism cannot distinguish between "dynamical" and "auxiliary" components of superfields).

In order to illustrate the use of this manifestly supersymmetric calculus of variations, we derive the field equations for a supersymmetric Φ^3 theory [11], the prototype of renormalizable models invariant under the WZ group. The action of any such model is of the form:

$$W = \int dV\,\Phi(y^\dagger,\bar{\theta})\,\Phi(y,\theta) + \int dV\,\delta(\bar{\theta})\,U(\Phi) + \int dV\,\delta(\theta)\,U^*(\Phi^\dagger) \qquad (2.10)$$

where $U(z)$ is a Hermitian, third degree polynomial of the complex vector z. The field equations for Φ are given by $\dfrac{\delta W}{\delta \Phi^\dagger} = 0$. The variation of the F^*-term in Eq. (2.10) is straightforward, viz.

$$\frac{\delta}{\delta \Phi^\dagger(P_1)} \int dV \, \delta(\theta) \, U^*(\Phi^\dagger) = \frac{\partial U^*(\Phi^\dagger(P_1))}{\partial \Phi^\dagger(P_1)}$$

cf. Eq. (2.7). In order to compute the functional derivative of the kinetic energy (the D-term in Eq. (2.10)), we use Eq. (2.2) to obtain:

$$\frac{\delta}{\delta \Phi^\dagger(P_1)} \int dV \, \Phi(y^\dagger, \bar\theta) \, \Phi(y, \theta) = \int dV \, \overline{\delta(P, P_1)} \, \Phi(y, \theta)$$

$$= \int dV \, \Phi(y, \theta) \, \delta^{(4)}(x - y_1^\dagger - i\theta\sigma\bar\theta_1) \, \delta(\bar\theta - \bar\theta_1)$$

$$= \int d^2\theta \, \Phi(y_1^\dagger + 2i\theta\sigma\bar\theta_1 \, , \, \theta) \, .$$

Putting these results together, we have the field equation:

$$\int d^2\theta \, \Phi(y_1^\dagger + 2i\theta\sigma\bar\theta_1 \, , \, \theta) + \frac{\partial U^*(\Phi^\dagger(y_1^\dagger, \bar\theta_1))}{\partial \Phi^\dagger(y_1^\dagger, \bar\theta_1)} = 0 \, . \tag{2.11}$$

This is indeed equivalent to the conventional form of the field equations; one just has to carry out the integration over θ. The result is the following:

$$\left(\frac{d}{d\theta} \frac{d}{d\theta}\right) \Phi(P) + \frac{\partial U^*}{\partial \Phi^\dagger(P)} = 0 \quad , \tag{2.12}$$

where

$$\frac{d}{d\theta^A} = \partial_A + 2i \, \partial_{A\dot A} \cdot \bar\theta^{\dot A} \, . \tag{2.13}$$

Upon decomposing Eq. (2.12) into component fields, the conventional form of the field equations is readily recovered.

3. EFFECTIVE ACTIONS AND WARD IDENTITIES

In a conventional theory (without supersymmetry), an effective action is defined as the Legendre transform of the generating functional of connected Green's functions. Having developed a variational calculus for chiral superfields, there stands nothing in the way of performing an analogous construction in terms of superfields. We define the generator, G, of connected Green's function by means of the usual formula, viz.

$$e^{iG} = \int D\Phi \; e^{iW[\Phi,\Phi^\dagger] + iS_1}$$

(3.1)

where W is a supersymmetric action and S_1 is a source term, viz.,

$$S_1 = \int dV \; \delta(\bar{\theta}) \; J(P) \; \Phi(P) \; + h.c.$$

(3.2)

In order to keep track of component fields, it is convenient to choose J as the *dual* of a chiral superfield,

$$J(y,\theta) = j_F + \sqrt{2} \; (\theta j_\psi) + (\theta\theta) \; j_A \; ,$$

(3.3)

where the subscripts $i = A$, ψ, F indicate the component field of which j_i serves as a source function.

Differentiation of G with respect to J yields the connected Green's functions. In particular, the *VEV* of the superfield Φ is given by:

$$\phi(P) = \frac{\delta G}{\delta J(P)} \quad , \quad \phi^\dagger(P) = \frac{\delta G}{\delta J^\dagger(P)} \quad .$$

(3.4)

We define an *effective action of the first kind* [9] as the Legendre transform,

$$A[\phi,\phi^\dagger] = G[J,J^\dagger] - \int dV \; \delta(\bar{\theta}) \; J \; \phi - \int dV \; \delta(\theta) \; J^\dagger \; \phi^\dagger \; .$$

(3.5)

By means of a straightforward application of the rules of the chiral variational calculus, we verify that A is indeed a functional of ϕ and $\phi^\dagger$ only and we derive the relationship reciprocal to Eq. (3.4),

$$\frac{\delta A}{\delta\phi(P)} = - J(P) \quad , \quad \frac{\delta A}{\delta\phi^\dagger(P)} = - J^\dagger(P)$$

(3.6)

which, of course, tells us that the effective action is stationary in the absence of external sources. One demonstrates in the usual way that higher derivatives of A give the irreducible vertex functions. In particular, the second derivatives yield the inverse two-point Green's functions

$$\frac{\delta^2 A}{\delta\phi(1)\delta\phi(2)} \equiv g^{-1}(1,2) = <T\Phi(1)\Phi(2)>^{-1}$$

$$\frac{\delta^2 A}{\delta\phi^\dagger(1)\delta\phi^\dagger(2)} \equiv g^{\dagger-1}(1,2) = <T\Phi^\dagger(1)\Phi^\dagger(2)>^{-1}$$

(3.7)

$$\frac{\delta^2 A}{\delta\phi(1)\delta\phi^\dagger(2)} \equiv h^{-1}(1,2) = <T\Phi(1)\Phi^\dagger(2)>^{-1} \; .$$

It is to be noted that at this point Eq. (3.7) is purely a formal relationship since all the two-point functions are elements of a Grassmann algebra. Consequently, the inverse of a Green's function has to be given a proper meaning, in particular when supersymmetry is broken. This question will be dealt with in the next section.

In complete analogy with the relationships Eq. (3.4) and Eq. (3.6), one can construct other kinds of effective actions. In particular, an effective action of the second kind is useful in order to investigate the dynamical breakdown of a symmetry, see e.g. Domokos and Suranyi, *loc. cit.*, Banks and Raby [12], Haymaker and Perez-Mercader [13]. A detailed investigation of the question of a dynamical breakdown of supersymmetry is deferred to a future publication; nevertheless, this is an appropriate place to outline the formal construction of a supersymmetric effective action of the second kind.

We modify the source term in Eq. (3.1) to read:

$$e^{H/2} \equiv \int D\Phi \, e^{iW + iS_2} \tag{3.8}$$

where

$$S_2 = \int dV_1 \, dV_2 \, \delta(\bar{\theta}_1) \, \delta(\bar{\theta}_2) \, k(P_1, P_2) \, \Phi(P_1) \, \Phi(P_2)$$

$$+ \int dV_1 \, dV_2 \, \delta(\theta_1) \, \delta(\theta_2) \, k^\dagger(P_1, P_2) \, \Phi^\dagger(P_1) \, \Phi^\dagger(P_2) \tag{3.9}$$

$$+ \int dV_1 \, dV_2 \, \delta(\theta_1) \, \delta(\bar{\theta}_2) \, l(P_1, P_2) \, \Phi(P_1) \, \Phi^\dagger(P_2)$$

and k, k^+, and l are chiral functions on the direct product of the supertranslation group with itself; the function l is Hermitian: $l(1,2) = l^\dagger(2,1)$. It is easily seen as before that the functional H generates the $2n$-point functions of the theory; in particular,

$$\frac{\delta H}{\delta k(1,2)} = g(2,1) \quad , \quad \frac{\delta H}{\delta k^\dagger(1,2)} = g^\dagger(2,1) \quad , \quad \frac{\delta H}{\delta l(1,2)} = h(2,1) . \tag{3.10}$$

An *effective action of the second kind* is the Legendre transform from k, k^+, l to g, g^+, h as independent variables, viz.

$$B = H - \int dV_1 \, dV_2 \, \delta(\bar{\theta}_1) \, \delta(\bar{\theta}_2) \, k(1,2) \, g(2,1)$$

$$- \int dV_1 \, dV_2 \, \delta(\theta_1) \, \delta(\theta_2) \, k^\dagger(1,2) \, g^\dagger(2,1) \tag{3.11}$$

$$- \int dV_1 \, dV_2 \, \delta(\theta_1) \, \delta(\bar{\theta}_2) \, l(1,2) \, h(2,1) .$$

We have the relations reciprocal to Eq. (3.10),

$$\frac{\delta B}{\delta g(1,2)} = -k(2,1) \quad , \quad \frac{\delta B}{\delta g^\dagger(1,2)} = -k^\dagger(2,1) \quad , \quad \frac{\delta B}{\delta h(1,2)} = -l(2,1) \,, \tag{3.12}$$

so that B *is stationary* if the sources k, k^+, and l are turned off.

The importance of the functional B lies in the fact that its higher functional derivatives are the inverse multiparticle Green's functions. To quote but one example,

$$\frac{\delta^2 B}{\delta g(1,2)\,\delta g(3,4)} = g^{-1}(2,1)\, g^{-1}(4,3) + K(1,3|2,4) \tag{3.13}$$

where K stands for the *exact* Bethe-Salpeter kernel in the scattering channel $(1,3) \rightarrow (2,4)$. As a consequence, vanishing Fredholm determinants of the operator $\delta^2 B$ at null total four-momentum in any given channel indicate the presence of massless collective modes (composite Goldstone particles).

Let us now turn to the derivation of the Ward identities satisfied by the effective actions. We subject Φ to an infinitesimal supertranslation of Grassmann parameters ξ^A, $\bar{\xi}^A$. The variation of the source term Eq. (3.1) is:

$$\delta S_1 = \int dV\, \bar{\xi}^{\dot A}\, \bar{\partial}_{\dot A}\, \delta(\bar\theta)\, J(P)\, \Phi(P) + \int dV\, \delta J(P)\, \Phi(P) + \int dV\, J(P)\, \delta\Phi(P) + h.\,c. \tag{3.14}$$

cf. Eq. (2.7) and Eq. (2.9). The first term in Eq. (3.14) is a total divergence and it can be omitted. Indeed, upon integration by parts,

$$\int dV\, J(P)\, \Phi(P)\, \bar{\xi}^{\dot A}\, \bar{\partial}_{\dot A}\, \delta(\bar\theta) = - \int dV\, \delta(\bar\theta)\, \bar{\xi}^{\dot A}\, \bar{\partial}_{\dot A}\, (J(P)\, \Phi(P)) \tag{3.15}$$

$$= -i \int dV\, \delta(\bar\theta)\, \partial_{A\dot A}\, (\bar{\xi}^{\dot A}\, \theta^A\, J(P)\, \Phi(P)) $$

because both J and Φ are chiral fields. In these equations,

$$\delta J(P) = \left(\xi^A \partial_A + 2i\, \theta^A \bar{\xi}^{\dot A}\, \partial_{A\dot A}\right) J(P) \tag{3.16}$$

and similarly for $\delta\Phi$. On introducing $\phi + \delta\Phi$ as a new functional argument and using the invariance of the action and functional measure in Eq. (3.1), we obtain the Ward identity:

$$\int dV\, \frac{\delta G}{\delta J(P)} \left\{ \xi^A \partial_A + 2i\, \theta^A \bar{\xi}^{\dot A}\, \partial_{A\dot A} \right\} J(P) + h.\,c. = 0\,. \tag{3.17}$$

Going over to the effective action Eq. (3.5) and using Eq. (3.4) and Eq. (3.6), we have finally after an integration by parts:

$$\int dV\, \frac{\delta A}{\delta\phi(P)} \left\{ \xi^A \partial_A + 2i\, \theta^A \bar{\xi}^{\dot A}\, \partial_{A\dot A} \right\} \phi(P) + h.\,c. = 0\,. \tag{3.18}$$

In a completely analogous way, we obtain Ward identities for the effective action of the second kind. We merely quote the result:

$$\int d(1)\, d(2) \left\{ \frac{\delta B}{\delta g(1,2)}\, \delta g(2,1) + \frac{\delta B}{\delta g^\dagger(1,2)}\, \delta g^\dagger(2,1) + \frac{\delta B}{\delta h(1,2)}\, \delta h(2,1) \right\} = 0 \qquad (3.19)$$

where

$$\delta g(1,2) = \left\{ \xi^A\, (\partial_{1A} + \partial_{2A}) + 2i\, \theta_1^A \bar{\xi}^{\dot{A}}\, \partial_{1A\dot{A}} + 2i\, \theta_2^A \bar{\xi}^{\dot{A}}\, \partial_{2A\dot{A}} \right\}\, g(1,2)$$

$$\delta h(1,2) = \left\{ \xi^A\, \partial_{1A} + \bar{\xi}^{\dot{A}} \partial_{2\dot{A}} + 2i\, \theta_1^A \bar{\xi}^{\dot{A}}\, \partial_{1A\dot{A}} - 2i\, \xi^A \bar{\theta}_2^{\dot{A}}\, \partial_{2A\dot{A}} \right\}\, h(1,2)$$

$$(3.20)$$

The simplicity of this manifestly covariant derivation of supersymmetric Ward identities is to be compared with the derivation using component field methods [14].

Ward identities for linearly realized *internal* symmetries can be solved immediately. In fact, the solution is provided by the first and second main theorems of the theory of invariants [15]: the effective action is a functional of a finite number of elementary invariants appropriate for the internal symmetry group and its representation spanned by the fields entering the effective action. In the case of induced representations, by contrast, we can construct an infinite number of independent invariants in terms of *arbitrary functions*. As a consequence, the solution to Eq. (3.16) can be, at best, given in the form of a Volterra series. We find:

$$A[\phi, \phi^\dagger] = a_0 + \int dV\, a_1(P)\, \phi(P)$$

$$+ \frac{1}{2} \int dV_1\, dV_2\, a_2(u_{12},\, \theta_1 - \theta_2)\, \delta(\bar{\theta}_1)\, \delta(\bar{\theta}_2)\, \phi(1)\, \phi(2)$$

$$+ \ldots$$

where $u_{12}^\mu = x_1^\mu - x_2^\mu + i\theta_2\, \sigma^\mu\, \bar{\theta}_1 - i\theta_1\, \sigma^\mu\, \bar{\theta}_2$, and the coefficients a_i are Lorentz invariant. (As a consequence, e.g. a_2 depends on $u^\mu u_\mu$ and $(\theta_1 - \theta_2)^2$ only.) Some of the terms contain irrelevant total divergences which must be isolated before proceeding further. On trying to proceed along this line, it soon becomes obvious that this approach is extremely cumbersome and the formulae become quite intractable. Fortunately, we can do without solving the Ward identities explicitly; this is the approach taken here.

4. APPLICATIONS OF THE FORMALISM: GOLDSTONE THEOREM AND DECOUPLING THEOREMS

We illustrate the usefulness of the effective action formalism in supersymmetric theories by rederiving the Goldstone theorem and various decoupling theorems [16] in this formalism for the case when the WZ group is broken by giving ϕ a non-vanishing VEV in the absence of external sources. First, however, we have to dispose of the problem of inverting Green's functions of superfields. We define the inverses of the Green's functions Eq. (3.7) by the covariant formulae respecting the chiral constraints:

$$\int dV_2\, \delta(\bar{\theta}_2)\, g(1,2)\, g^{-1}(2,3) = \delta(P_1\,,\, P_3) \tag{4.1.a}$$

$$\int dV_2\, \delta(\theta_2)\, h(1,\bar{2})\, h^{-1}(\bar{2},3) = \delta(P_1\,,\, P_3) \tag{4.1.b}$$

$$\int dV_3\, \delta(\bar{\theta}_3)\, h^{-1}(\bar{2},3)\, h(3,\bar{1}) = \overline{\delta(P_1\,,\, P_3)} \tag{4.1.c}$$

where bars above the arguments indicate that the corresponding (right-handed) chiral field depends on $y^\dagger$ rather than y; $\delta(P_1\,,P_2)$ stands for the chiral δ-function, cf. Eq. (2.3). If supersymmetry is broken, the Green's functions cannot be expected to obey supersymmetry constraints either; they are general functions on the Grassmann algebra corresponding to their arguments. Let us illustrate the procedure of computing the inverse of $g(P_1\,,P_2) = <T(\Phi(P_1)\,(P_2\,))>$. We decompose g into its components; the decomposition is constrained by Lorentz invariance (which we want to maintain, see Section 1) and symmetry. We find:

$$g(1,2) = g_0 + (\theta_1\, \theta_2\,)\, g_1 + (\theta_1\, \theta_1 + \theta_2\, \theta_2\,)\, g_2 + (\theta_1\, \theta_1\,)\, (\theta_2\, \theta_2\,)\, g_3\,, \tag{4.2}$$

where, in terms of the propagators of the component fields,

$$g_0 = <T(A(1)\, A(2))>\,, \qquad\qquad g_1 = <T(\psi(1)^A\, \psi(2)_A)>\,,$$

$$g_2 = <T(A(1)\, F(2))>\,, \qquad\qquad g_3 = <T(F(1)\, F(2))>\,, \tag{4.3}$$

all arguments in Eq. (4.3) being y_1^μ and y_2^μ respectively. (We know, of course, that $g_0 = g_3 = 0$ in the supersymmetric limit.) We now make a similar Ansatz for the inverse Green's function:

$$g^{-1}(2,3) = X_0 + (\theta_2\, \theta_3\,)\, X_1 + (\theta_2\, \theta_2 + \theta_3\, \theta_3\,)\, X_2 + (\theta_2\, \theta_2\,)\, (\theta_3\, \theta_3\,)\, X_3\,. \tag{4.4}$$

On inserting Eq. (4.2) and Eq. (4.4) into Eq. (4.1.a), we obtain relationships between the g_i and X_i in terms of ordinary space-time integrals. We discover that unless $g_0 = g_3 = 0$, the two-point function is not invertible! In other words, no matter how super-symmetry is broken, the (AA) and (FF) propagators must remain zero. With this observation, we easily obtain:

$$X_0 = X_3 = 0 \ , \quad X_1 = 4g_1^{-1} \ , \quad X_2 = g_2^{-1} \ . \tag{4.5}$$

We merely quote the appropriate relationships for the Green's function, h. We find:

$$h(1,\bar{2}) = h_0 - \theta_1 \, \sigma^\mu \, \bar{\theta}_2 \, h_{1\mu} + (\theta_1 \, \theta_1)\,(\bar{\theta}_2 \, \bar{\theta}_2)\, h_2 \ , \tag{4.6}$$

where, in terms of component fields,

$$h_0 = \langle T(A(1)A(\bar{2})^*) \rangle \ , \quad h_1^\mu = \langle T(\psi(1)\sigma^\mu\bar{\psi}(\bar{2})) \rangle \ , \quad h_2 = \langle T(F(1)F(\bar{2})^*) \rangle \tag{4.7}$$

(Again, (AF^*) -propagators must be absent; otherwise h is not invertible). The component h_1^μ can be written in terms of the derivative of a scalar function, viz. $h_{1\mu} = \partial_\mu S$. The inverse of h now reads:

$$h^{-1}(\bar{2},1) = h_2^{-1} + \bar{\theta}_2 \, \bar{\sigma}^\mu \, \theta_1 \, \partial_\mu \, (\ ^{-1} S^{-1}) + (\theta_1 \, \theta_1)\,(\bar{\theta}_2 \, \bar{\theta}_2)\, h_0^{-1} \ . \tag{4.8}$$

We are now ready to exploit the consequences of the Ward identity Eq. (3.18) for the case when supersymmetry is spontaneously broken. In the absence of external sources, A is stationary, $\delta A/\delta\phi = \delta A/\delta\phi^\dagger = 0$ and we also know that ϕ can be chosen to be of the form: $\phi = a + (\theta\theta)f$, a and f being constants, cf. Section 1. We differentiate Eq. (3.18) with respect to ϕ and put the coefficients of ξ and $\bar{\xi}$ equal to zero separately. In this way we obtain, after an integration by parts and dropping first derivatives of A:

$$\int dV_1 \left\{ \phi(1) \left(\xi \frac{\partial}{\partial\theta_1} \right) g^{-1}(2,1) + 2i \, \xi \, \sigma^\mu \, \bar{\theta}_1 \, \partial_\mu \phi(1) \right\} = 0$$

$$\int dV_1 \left\{ \phi^\dagger(1) \left(\bar{\xi} \frac{\partial}{\partial\theta_1} \right) h^{-1}(1,2) - 2i \, \theta_1 \, \sigma^\mu \, \bar{\xi} \, \partial_\mu \phi(1) \right\} = 0 \tag{4.9}$$

If the *VEV* is constant, the space-time derivatives vanish, so that Eq. (4.9) simplifies to

$$\int dV_1 \, \phi(1) \left(\xi \frac{\partial}{\partial\theta_1} \right) g^{-1}(2,1) = 4f \int d^4x_1 \, d^2\bar{\theta}_1 \, (\xi \, \theta_2)\, g_1^{-1}(2,1) = 0 \tag{4.10}$$

$$f^* \int d^4x_1 \, d\theta_1 \, \partial_\mu \left(\bar{\xi} \, \bar{\sigma}^\mu \, \theta_2 \quad^{-1} S^{-1} \right) = 0 \, . \tag{4.11}$$

Eq. (4.10) expresses the Goldstone alternative. We notice that integration over $\bar{\theta}_1$ picks out an F^*-term from g_1^{-1}, while integration over x_1 selects the zeroth Fourier coefficient: hence, the coefficient of f is just the exact mass operator of the Fermi component of Φ. Consequently, $f \neq 0$ leads to a vanishing Fermion mass term: ψ is the Goldstone Fermion.

Eq. (4.11), in essence, asserts the absence of infrared singularities in S^{-1}. Indeed, on going over to Fourier components and carrying out the integrations, Eq. (4.11) leads to the equation:

$$\int d^4k \, k_\mu \, \delta^{(4)}(k) \, S^{-1}(k^2) = 0 \, .$$

This is automatically satisfied unless S^{-1} is infrared singular. (In that case the integral would be ambiguous).

Let us now turn to a derivation of various decoupling theorems. Differentiating Eq. (3.18) once more, integrating by parts and assuming ϕ to be constant in space-time as before, leads to the identities:

$$\left(\xi \frac{\partial}{\partial \theta_2} + \xi \frac{\partial}{\partial \theta_3} \right) g^{-1}(2,3) + \int dV_1 \, \phi(1) \left(\xi \frac{\partial}{\partial \theta_1} \right) V(1, 2, 3) = 0 \, , \tag{4.12}$$

$$\int dV_1 \, \phi^\dagger(\bar{1}) \left(\bar{\xi} \frac{\partial}{\partial \bar{\theta}_1} \right) W(\bar{1}, 2, 3) = 0 \, , \tag{4.13}$$

where V and W are irreducible $(\Phi\Phi\Phi)$ and $(\Phi\Phi\Phi^\dagger)$ vertices, respectively. We now decompose V and W similarly to the two-point functions, viz.

$$
\begin{aligned}
V(1,2,3) = V_0 &+ (\theta_1 \theta_2 + \theta_1 \theta_3 + \theta_2 \theta_3) V_1 \\
&+ (\theta_1 \theta_1 + \theta_2 \theta_2 + \theta_3 \theta_3) V_2 \\
&+ \left\{ (\theta_1 \theta_2)(\theta_3 \theta_3) + (\theta_1 \theta_3)(\theta_2 \theta_2) + (\theta_2 \theta_3)(\theta_1 \theta_1) \right\} V_3 + \ldots
\end{aligned}
\tag{4.14}
$$

$$
W(\bar{1},2,3) = \begin{aligned} \theta_2 \sigma^\mu \bar{\theta}_1 &\left\{ W_{1\mu} + (\theta_2 \theta_3) \, W_{2\mu} \right\} + \\ \theta_3 \sigma^\mu \bar{\theta}_1 &\left\{ W_{1\mu} + (\theta_1 \theta_2) \, W_{2\mu} \right\} + \ldots \end{aligned}
\tag{4.15}
$$

Here the interpretation of the various components is again obvious; for instance, V_0 is an (AAA) vertex, V_1 is a $(\psi\psi A)$ vertex, and so on. Now inserting Eq. (4.14) and Eq. (4.15) into Eq. (4.12) and Eq. (4.13) respectively and using Eq. (4.4) and Eq. (4.5) we get the component field identities:

$$2\, g_1^{-1} + g_2^{-1} = \tfrac{f}{2} \int d^4x_1\, d\bar\theta_1\, V_1 \tag{4.16}$$

$$f \int d^4x_1\, d\bar\theta_1\, V_3 = 0 \tag{4.17}$$

$$f^* \int d\theta_1\, d^4x_1 \left[\, \theta_2\, \sigma^\mu\, \bar\xi \left\{ W_{1\mu} + (\theta_2\, \theta_3)\, W_{2\mu} \right\} \right.$$
$$\left. + \theta_3\, \sigma^\mu\, \bar\xi \left\{ W_{1\mu} + (\theta_3\, \theta_2)\, W_{3\mu} \right\} \right] = 0\,. \tag{4.18}$$

If supersymmetry is unbroken ($f = 0$), Eq. (4.16) simply asserts the well-known (at least, *perturbatively* well-known) relationship between the ($\psi\psi$) and (FA) propagators. If $f \neq 0$, this relationship becomes a "tadpole-insertion theorem" relating tadpole insertions into the propagators to the ($\psi\psi A$) vertex. Equations Eq. (4.17) and Eq. (4.18) are genuine "decoupling theorems": Eq. (4.17) asserts (for $f \neq 0$) the vanishing of the zeroth Fourier component of the ($F\psi\psi$) vertex; this means that the Goldstone Fermion must couple to the auxiliary scalar field, ($F-f$), *via* a derivative coupling. Likewise, Eq. (4.18) tells us after Fierz rearrangement that the effective ($\psi\bar\psi A$) and ($\psi\bar\psi F$) couplings are also of the derivative type if supersymmetry is spontaneously broken.

5. DISCUSSION

We believe that the most important result of this paper is the development of a manifestly supersymmetric formalism allowing us to undertake a systematic algebraic study of the general properties of theories with a spontaneously broken supersymmetry. The algebraic relationships derived from the supersymmetric Ward identities are independent of specific models and/or approximation schemes, provided the infinities of a given theory can be removed in a way consistent with WZ invariance. Despite valiant efforts of many investigators, it is still unclear whether the concept of a spontaneously broken supersymmetry is of any relevance for the description of Nature or if it will go down in the history of physics as one of the many ingenious schemes which *might* describe the observed phenomena in particle physics but they do not do so in our world. We hope that this paper represents a small contribution towards clarifying this important question.

ACKNOWLEDGEMENTS

This research has been supported by the U. S. Department of Energy under Contract No. DE-AC02-76ER03285.

REFERENCES

[1]. F. Gürsey, *Proceedings of the Conference on Gauge Theories and Modern Field Theory*, ed. P. Nath and R. Arnowitt. MIT Press, Cambridge, MA. 1976. F. Gürsey and L. Marchildon, J. Math. Phys. 19 (1978) 942. F. Gürsey, *Proceedings of the Second Johns Hopkins Workshop on Current Problems in High Energy Particle Theory*, ed. G. Domokos and S. Kövesi-Domokos. Johns Hopkins University, Baltimore, MD, 1978.

[2]. See, for instance, F. Gürsey and T. D. Lee, Proc. Nat. Acad. Sci. 49 (1963) 179. F. Gürsey, *Group Theoretical Concepts and Methods in Elementary Particle Physics*, ed. F. Gürsey. Gordon and Breach, New York 1964.

[3]. F. Gürsey, Ann. Phys. 24 (1963) 211. F. Gürsey and H. C. Tze, Ann. Phys. 128 (1980) 29.

[4]. J. Wess and B. Zumino, Nucl. Phys. B70 (1974) 39. For recent reviews, see P. Fayet and S. Ferrara, Phys. Lett. C32 (1977) 249. J. Wess, *Supersymmetry and Supergravity*, Princeton University Press (to be published).

[5]. A. Salam and J. Strathdee, Nucl. Phys. B76 (1974) 477.

[6]. See, for instance, H. Georgi, Harvard University preprint HUTP-81/A039 (1981). S. Dimopoulos and H. Georgi, Harvard University preprint, HUTP-81/A022 (1981). E. Witten, ICTP Trieste preprint (1981). S. Weinberg, Harvard University preprint HUTP-81/A047 (1981). J. Ellis, M. K. Gaillard and B. Zumino, LAPP (Annecy) preprint LAPP-TH-44 (1981).

[7]. It is amusing to watch how Michel's conjecture (L. Michel, *Lecture Notes in Physics*, Vol. 6 Springer, New York, 1970) works even in the case of symmetry groups involving space-time transformations. Note that the homogeneous Lorentz group is a maximal subgroup of the graded De Sitter group from which the WZ group is obtained by contraction. It remains the unbroken symmetry after contraction too.

[8]. L. O'Raifeartaigh, Nucl. Phys. B96 (1975) 331.

[9]. G. Jona-Lasinio, Nuovo Cim. 34 (1964) 1970. G. Domokos and P. Suranyi, J. Nucl. Phys. (USSR) 2 (1965) 501. English translation: Sov. Journ. Nucl. Phys. 2 (1966) 361. B. Zumino, *Brandeis Lectures in Elementary Particles and Quantum Field Theory*, ed. S. Deser and H. Pendleton. (MIT Press, Cambridge, MA, 1970).

[10]. The formalism developed here is essentially identical with the one described in the review article of Fayet and Ferrara, see Ref. 4. However, instead of Majorana spinors, we use Weyl spinors throughout; this seems to be somewhat better suited to our purposes.

[11]. J. Wess and B. Zumino, Phys. Lett. 49B (1974) 52.

[12]. T. Banks and S. Raby, Phys. Rev. D14 (1976) 2182.

[13]. R. Haymaker and J. Perez-Mercader, Phys. Lett. 106B (1981) 201.

[14]. H. S. Tsao, Phys. Lett. 53B (1974) 381. O. Piguet and M. Schweda, Nucl. Phys. B92 (1975) 334.

[15]. Cf. H. Weyl, *The Classical Groups* (Princeton University Press, 1946).

[16]. Decoupling theorems in supersymmetric theories were first derived by B. deWit and D. Z. Freedman, Phys. Rev. D12 (1975) 2286.

AN ACTION IN SUPERSPACE
FOR
SO(N)-SUPERGRAVITY

S. W. MacDowell

Yale University
Physics Department
J. W. Gibbs Laboratory
New Haven, Connecticut 06511

Abstract

We review the formulation of supergravity as the geometry of a bundle of frames with structure group $SL(2,C) \otimes SO(N)$ and base supermanifold $\mathcal{M}$ with 4 even and $4N$ odd dimensions. The action principle is discussed and a construction is given of a superspace action $\mathcal{S}$ that becomes stationary when the constraints and equations of motion of conventional $SO(N)$-models are imposed. One attractive feature of this action is that it contains Einstein's action plus other terms which depend on the torsion and the Yang-Mills curvature.

INTRODUCTION

The concept of superspace was introduced by Salam and Strathdee [1] as a framework for the formulation of supersymmetric theories [2]. It has been an invaluable tool not only for the construction of a variety of supersymmetric models, but also in understanding the meaning of supersymmetry and its role in the cancellation of ultraviolet quantum divergences.

In locally supersymmetric models, generically called supergravity, the superspace may be given a deeper, more intrinsic, geometric meaning. In this article the geometric formulation of supergravity is reviewed and new results are presented on the construction of an action in superspace for $SO(N)$-supergravity.

The paper is organized as follows. The first section consists of a geometric presentation of the superspace approach to $SO(N)$-supergravity in terms of a bundle of frames with structure group $SL(2,C) \otimes SO(N)$ and base supermanifold $\mathcal{M}$ with 4 even and $4N$ odd dimensions. In the second section we derive by a somewhat different method general results concerning the action principle in superspace. In the third section we develop a method for the construction of invariant actions in superspace for the known models of $SO(N)$-supergravity. In particular, we arrive at an action, without *a priori* constraints, which generalizes Einstein's action. The equations of motion of $SO(N)$-supergravity are compatible with the Euler-Lagrange equations derived from this superaction.

1. BUNDLE OF FRAMES FOR SUPERGRAVITY

Since the commutator of two supersymmetry transformations is a translation, it follows that a theory invariant under *local* supersymmetry is also invariant under local translations, that is under general coordinate transformations. Hence, any such theory is a generalization of Einstein's theory of gravitation wherefore it is called Supergravity. A fundamental difference between a globally supersymmetric theory and supergravity is that the former is formulated on a flat superspace whereas the latter requires a curved manifold.

We have derived a geometrical approach to supergravity based on the formalism of fibre bundles [3]. In its essence this formalism was first used by Wess and Zumino [4] and followed by many others [5, 6, 7, 11].

The natural framework for the formulation of a physical theory is the bundle of frames in which at each point on the base manifold one associates a local Lorentz frame. As emphasized by Einstein the laws of nature are invariant under *local* Lorentz transformations of the inertial frames. Furthermore, one needs to give a prescription on how to compare measurements performed at distant points in the space-time manifold. This is done by introducing a Lorentz connection which is equivalent to the notion of parallel transport of vectors along a path.

In Einstein's General Relativity, the bundle of frames B consists of a four-dimensional manifold $\mathcal{M}$ covered by a complete set of coordinate neighborhoods $\{U_I\}$; in each neighborhood U_I is defined a tetrad of vector fields $\{e^I_{,i}\}$ and a scalar product $\{e^I_i, e^I_j\} = \eta_{ij}$, where $\eta_{ij} = (-1, -1, -1, 1)$ is the Minkowski metric [8]. At each point $m \in U_I$, the fibre over $\mathcal{M}$ is the set $\{p = (m, \{e_i\})\}$, where $\{e_i\}$ is obtained from

$\{e^I_{\;j}\}$ by means of a Lorentz transformation: $e_i = (g_I)^j_{\;i}\, e^I_{\;j}$. Then the structure group in B is the metric preserving group $H = SO(3,1)$ and the projection π is just $\pi(p) = m$. The bundle space B is locally homeomorphic to $\mathcal{M} \otimes H$, that is, for every neighborhood U_I there is a mapping $\Phi_I\colon \pi^{-1}(U_I) \twoheadrightarrow (U_I \otimes H)$. In the bundle of frames this mapping is specified by $p \twoheadrightarrow (m, g_I)$. In the intersection $U_I \cap U_{II}$ of two neighborhoods, the tetrads $\{e^I_{\;i}\}$ and $\{e^{II}_{\;i}\}$ are related by: $e^I_{\;i} = (g_{I,II}(m))^j_{\;i}\, e^{II}_{\;j}$, where $g_{I,II}(m) \overset{\cdot}{=} g_I(m)^{-1}\, g_{II}(m) \in H$.

The displacements along the fibres in B are generated by a set of tangent vectors $\{D_{[ij]}\}$ called *vertical* vectors which can be so chosen as to define, under commutation, a Lie algebra isomorphic to the Lie algebra of H, whose generators will be denoted by $\{X_{[ij]}\}$.

The set of *horizontal* one-forms $\{h^i\}$ dual to the basis of vectors $\{e_i\}$ is called the solder form [9]; then we have[1]: $h^i(e_j) = \delta^i_j$, $h^i(D_{[jk]}) = 0$. In order to define the parallel transport of vectors along a curve in $\mathcal{M}$, one introduces a connection in B, which is an equivariant *vertical* one-form taking values in the Lie algebra of H, $\omega = \omega^{[ij]} X_{[ij]}$ and dual to the basis of vectors $\{D_{[ij]}\}$ so that $\omega(D_{[ij]}) = X_{[ij]}$. It can be shown that the invariance of the metric under the structure group H in the bundle of frames insures the invariance of the scalar product of vectors parallel translated along a curve in $\mathcal{M}$ [9]. The connection ω defines a horizontal tangent space by $\omega(t) = 0$, for every horizontal vector t. One can choose a basis $\{D_{[ij]}, D_k\}$ in the tangent space which is dual to the basis of forms $\{\omega^{[ij]}, h^k\}$, so that;

$$\omega^{[ij]}\,(D_{[kl]}) = \delta^{[ij]}_{[kl]} \ , \qquad\qquad\qquad \omega^{[ij]}\,(D_k) = 0 \ , \qquad\qquad (1.1)$$

$$h^i\,(D_{[kl]}) = 0 \quad , \qquad\qquad\qquad h^i\,(D_k) = \delta^i_k \ . \qquad\qquad (1.2)$$

The curvature and the torsion are horizontal two-forms defined by:

$$\mathcal{R}^{[ij]} = d\omega^{[ij]} \circ Hr \quad , \qquad\qquad\qquad\qquad\qquad\qquad\qquad (1.3)$$

$$\mathcal{T}^k = dh^k \circ Hr \qquad\qquad\qquad\qquad\qquad\qquad\qquad\qquad (1.4)$$

where d is the exterior derivative in B and Hr means horizontal projection.

---

[1] We use the notation $\omega(t_A, t_B, \ldots)$ to denote the contraction of a form ω with the vectors $t_A, t_B, \ldots$

The bundle space of general relativity is such that its tangent space at each point is homeomorphic to the tangent space of either the Poincaré group or one of the de Sitter groups, for non-vanishing cosmological constant. Let us denote this group by G. The vertical tangent space at a point $p \in B$ is homeomorphic to the tangent space of the subgroup H of G at the identity, whereas the horizontal tangent space is homeomorphic to the coset space G/H.

Now we want to extend this structure to supergravity. Since the physical space appears to be four-dimensional (namely three space and one time dimensions) one can change the base manifold only by the addition of extra dimensions which are not parametrized by real numbers, but instead, by anti-commuting Grassmann numbers. The internal symmetry gauge group can be incorporated in a unified way within the formalism of the bundle of frames, by letting the parameters associated with the extra dimensions carry an internal symmetry index. Then one should look for supergroups G which contain a Lie subgroup $H = SL(2,C) \otimes K$ where K is a compact internal symmetry group [3]. Furthermore, the coset space G/H should have only four generators belonging to the even sector of the Lie super-algebra of G; in the case of Poincaré supergravity they are associated with translations. The simplest such supergroups are the orthosymplectic supergroups $OSp(N|4)$ or contractions thereof.

We shall denote by $\{X_A\}$ the set of generators of the Lie superalgebra of G, with the following Lie brackets:

$$[X_A, X_B\} = -(-1)^{\sigma_A \sigma_B}[X_B, X_A\} = f_{AB}{}^C X_C \qquad (1.5)$$

where $\sigma_A = 0, 1$ is the signature of the generator X_A (even or odd) and $\{f_{AB}{}^C\}$ is the set of structure constants.

The bundle of frames B for $SO(N)$-supergravity has $H = SL(2,C) \otimes SO(N)$ as its structure group, and a base supermanifold $\mathcal{M}$ which is locally homeomorphic to G/H, where $G = OSp(N|4)$ for de Sitter supergravity or its Wigner-Inönu contraction for Poincaré supergravity [8]. The generators of the subgroup H of G will be denoted by $\{X_{A_0}\}$ and the generators of the coset space G/H will be denoted by $\{X_{A_1}\}$. More generally, the indices A_0 and A_1 will be used for the elements of a set which transform under H as $\{X_{A_0}\}$ and $\{X_{A_1}\}$ respectively. For the supergroup $OSp(N|4)$ the set $\{X_{A_0}\}$ consists of the generators $\{X_{[ij]}\}$ of $SL(2,C)$ and the generators $\{X_{[ab]}\}$ of $SO(N)$; the set $\{X_{A_1}\}$ consists of two irreducible sets $\{X_i\}$ and $\{X_{aa}\}$, the former transforming as a basis of vectors under $SL(2,C)$ and invariant under $SO(N)$, the latter transforming as a basis for Majorana spinors (with index a) under $SL(2,C)$ and vectors (with index

a) under $SO(N)$.[2]

Let η_{AB} be the Killing metric[3] of the superalgebra of G; it has the following symmetry:

$$\eta_{AB} = (-1)^{\sigma_A \sigma_B} \eta_{BA} \tag{1.6}$$

and invariance property

$$f_{AB}{}^D \eta_{CD} \doteq f_{CA}{}^D \eta_{DB} = 0 . \tag{1.7}$$

For a simple supergroup such as $OSp(N|4)$, η_{AB} is non-singular. Then we define η^{AB} by:

$$\eta_{AC} \, \eta^{BC} = \delta_A{}^B . \tag{1.8}$$

The matrices η_{AB} and η^{AB} are used to lower and raise indices according to the rule $\phi^A \eta_{AB} = \phi_B$, $\eta^{AB}\phi_B = \phi^A$.

A metric in the base supermanifold $\mathcal{M}$ is defined by specifying the scalar product of the elements of the basis vectors $\{e^I_{A_I}\}$, in each neighborhood U_I of $\mathcal{M}$, to be:

$$(e^I_{A_I}, e^I_{B_I}) = \eta_{A_I B_I} . \tag{1.9}$$

An element $p \in B$, consists of a point $m \in \mathcal{M}$ and a basis of vectors $\{e_{A_I} = g_{IA_I}^{B_I} e^I_{B_I}\}$, where $g_I \in H$ and $e^I_{B_I}$ is taken at m.

The set of vertical vectors $\{D_{A_0} = D_{[ij]}, D_{[ab]}\}$, which generate displacements along a fibre may be defined by:

---

[2] The structure constants $f_{aabb}{}^i = f_a{}^i \eta_{ab}$, $f_{ab}{}^{[ij]} = f_{ab}{}^{[ij]} \eta_{ab}$ are proportional to $(\gamma^i C)_{ab} \delta_{ab}$, $(\sigma^{[ij]}C)_{ab}\delta_{ab}$, where $(\gamma^i)_a{}^b$ is a Dirac matrix, $\sigma^{[ij]} \doteq 1/2 [\gamma^i, \gamma^j]$ and $C_{ab} = -C_{ba}$ is the charge conjugation matrix, all in the Majorana representation where they are real. We define C^{ab} by $C_{ab}C^{cb} = I_a{}^c$ where I is the unit matrix.

The matrices C_{ab} and C^{ab} are used to lower and raise spinor indices: $u^a C_{ab} = u_b$, $u_b C^{ab} = u^a$; this justifies the notation $C_a{}^b$ for the unit matrix.

[3] If $(X'_A)_{A'}{}^{B'}$ are the generators of the Lie algebra of G in the representation R', then:

Sup. Tr. $(X'_A, X'_B) \equiv (-1)^{\sigma_{D'}} (X'_A)_{C'}{}^{D'} (X'_B)_{D'}{}^{C'} = -f(R') \eta_{AB}$

where η_{AB} is the Killing metric (independent of the representation) and $f(R')$ is the index of the representation R'. For simple Lie groups one sets $f(R') = 1$ for the adjoint representation; for simple supergroups, in some instances $f(R') = 0$ for the adjoint representation and the metric has to be defined in terms of the fundamental representation.

$$dg_{1B_1}^{C_1}(D_{A_0}) = D_{A_0} g_{1B_1}^{C_1} = -f_{A_0B_1}{}^{D_1} g_{1D_1}^{C_1} \tag{1.10}$$

Because of the invariance of $\eta_{A_1B_1}$ under H, the metric η in $\mathcal{M}$ induces a metric in B which at $p = (m, \{e_{A_1}\})$ is block diagonal in the basis $\{D_{A_0}, e_{A_1}\}$ and is given by:

$$\eta(D_{A_0}, D_{B_0}) = \eta_{A_0B_0},$$

$$\eta(e_{A_1}, e_{B_1}) = \eta_{A_1B_1}.$$

The solder form $h = h^{A_1} X_{A_1}$ is defined as a set of horizontal one-forms dual to the basis $\{e_{A_1}\}$ at $p = (m, \{e_{A_1}\})$, that is:

$$h^{A_1}(D_{A_0}) = 0 \quad , \quad h^{B_1}(e_{A_1}) = \delta_{A_1}{}^{B_1}. \tag{1.11}$$

The definitions of the metric and of the solder form are clearly independent of the existence of a connection. We now introduce in B a connection $\omega = \omega^{A_0} X_{A_0}$, which is an equivariant vertical one-form, taking values in the Lie algebra of H and dual to the set of vertical vectors $\{D_{A_0}\}$: $\omega(D_{A_0}) = X_{A_0}$ and $\omega_{gp} = g\, \omega_p\, g^{-1}$.

It is convenient to introduce the one-form Φ, taking values in the Lie algebra of G, defined by [8,10]:

$$\Phi = \Phi^A X_A = \omega + h \tag{1.12}$$

and the basis of vectors $\{D_A = D_{A_0}, D_{A_1}\}$ dual to Φ, where the set $\{D_{A_1}\}$ forms a basis for the horizontal tangent space.

We also define the two-form [8,10]:

$$\mathcal{F} = \mathcal{F}^A X_A = d\,\Phi. \tag{1.13}$$

Then we have:

$$[D_A, D_B\} = -F_{AB}{}^C D_C \tag{1.14}$$

where $F_{AB}{}^C = \mathcal{F}^C(D_A, D_B)$. Any vertical component of $\mathcal{F}^C$ coincides with a structure constant, $F_{A_0B}{}^C = f_{A_0B}{}^C$. The horizontal components of $\mathcal{F}$ are the curvature and torsion forms:

$$\mathcal{F} \circ Hr = \mathcal{R} + \mathcal{T} = \mathcal{R}^{A_0} X_{A_0} + \mathcal{T}^{A_1} X_{A_1} \tag{1.15}$$

where

$$\mathscr{R} = d\omega \circ Hr \ ; \qquad\qquad\qquad \mathscr{T} = dh \circ Hr \ . \qquad\qquad (1.16)$$

From the definition (1.13) it follows that:

$$d\mathscr{F} = 0 \qquad\qquad\qquad\qquad (1.17)$$

which in component form in the basis $\{D_A\}$ gives the following set of Bianchi and Jacobi identities [8].

$$(ABC)^D \equiv \sum_{(ABC)} (-1)^{\sigma_A \, \sigma_C} [D_A F_{BC}{}^D + F_{AB}{}^E F_{EC}{}^D] = 0 \qquad\qquad (1.18)$$

where the sum is over cyclic permutations of the indices (ABC).

Let $\{U_I\}$ be a complete set of neighborhoods in $\mathcal{M}$. For each U_I a coordinate basis is a set of vector fields $\{\partial_\Lambda\}$ satisfying the conditions

$$[\partial_\Lambda, \partial_\Sigma\} \equiv (\partial_\Lambda \partial_\Sigma - (-1)^{\sigma_\Lambda \, \sigma_\Sigma} \partial_\Sigma \partial_\Lambda) = 0 \qquad\qquad (1.19)$$

and spanning the tangent space $\mathcal{M}_m$ at every $m \in U_I$. The dual of the basis $\{\partial_\Lambda\}$ is the set of one-forms $\{dx^\Lambda\}$ such that $dx^\Lambda(\partial_\Sigma) = \delta_\Sigma{}^\Lambda$.

The introduction of a connection ω is equivalent to the definition of the horizontal lifts $\{D_\Lambda\}$ of the vectors $\{\partial_\Lambda\}$, with the following properties [9]:

(i)They are horizontal ($\omega(D_\Lambda) = 0$).

(ii)The projection of D_Λ onto the base is ∂_Λ.

(iii)D_Λ is invariant, that is $[D_{A_0}, D_\Lambda] = 0$.

Therefore, we have:

$$D_\Lambda = \partial_\Lambda - h_\Lambda^{A_0} D_{A_0} = h_\Lambda^{A_1} D_{A_1} \qquad\qquad (1.20)$$

hence:

$$\partial_\Lambda = h_\Lambda^A D_A \qquad\qquad\qquad (1.21)$$

where

$$h_\Lambda^{A_0} = \omega^{A_0}(\partial_\Lambda) \ , \qquad\qquad\qquad (1.22)$$

$$h_\Lambda^{A_1} = h^{A_1}(D_\Lambda) = h^{A_1}(\partial_\Lambda) \ . \qquad\qquad (1.23)$$

The equivariance of ω and h implies that:

$$D_{A_0} h_\Lambda^C = f_{A_0 B}{}^C h_\Lambda^B .$$

(1.24)

The matrix $h_\Lambda^{A_1}$ which relates the two horizontal bases $\{D_\Lambda\}$ and $\{D_{A_1}\}$ must be non-singular. Let us denote its inverse by $h^{-1}{}^\Lambda_{A_1}$ so that

$$h_\Lambda^{A_1} h^{-1}{}^\Sigma_{A_1} = \delta_\Lambda^\Sigma ; \qquad\qquad h^{-1}{}^\Lambda_{A_1} h_\Lambda^{B_1} = \delta_{A_1}^{B_1} .$$

(1.25)

The components of the curvature and torsion in the basis D_Λ are given by:

$$\mathcal{R}_{\Lambda\Sigma}^{A_0} D_{A_0} = - [D_\Lambda, D_\Sigma] = \{ \partial_\Lambda h_\Sigma^{A_0} - (-1)^{\sigma_\Lambda \sigma_\Sigma} \partial_\Sigma h_\Lambda^{A_0} - h_\Lambda^{B_0} h_\Sigma^{C_0} f_{B_0 C_0}{}^{A_0} \} D_{A_0} ,$$

(1.26)

$$\mathcal{T}_{\Lambda\Sigma}^{A_1} = D_\Lambda h_\Sigma^{A_1} - (-1)^{\sigma_\Lambda \sigma_\Sigma} D_\Sigma h_\Lambda^{A_1}$$

$$= \partial_\Lambda h_\Sigma^{A_1} - (-1)^{\sigma_\Lambda \sigma_\Sigma} \partial_\Sigma h_\Lambda^{A_1} - \{h_\Lambda^{B_0} h_\Sigma^{C_1} - (-1)^{\sigma_\Lambda \sigma_\Sigma} h_\Sigma^{B_0} h_\Lambda^{C_1} \} f_{B_0 C_1}{}^{A_1} .$$

(1.27)

They are related to the components in the basis $\{D_{A_1}\}$ by:

$$F_{\Lambda\Sigma}{}^A = (-1)^{\sigma_{B_1} (\sigma_{C_1} + \sigma_\Sigma)} h_\Lambda^{B_1} h_\Sigma^{C_1} F_{B_1 C_1}{}^A .$$

(1.28)

Finally, we consider transformations of the connection and solder form, defined in terms of the following transformations of the basis $\{D_{A_1}\}$, which leave invariant the basis $\{D_{A_0}, \partial_\Lambda\}$ [3, 7, 8]:

$$D'_A = \exp\{\alpha^B(p)D_B\} D_A \exp\{-\alpha^C(p)D_C\}$$

(1.29)

where $\alpha^C(p)$ satisfies the condition

$$D_{A_0} \alpha^C(p) = f_{A_0 B}{}^C \alpha^B(p) .$$

(1.30)

For an infinitesimal transformation, we have

$$\delta D_A = [\epsilon^C D_C, D_A] = - (\epsilon^C F_{CA}{}^B + D_A \epsilon^B) D_B .$$

(1.31)

Therefore, from Eq. (1.21) with $\delta\partial_\Lambda = 0$, one obtains:

$$\delta h_\Lambda^B D_B + h_\Lambda^B \delta D_B = 0$$

(1.32)

which gives:

$$\delta h_\Lambda^A = \partial_\Lambda \epsilon^A + h_\Lambda^C \epsilon^B F_{BC}{}^A$$

(1.33)

and from Eq. (1.14)

$$\delta F_{BC}{}^A = \epsilon^D D_D F_{BC}{}^A . \tag{1.34}$$

These transformations close under commutation, $[\delta_1, \delta_2] = \delta_3$, with the composition law [8]:

$$\epsilon_3^A = \epsilon_1^B D_B \epsilon_2^A - \epsilon_2^B D_B \epsilon_1^A + \epsilon_1^B \epsilon_2^C F_{CB}{}^A . \tag{1.35}$$

Transformations with $\alpha^{B_1}(p) = 0$ are equivalent to gauge transformations. On the other hand, an infinitesimal transformation with parameters [3, 7]

$$\epsilon^A = \epsilon^\Lambda h_\Lambda^A \tag{1.36}$$

is equivalent to a general coordinate transformation, $\{x^\Lambda\} \rightarrow \{x'^\Lambda = x^\Lambda + \epsilon^\Lambda(x)\}$.

2. INVARIANT ACTION AND THE ACTION PRINCIPLE

The action for supergravity in superspace is of the form:

$$\mathscr{S} = \int \mathscr{L} \, \Omega \tag{2.1}$$

where the Lagrangian $\mathscr{L}$ is assumed to depend explicitly only on the covariant fields $F_{A_1 B_1}{}^C$, and the invariant measure Ω depends on the metric, therefore on the components of the solder form, the "vielbein" $h_\Lambda^{A_1}$.

Let us consider the horizontal form $h^A = h_\Lambda^A \, dx^\Lambda$ and let $\delta h^A = \delta h_\Lambda^A \, dx^\Lambda$ be an infinitesimal variation of h^A; define also $\delta h_{B_1}^A$ by:

$$\delta h_{B_1}^A = \delta h^A (D_{B_1}) = h^{-1\Lambda}{}_{B_1} \, \delta h_\Lambda^A . \tag{2.2}$$

The corresponding variation of Ω must fulfill the following conditions:

  (i)It is a measure.

  (ii)It is linear and homogenous in δh^{A_1}.

  (iii)It is invariant under the gauge group H.

Therefore it must be of the form [4,5]:

$$\delta\Omega = (-1)^{\sigma_{A_1}} \, \delta h_{A_1}^{A_1} \, \Omega \tag{2.3}$$

up to a numerical constant; this constant may be determined from the known expressions for the invariant measure for ordinary manifolds and turns out to be one. From the expression Eq. (2.3) for the variation of Ω one deduces that:

$$D_\Lambda \, \Omega = (D_\Lambda \, h^{B_1}_\Sigma) \, h^{-1\Sigma}_{B_1} \, (-1)^{\sigma_\Sigma} \, \Omega \tag{2.4}$$

or

$$D_{A_1} \, \Omega = h^{-1\Lambda}_{A_1} \, D_\Lambda \, \Omega = (D_{A_1} \, h^{B_1}_\Sigma) \, h^{-1\Sigma}_{B_1} \, (-1)^{\sigma_\Sigma} \, \Omega \tag{2.5}$$

and

$$D_{A_0} \, \Omega = (-1)^{\sigma_{A_1}} \, h^{-1\Lambda}_{A_1} \, D_{A_0} \, h^{A_1}_\Lambda \Omega = (-1)^{\sigma_{A_1}} \, h^{-1\Lambda}_{A_1} \, h^{B_1}_\Lambda \, f_{A_0 B_1}{}^{A_1} \, \Omega$$

$$= (-1)^{\sigma_{A_1}} \, f_{A_0 A_1}{}^{A_1} \, \Omega = 0 \, . \tag{2.6}$$

Let us now demand that the action $\mathscr{S}$ be invariant under local gauge transformations and general coordinate transformations in the base supermanifold or, equivalently, under the set of infinitesimal transformations of the connection and the "vielbein" as expressed by Eq. (1.33) and Eq. (1.34). We have:

$$\delta\mathscr{S} = \int \{\delta\mathscr{L} \, \Omega + \mathscr{L} \, \delta\Omega\} = \int \{\epsilon^A \, D_A \, \mathscr{L} + \mathscr{L} \, (-1)^{\sigma_{A_1}} \, \delta h^{A_1}_{A_1}\} \, \Omega$$

$$= \int [\epsilon^{A_0} \, D_{A_0} \, \mathscr{L} + \epsilon^{A_1} \, D_{A_1} \, \mathscr{L} + (-1)^{\sigma_{A_1}} \, \{D_{A_1} \, \epsilon^{A_1} + \epsilon^{B_1} \, F_{B_1 A_1}{}^{A_1}\} \, \mathscr{L}] \, \Omega$$

$$= \int [\epsilon^{A_0} \, D_{A_0} \, \mathscr{L} + (-1)^{\sigma_{A_1}} \, \{D_{A_1}(\epsilon^{A_1} \, \mathscr{L}) + (-1)^{\sigma_{B_1}} \, F_{A_1 B_1}{}^{B_1} \, (\epsilon^{A_1} \, \mathscr{L})\}] \, \Omega \, . \tag{2.7}$$

Therefore the invariance of the action integral requires:

1. Invariance of the Lagrangian under the gauge group H.

$$D_{A_0} \, \mathscr{L} = 0 \, . \tag{2.8}$$

2. The following rule of integration by parts:

$$\int (D_{A_1} \, U^{A_1}) \, \Omega = - \int (-1)^{\sigma_{B_1}} \, T_{A_1 B_1}{}^{B_1} \, U^{A_1} \, \Omega \, . \tag{2.9}$$

We shall now show that the rule of integration by parts follows from the condition

$$D_{A_0} \, U^{A_1} = f_{A_0 B_1}{}^{A_1} \, U^{B_1} \tag{2.10}$$

that is that $\{U^{A_1}\}$ transform under the group H as the solder form $\{h^{A_1}\}$.

Indeed we have:

$$\int (D_{A_1} U^{A_1}) \, \Omega = \int h^{-1\Lambda}_{A_1} (D_\Lambda U^{A_1}) \, \Omega \tag{2.11}$$

$$= \int (-1)^{\sigma_\Lambda (\sigma_\Lambda + \sigma_{A_1})} D_\Lambda \, (h^{-1\Lambda}_{A_1} U^{A_1} \Omega)$$

$$- \int [(-1)^{\sigma_\Lambda(\sigma_\Lambda + \sigma_{A_1})} (D_\Lambda h^{-1\Lambda}_{A_1}) U^{A_1} \Omega + (-1)^{\sigma_{A_1}} U^{A_1} D_{A_1} \Omega] \, .$$

The first integral vanishes because of Eq. (2.10). Then:

$$\int(D_{A_1} U^{A_1})\Omega = -\int (-1)^{\sigma_{B_1}}[(-1)^{\sigma_{B_1}\sigma_{A_1}}(D_{B_1} h^{-1\Lambda}_{.\ A_1}) - (D_{A_1} h^{-1\Lambda}_{B_1})]h^{B_1}_\Lambda U^{A_1}\Omega \, . \tag{2.12}$$

But from Eq. (1.14) we have:

$$F_{A_1 B_1}{}^C D_C = - [D_{A_1}, D_{B_1}\} = - [h^{-1\Lambda}_{A_1} D_\Lambda, h^{-1\Sigma}_{B_1} D_\Sigma\}$$

$$= - (-1)^{\sigma_\Lambda(\sigma_\Sigma + \sigma_{B_1})} h^{-1\Lambda}_{A_1} h^{-1\Sigma}_{B_1} [D_\Lambda, D_\Sigma\}$$

$$- [D_{A_1} h^{-1\Lambda}_{B_1} - (-1)^{\sigma_{A_1}\sigma_{B_1}} (D_{B_1} h^{-1\Lambda}_{A_1})] h^{C_1}_\Lambda D_{C_1} \, . \tag{2.13}$$

Since the first term is vertical it follows that:

$$T_{A_1 B_1}{}^{C_1} = - [D_{A_1} h^{-1\Lambda}_{B_1} - (-1)^{\sigma_{A_1} \sigma_{B_1}} (D_{B_1} h^{-1\Lambda}_{A_1})] \, h^{C_1}_\Lambda \, . \tag{2.14}$$

Using this relation in Eq. (2.12) one obtains the desired result Eq. (2.9). For $U^{A_1} = \epsilon^{A_1} \mathcal{L}$, taking into account Eq. (1.30), the condition Eq. (2.10) will follow from the condition 1. We come then to the following result[4].

"If the Lagrangian $\mathcal{L}$ is a gauge invariant function of the components $F_{A_1 B_1}{}^C$ of the curvature and the torsion, the action Eq. (2.1), where Ω is an invariant measure, is gauge invariant and independent of the choice of coordinates in the base superman-ifold."

Let us now derive the Euler-Lagrange equations for the variational principle $\delta\mathcal{S} = 0$. We have:

$$\delta\mathcal{S} = \int \{\delta\mathcal{L} \, \Omega + \mathcal{L} \, \delta\Omega\} = \int \{ \, \delta F_{A_1 B_1}{}^C \frac{\partial \mathcal{L}}{\partial F_{A_1 B_1}{}^C} \Omega + \mathcal{L} \, (-1)^{\sigma_{A_1}} \delta h^{A_1}_{A_1} \, \Omega\} \, . \tag{2.15}$$

The variation of $F_{A_1 B_1}{}^C$ in terms of $\delta h^B_{A_1}$ may be obtained from Eq. (1.14) which gives:

---

[4]We have here derived this result without reference to the explicit form of the invariant measure Ω. In the literature it has been obtained using the expression $\Omega = \exp \, Sup \, Tr \ln h^{A_1}_\Lambda \, d^{4+4N}x$.

$$[\delta D_{A_1}, D_{B_1}\} + [D_{A_1}, \delta D_{B_1}\} = -\,\delta F_{A_1 B_1}{}^C D_C - F_{A_1 B_1}{}^C \delta D_C \qquad (2.16)$$

where:

$$\delta D_{A_0} = 0 \,. \qquad (2.17)$$

But,

$$\delta \partial_\Lambda = \delta h_\Lambda^A D_A + h_\Lambda^A \delta D_A = 0 \,. \qquad (2.18)$$

Hence:

$$\delta D_{A_1} = -\,\delta h_{A_1}^A D_A \qquad (2.19)$$

and:

$$(\delta F_{A_1 B_1}{}^C) D_C = F_{A_1 B_1}{}^{C_1} \delta h_{C_1}^D D_D + [\delta h_{A_1}^D D_D, D_{B_1}\} + [D_{A_1}, \delta h_{B_1}^D D_D\} \qquad (2.20)$$

wherefrom one obtains:

$$\delta F_{A_1 B_1}{}^C = F_{A_1 B_1}{}^{C_1} \delta h_{C_1}^C - \left\{ \delta h_{A_1}^D F_{DB_1}{}^C - (-1)^{\sigma_{A_1}\,\sigma_{B_1}} \delta h_{B_1}^D F_{DA_1}{}^C \right\}$$

$$+ D_{A_1} \delta h_{B_1}^C - (-1)^{\sigma_{A_1}\,\sigma_{B_1}} D_{B_1} \delta h_{A_1}^C \,. \qquad (2.21)$$

Then:

$$\delta \mathscr{S} = \int \delta h_{C_1}^D \left\{ (-1)^{\sigma_{C_1}} \delta_D^{C_1} \mathscr{L} + \left[\, (-1)^{(\sigma_D + \sigma_{C_1})(\sigma_{A_1} + \sigma_{B_1} + \sigma_{C_1})} \delta_D^C F_{A_1 B_1}{}^{C_1} \right.\right.$$

$$\left.\left. -\,(\,\delta_{A_1}^{C_1} F_{DB_1}{}^C - (-1)^{\sigma_{A_1}\,\sigma_{B_1}} \delta_{B_1}^{C_1} F_{DA_1}{}^C)\,\right] \frac{\partial \mathscr{L}}{\partial F_{A_1 B_1}{}^C} \right\} \Omega$$

$$+ \int [\, D_{A_1} \delta h_{B_1}^C - (-1)^{\sigma_{A_1}\,\sigma_{B_1}} D_{B_1} \delta h_{A_1}^C \,] \frac{\partial \mathscr{L}}{\partial F_{A_1 B_1}{}^C} \Omega \,. \qquad (2.22)$$

The last integral, integrated by parts, gives:

$$\int \delta h_{C_1}^D \delta_D^C \left[\, (-1)^{\sigma_{E_1}} \left\{ (-1)^{\sigma_{B_1}\sigma_D} \delta_{A_1}^{C_1} T_{B_1 E_1}{}^{E_1} - (-1)^{\sigma_{A_1}(\sigma_{B_1} + \sigma_D)} \delta_{B_1}^{C_1} T_{A_1 E_1}{}^{E_1} \right\} \right.$$

$$\left. +\, \left\{ (-1)^{\sigma_{B_1}\,\sigma_D} \delta_{A_1}^{C_1} D_{B_1} - (-1)^{\sigma_{A_1}(\sigma_{B_1} + \sigma_D)} \delta_{B_1}^{C_1} D_{A_1} \right\} \right] \frac{\partial \mathscr{L}}{\partial F_{A_1 B_1}{}^C} \Omega \,. \qquad (2.23)$$

Therefore the Euler-Lagrange equations for the variational principle $\delta \mathscr{S} = 0$ are:

$$(-1)^{\sigma_{C_1}} \delta_D^{C_1} \mathcal{L} + \Big\{ (-1)^{(\sigma_D + \sigma_{C_1})(\sigma_{A_1} + \sigma_{B_1} + \sigma_{C_1})} \delta_D^C F_{A_1 B_1}{}^{C_1}$$

$$- (\delta_{A_1}^{C_1} F_{DB_1}{}^C - (-1)^{\sigma_{A_1}\sigma_{B_1}} \delta_{B_1}^{C_1} F_{DA_1}{}^C) \tag{2.24}$$

$$+ \delta_D^C [(-1)^{\sigma_{B_1}\,\sigma_D} \delta_{A_1}^{C_1} (D_{B_1} + (-1)^{\sigma_{E_1}} T_{B_1 E_1}{}^{E_1})$$

$$- (-1)^{\sigma_{A_1}(\sigma_{B_1} + \sigma_D)} \delta_{B_1}^{C_1} (D_{A_1} + (-1)^{\sigma_{E_1}} T_{A_1 E_1}{}^{E_1})]\Big\} \frac{\partial \mathcal{L}}{\partial F_{A_1 B_1}{}^C} = 0 \, .$$

3. SUPERSPACE ACTION FOR $SO(N)$-SUPERGRAVITY

We want to construct an action in superspace subject to the following criteria: first that it is stationary when the constraints and equations of motion of $SO(N)$-supergravity are imposed[5]; second that it contains the Einstein action.

An action for $SO(N)$-supergravity as an integral over space-time may be obtained, at least for $N \le 3$, by functional integration of the equations of motion for the "vierbein" h_μ^i [8]. If appropriate constraints are imposed on the torsion this equation is derived from the Bianchi identity:

$$(i, j, aa)^{bb} = 0 \, . \tag{3.1}$$

We take this identity as the starting point for our construction of the superspace action satisfying the two criteria given above. Multiplying it by $\frac{1}{4} [f^i, f^j]_b^a \delta_b^a$ and integrating with the invariant measure Ω, we have:

$$\int \Big\{ D_{[i} T_{j]aa}{}^{bb} + D_{aa} T_{ij}{}^{bb} + R_{ij}{}^{[rs]} f_{[rs]a}{}^b \delta_a^b + R_{ij}{}^{[gh]} f_{[gh]a}{}^b C_a^b + T_{ij}{}^k T_{kaa}{}^{bb} + T_{ij}{}^{cc} T_{aacc}{}^{bb}$$

$$- T_{aa[i}{}^k T_{j]k}{}^{bb} + T_{[iaa}{}^{cc} T_{j]cc}{}^{bb} \Big\} \times \frac{1}{4} [f^i, f^j]_b^a \delta_b^a \, \Omega = 0 \, . \tag{3.2}$$

Integrating by parts the derivative terms, one obtains the following topologically invariant integral:

$$\mathcal{I} = \int \Big\{ (T_{[icc}{}^{cc} - T_{[ik}{}^k) T_{j]aa}{}^{bb} + (T_{aacc}{}^{cc} - T_{aak}{}^k) T_{ij}{}^{bb} + R_{ij}{}^{[rs]} f_{[rs]a}{}^b \delta_a^b +$$

$$T_{ij}{}^{cc} T_{aacc}{}^{bb} + T_{ij}{}^k T_{kaa}{}^{bb} + T_{[iaa}{}^{cc} T_{j]cc}{}^{bb} - T_{aa[i}{}^k T_{j]k}{}^{bb} \Big\} \times \frac{1}{4} [f^i, f^j]_b^a \delta_b^a \, \Omega \, . \tag{3.3}$$

---

[5]Here, when we refer to $SO(N)$-supergravity we mean the known models for which an action has been given in ordinary space-time or the constraints and equations of motion have been established.

In order to obtain an action we expand the torsion components around their on-shell values (their values as constrained by the equations of motion) and drop the quadratic terms in this expansion. Let us work out this procedure for $SO(N)$-supergravity with $N \le 3$. The on-shell constraints are [8]:

$$T^{\circ}{}_{aabb}{}^{i} = f_{ab}^{\,i} \eta_{ab} \, , \tag{3.4}$$

$$T^{\circ}{}_{aaj}{}^{i} = 0 \, , \tag{3.5}$$

$$T^{\circ}{}_{aabb}{}^{cc} = T^{d}{}_{[abd]} \left(C_{ab} C_{d}^{\,c} + \gamma_{5ab} \gamma_{5d}^{\,c} \right) \eta^{cd} \, , \tag{3.6}$$

$$T^{\circ}{}_{iaa}^{\,bb} = f_{ia}^{\,b} \delta_{a}^{b} + T_{5j} \left(\gamma_{5} f_{i} f^{j} \right)_{a}^{\,b} \delta_{a}^{b} - \frac{1}{8} R_{kl}^{\,[gh]} f_{[gh]a}^{\quad b} \left(f_{i} f^{k} f^{l} \right)_{a}^{\,b} \, , \tag{3.7}$$

$$T^{\circ}{}_{ij}^{\,cc} = - \frac{1}{16} \left\{ T_{kl}^{\,dc} \left([f_{i} , f_{j}] f^{k} f^{l} + \frac{1}{3} f^{k} f^{l} [f_{i} , f_{j}] \right)_{d}^{\,c} \right.$$
$$\left. + \frac{1}{6} T^{d[cde]} \left(f_{[i} f^{k} f^{l} f_{j]} \right)_{d}^{\,c} R_{kl}^{\,[gh]} f_{[gh]e} \eta_{bd} \right\} \, , \tag{3.8}$$

$$T^{\circ}{}_{ij}^{\,k} = \frac{1}{2} \mathrm{Tr}(\gamma_{5} f_{i} f_{j} f^{k} f^{l}) T_{5l} = - 2 \epsilon_{ij}^{\,kl} T_{5l} \, , \tag{3.9}$$

where:

$$T_{5j} = - \frac{1}{12} T^{c[abc]} T^{d}{}_{[abc]} \left(\gamma_{5} f_{j} \right)_{cd} \, , \tag{3.10}$$

$$T^{d}{}_{[abc]} = \frac{1}{24} \sum_{(abc)} T_{aabb}^{\,cd} \left(C^{ab} C_{c}^{\,d} + \gamma_{5}^{\,ab} \gamma_{5c}^{\,d} \right) \eta_{cd} \, . \tag{3.11}$$

Therefore the action integral will be:

$$\mathcal{S} = \int \left\{ N \left(R_{ij}^{\,[rs]} f_{[rs]k}^{\quad i} - 12 \eta_{jk} \right) \eta^{jk} + \left[\left(T_{aacc}^{\,cc} - T_{aak}^{\,k} \right) T^{\circ}{}_{ij}^{\,bb} \right. \right.$$

$$+ T^{\circ}{}_{ij}^{\,cc} T^{\circ}{}_{aacc}^{\,bb} + \left(T_{ij}^{\,cc} - T^{\circ}{}_{ij}^{\,cc} \right) T_{aacc}^{\,bb} + T^{\circ}{}_{ij}^{\,cc} \left(T_{aacc}^{\,bb} - T^{\circ}{}_{aacc}^{\,bb} \right)$$

$$+ \left(T_{ij}^{\,k} - T^{\circ}{}_{ij}^{\,k} \right) T^{\circ}{}_{kaa}^{\,bb} + T^{\circ}{}_{ij}^{\,k} \left(T_{kaa}^{\,bb} - T^{\circ}{}_{kaa}^{\,bb} \right)$$

$$+ \left(T_{[iaa}^{\,cc} - T^{\circ}{}_{[iaa}^{\,cc} \right) T^{\circ}{}_{j]cc}^{\,bb} + T^{\circ}{}_{[iaa}^{\,cc} \left(T_{j]cc}^{\,bb} - T^{\circ}{}_{j]cc}^{\,bb} \right)$$

$$\left. \left. - T_{aa[i}^{\,k} T^{\circ}{}_{j]k}^{\,bb} \right] \times \frac{1}{4} [f^{\,i}, f^{j}]_{b}^{\,a} \delta_{b}^{\,a} \right\} \Omega \, . \tag{3.12}$$

By construction this action is such that the on-shell constraints and equations of motion of $SO(N)$-supergravity are compatible with the Euler-Lagrange equations for the variational principle $\delta \mathcal{S} = 0$. Moreover $\mathcal{S}$ is a superspace generalization of Einstein's action since it contains this action as its first term. This is, therefore, the desired result. For $N = 1$, $\mathcal{S}$ reduces to:

$$\mathcal{S}_1 = \int \left\{ \left[R_{ij}^{[rs]} f_{[rs]k}{}^i + 12\,\eta_{jk} - 6\,T_{ja}^{\ b} f_{kb}^{\ a} \right] \eta^{jk} + \left[\left(T_{ac}^{\ c} - T_{ak}^{\ k} \right) T^o{}_{ij}^{\ b} \right. \right.$$

$$\left. \left. + T^o{}_{ij}^{\ c} T_{ac}^{\ b} - T_{a[i}^{\ k} T^o{}_{j]k}^{\ b} \right] \times \tfrac{1}{4} \left[f^i, f^j \right]_b^a \right\} \Omega \,. \tag{3.13}$$

The only other choice of a Bianchi identity from which an action, fulfilling the first criterion, can be obtained by this method is

$$(aa,\ bb,\ i)^j = 0 \,. \tag{3.14}$$

Multiplying by $\tfrac{1}{4} \left[f^i, f_j \right]^{ab} \eta^{ab}$ and integrating one obtains:

$$\int \left\{ D_i\, T_{aabb}{}^j + D_{(aa}\, T_{bb)i}{}^j + R_{aabb}^{[rs]} f_{[rs]i}{}^j + T_{aabb}^{\ \ k} T_{ki}{}^j + T_{aabb}^{\ \ cc} T_{cci}{}^j \right.$$

$$\left. + T_{(aai}^{\ \ k} T_{bb)k}{}^j + T_{i(aa}^{\ \ cc} T_{bb)cc}{}^j \right\} \times \tfrac{1}{4} \left[f^i, f_j \right]^{ab} \eta^{ab}\, \Omega = 0 \,. \tag{3.15}$$

Integrating by parts the derivative terms one obtains a topological invariant:

$$\mathcal{S}' = \int \left\{ \left(T_{icc}^{\ \ cc} - T_{ik}^{\ k} \right) T_{aabb}{}^j + \left(T_{(aacc}^{\ \ cc} - T_{(aak}^{\ \ k} \right) T_{bb)i}{}^j + R_{aabb}^{[rs]} f_{[rs]i}{}^j + T_{aabb}^{\ \ k} T_{ki}{}^j \right.$$

$$\left. + T_{aabb}^{\ \ cc} T_{cci}{}^j + T_{(aai}^{\ \ k} T_{bb)k}{}^j + T_{i(aa}^{\ \ cc} T_{bb)cc}{}^j \right\} \times \tfrac{1}{4} \left[f^i, f_j \right]^{ab} \eta^{ab}\, \Omega \,. \tag{3.16}$$

Again expanding the torsion components around their on-shell values and dropping the quadratic terms one obtains the action:

$$\mathcal{S}' = \int \left\{ R_{aabb}^{[rs]} f_{[rs]i}{}^j - \left(T^o{}_{iaa}^{\ \ cc} f_{cb}^{\ j} \eta_{cb} + T^o{}_{ibb}^{\ \ cc} f_{ca}^{\ j} \eta_{ca} \right) + T_{aabb}^{\ \ k} T^o{}_{ki}^{\ j} + T^o{}_{aabb}^{\ \ cc} T_{cci}{}^j \right.$$

$$\left. + T^o{}_{i(aa}^{\ \ cc} T_{bb)cc}^{\ \ j} + \left(T_{iaa}^{\ cc} f_{cb}^{\ j} \eta_{cb} + T_{ibb}^{\ cc} f_{ca}^{\ j} \eta_{ca} \right) \right\} \times \tfrac{1}{4} \left[f^i, f_j \right]^{ab} \eta^{ab} \Omega \,. \tag{3.17}$$

For $N = 1$ this action becomes linear in the curvature and torsion components as it reduces to:

$$\mathscr{S}'_1 = \int \left\{ 2\, R_{ac}{}^{[rs]}\, f_{[rs]b}{}^c + 3\left[-4\, C_{ab} + f_{ia}{}^c\, T_{cb}{}^i + T_{ia}{}^c\, f_{cb}{}^i \right] \right\} C^{ab}\, \Omega \ . \tag{3.18}$$

From an inspection of the Euler-Lagrange equations for the variational principle $\delta\mathscr{S}'_1 = 0$ one infers that, for any linear action of the form:

$$\mathscr{S}''_1 = \int \left\{ 2\, R_{ac}{}^{[rs]}\, f_{[rs]b}{}^c + 3\left[-4\alpha\, C_{ab} + \beta\, f_{ia}{}^c\, T_{cb}{}^i + \gamma\, T_{ia}{}^c\, f_{cb}{}^i \right] C^{ab} \right\} \Omega \tag{3.19}$$

the Euler-Lagrange equations for the variational principle $\delta\mathscr{S}''_1 = 0$ are compatible with the equations of simple supergravity provided that:

$$\gamma = 1 \quad , \qquad\qquad\qquad \beta = \tfrac{1}{2}\,(1 + \alpha) \ . \tag{3.20}$$

The linear Lagrangian proposed by Brink et. al. [5] belongs to this class with $\alpha = 0$, $\beta = \tfrac{1}{2}$, $\gamma = 1$. The linear Lagrangian Eq. (3.18) is obtained by imposing the additional constraint that $\mathscr{S}$ vanish on-shell which implies that $\beta + \gamma - \alpha = 1$, hence $\alpha = \beta = \gamma = 1$.

An advantage of our method is that it applies to any model of $SO(N)$-supergravity for which the on-shell constraints on the torsion have been established. However, it follows from its construction that the action always vanishes when the torsion and curvature components are restricted by the on-shell constraints and equations of motion. This is in contradistinction with the space-time action which, except in the case of simple Poincaré supergravity, does not vanish on-shell. Therefore we are led to conclude that the space-time actions of simple and extended supergravity are not obtained from these superspace actions upon integration of the Grassmann variables. Then the Euler-Lagrange equations derived from the variational principle in superspace are not equivalent to the on-shell equations of $SO(N)$-supergravity but should admit more general solutions. Whether these solutions have a satisfactory physical interpretation is yet to be investigated.

In conclusion we remark that the relevance of this superspace formulation of the classical theory rests on the possibility of quantization directly in terms of the superspace action. It is generally believed that finiteness of quantum supergravity may only be established if such a quantization procedure is achieved.

REFERENCES

[1]. A. Salam and J. Strathdee, Nucl. Phys. B76 (1974) 477.

[2]. J. Wess and B. Zumino, Nucl. Phys. B70 (1974) 39.

[3]. S. W. MacDowell, Phys. Letters 80B, (1979) 212.

[4]. J. Wess and B. Zumino, Phys. Letters 66B (1977) 361; Phys. Letters 74B (1978) 51. J. Wess, R. Grimm and B. Zumino, Phys. Letters 73B (1978) 415.

[5]. L. Brink, M. Gell-Mann, P. Ramond and J. Schwarz, Phys. Letters 74B (1978) 336; Phys. Letters 76B (1978) 417.

[6]. R. Arnowitt and P. Nath, Phys. Letters 78B (1978) 581.

[7]. Y. Ne'emann and T. Regge, Phys. Letters 74B (1978) 31; Rivista del Nuovo Cimento, Vol. 1 Sec. 5 (1978) 1.

[8]. S. W. MacDowell, Proceedings of the VI Brazilian Symposium of Theoretical Physics, Rio de Janeiro, (1980). The notation and conventions used in the present article are the same as in this reference.

[9]. L. Bishop and R. J. Crittenden, *Geometry of Manifolds*, Academic Press; New York, London (1964).

[10]. S. W. MacDowell, *Supergravity*, pg. 77; P. van Nieuwenhuizen and D. Z. Freedman, Editors; North Holland Publishing Company (1979).

[11]. W. Siegel and S. J. Gates, Jr., Nucl. Phys. B147 (1979) 77.

INTRINSIC GEOMETRY
OF SUPERGRAVITY

V. I. Ogievetsky

Joint Institute for Nuclear Research
Laboratory of Theoretical Physics
Dubna, USSR

1. INTRODUCTION

Supersymmetry and supergravity provide a considerable extension of possibilities and a new understanding of the unification tendency. It is appropriate to recall here the early attempts of Feza Gürsey and others [1,2] in the sixties to unify the space-time symmetries with the internal ones. Due to the analysis of these attempts there appear important no-go theorems concerning impossibility of such a merging in the framework of standard groups. At the same time internal symmetry can be nontrivially merged with the space-time one in the framework of supergroups, i.e., in supersymmetric models. After a remarkable success of the unified electroweak theory public opinion has been consolidated that this theory together with quantum chromodynamics will amalgamate into some gauge theory of grand unification of weak electromagnetic and strong interactions. In such a theory all the interacting fields (photon, intermediate bosons, gluons, etc.) come out on at least formally equal footings as gauge fields. At this stage there remains still an essential difference between them and the matter fields of leptons and quarks. The introduction of supersymmetry can help to eliminate this difference. Both the fermionic and bosonic fields can be equally the gauge ones in supersymmetric theories. These fields are functions of coordinates. One can try to eliminate this last difference and to equate in rights fields with coordinates. It is conceivable that supergravity will be useful to this end. In any case supergravity makes the next step and it has a claim to unify all the basic interactions including gravitation.

In the present paper it will be demonstrated that this equality between fields and coordinates is realized in fact at least in the simplest of supergravities, just in the $N = 1$ supergravity. We shall review without details our geometric approach to it [3] resulting from our dynamical superfield consideration of 76 [4] with the axial superfield generated by the supercurrent. Now this geometrical approach is practically completed [5,6]. It is based on the complex geometry where the gravitational axial superfield comes out as an imaginary part of a complexified coordinate and it defines some hypersurface. The action principle says that this hypersurface has to have the extremal supervolume. It gives straightforwardly the equations of this hypersurface which are just the supergravity equations of motion. We shall also discuss briefly an alternative method of deriving our action by Schwarz [7] which is based on a superanalog of Levy form well known in complex geometry.

We would like to emphasize the absence of any constraints of non-geometrical origin in our approach. For this reason the geometry proposed can be characterized as the intrinsic one. The results are consistent with those of the known powerful superfield approach by Wess and Zumino [8,9] which is connected with the very broad real superspace geometry. In the latter approach the geometric considerations and the action principle do not exhaust the content of the theory. Of paramount importance there are non-geometric constraints on the components of torsion. In the framework of the same geometric pattern one can impose different sets of constraints to obtain theories which differ from one another, e.g., by the auxiliary field contents. One such set corresponds to our case while the other one to the Siegel-Gates case [10] etc. A very important analysis of constraints was performed by Gates, Stelle and West [11]. At present there are also some sets of constraints known for the $N = 2$ supergravity [12, 13]. In any case a minimal geometric pattern seems to be of importance in which the whole content of the theory is exhausted by the geometry and by the action principle.

Such a minimal pattern could arise naturally (as in our case) or may result from a solution of some set of constraints. The geometry of just this pattern is the intrinsic one and it is most relevant.

It is very tempting to have the future superunified theory in the following form. At the beginning there is a multidimensional space some dimensions of which are of fermionic nature. Then some hypersurface with a smaller number of dimensions is defined in this superspace and some of the coordinates become physical fields. The variational equations for this hypersurface exhaust all the equations of motion for fields in the superunified theory thus obtained. It would be very tempting to attach such a form to the $N = 8$ supergravity, but it is difficult for the time being. However, the program is already completed for the $N = 1$ supergravity and now we pass to the discussion of this case. Let us begin with a brief list of the necessary facts from supersymmetry.

2.

The supersymmetry transformations intermix bosons and fermions. The Poincaré N-extended superalgebra contains the generators of translations P^a and of 4-rotations L_{ab} as well as the odd generators of supertranslations $Q_{\alpha i}$ and $Q_{\dot\alpha}{}^i$ (α, $\dot\alpha$ = 1, 2, i = 1, 2, ..., N). The latter are two component spinors. In what follows we shall be mainly interested in the $N = 1$ superalgebra and therefore the index i can be often omitted. The superalgebra contains anticommutation relations together with the commutation ones, in particular,

$$\{\, Q_{\alpha i} \,,\, \bar Q_{\dot\alpha}{}^j \,\} = 2\, \delta_i^j\, \sigma^a{}_{\alpha\dot\alpha}\, P_a \,, \qquad [\, P_a \,,\, Q_{\alpha i} \,] = [\, P_a \,,\, \bar Q_{\dot\alpha}{}^i \,] = 0 \,, \qquad etc \tag{2.1}$$

where $\sigma^a = (I \,,\, \vec\sigma)$, and $\vec\sigma$ are the 2×2 Pauli matrices. The supersymmetry transformations $\eta^\alpha Q_\alpha + \bar\eta_{\dot\alpha} \bar Q^{\dot\alpha}$ have parameters η which mix fields with different statistics. For this reason these parameters must be Grassmann numbers,

$$\{\eta_\alpha \,,\, \eta_\beta \} = \{\eta_\alpha \,,\, \bar\eta_{\dot\beta} \} = \ldots = 0 \,. \tag{2.2}$$

Special relativity (the $N = 0$ supersymmetry) describes the phenomena in the 4-dimensional space-time which can be called the (4, 0) real manifold (4 even and 0 odd coordinates). The extended N-supersymmetry can be easily realized in the (4, 4N) real supermanifold or superspace with 4 real and 4N odd real coordinates. For example, $N = 1$ supersymmetry can be realized in the (4, 4) real superspace having as real coordinates the 4-vector X^m and the Majorana spinor

$$\Theta^\alpha = \begin{pmatrix} \Theta_\alpha \\ \bar\Theta^{\dot\alpha} \end{pmatrix}$$

The latter is real in the sense that it transforms into itself under charge conjugation

$$\Theta^C = C\, \bar\Theta^T = \Theta \,.$$

Odd coordinates are Grassmann numbers $\{\Theta^\alpha \,,\, \Theta^\beta \} = 0$.

In this superspace it is easy to write down the supersymmetry transformations with the generators (2.1)

$$X'^m = X^m + i\, (\Theta\sigma^m\bar\eta - \eta\sigma^m\bar\Theta) \,, \quad \Theta'^\mu = \Theta^\mu + \eta^\mu \,, \quad \bar\Theta'^{\dot\mu} = \bar\Theta^{\dot\mu} + \bar\eta^{\dot\mu} \,. \tag{2.3}$$

The superfield $\phi(X,\Theta,\bar\Theta)$ is a set of $2^4 = 16$ fields because the number of Grassmann variables is four and, correspondingly, the decomposition of the superfield stops at $\Theta\Theta\bar\Theta\bar\Theta$. The transformation law for the superfield is as follows

$$\phi(X',\Theta',\bar{\Theta}') = \phi(X,\Theta,\bar{\Theta}) . \tag{2.4}$$

Of course a superfield can have Lorentz indices or internal symmetry indices. For extended N-supersymmetries the situation is analogous. In this case, however, there are $4N$ Grassmann variables and the number of fields in a given superfield is 2^{4N}. This number is very big, for $N = 8$ it takes the huge value of 2^{32}.

In supergravity the general coordinate transformation group

$$X'^m = X^m(X,\Theta,\bar{\Theta}) \quad , \quad \Theta'^\mu = \Theta^\mu(X,\Theta,\bar{\Theta}) \quad , \quad \bar{\Theta}'^{\dot\mu} = \bar{\Theta}^{\dot\mu}(X,\Theta,\bar{\Theta}) \tag{2.5}$$

is used instead of special "superrelativity" group (3). The superanalog of the vierbein $E_A{}^M(X,\Theta,\bar{\Theta})$ is introduced [8,9] where $M = (m, \mu, \dot\mu)$ is a set of $4 + 4N$ world vector and spinor indices and $A = (a, \alpha, \dot\alpha)$ are $4 + 4N$ indices in a tangent flat superspace. The superfield $E_A{}^M(X,\Theta,\bar{\Theta})$ contains $(4 + 4N)^2$ fields in its decomposition. Due to the gauge supergroup (2.5), $(4 + 2N + 2N)\cdot 2^{4N}$ fields become gauge degrees of freedom. There remain $(N + 1)(4N + 3)\cdot 2^{4N+2}$ physical and auxiliary degrees of freedom. Even at $N = 1$ one has to deal with 896 fields to describe only $(12 + 12)$ physical and auxiliary degrees of freedom (graviton e_a^m gives 6, the auxiliary fields $A^m(X)$, $s(X)$ and $p(X)$ give another 6, and the gravitino Ψ_α^μ another 12). At $N = 8$ there are 5×10^{12} and 4×10^{12} physical and auxiliary fields, respectively. In the cases already known, $N = 1$ and $N = 2$, external constraints are imposed on the torsion components for the elimination of superfluous fields. As was said above, these constraints are of the non-geometric origin. Thus, even being natural, the real approach is not pure geometrical and it deals with too many superfluous fields. At the same time at least for $N = 1$ there is a pure geometric complex approach containing, moreover, a minimal number of fields.

3.

This *complex approach is based on complex superspace*. It is instructive to start again with the flat case. A large number of fields in the real approach was connected with a large number of spinor coordinates. Can one reduce this number? The answer is yes. To this aim one has to pass in the $N = 1$ case from the (4, 4) real superspace to the complex superspace of complex dimensionality (4, 2), i.e., of the real one (8, 4). Introducing a complex coordinate $X_L{}^m$ ($X_R{}^m = (X_L{}^m)^\dagger$) one can realize the superalgebra 2.1 in complex superspace $Z^M = \{X_L{}^m, \Theta_L{}^\mu\}$ as follows

$$X_L'^m = X_L^m + 2i\,\Theta_L\sigma^m\bar\eta \;, \qquad\qquad \Theta_L'^\mu = \Theta_L^\mu + \eta^\mu \;. \tag{3.1}$$

Now the (4, 4) real superspace with transformations (2.3) is a (4, 4) real hypersurface whose equations have the form

$$\text{Re}\,X_L^m = X^m \;,\quad \Theta_L^\mu = \Theta^\mu \;,\quad \bar\Theta_R^{\dot\mu} = \bar\Theta^{\dot\mu} \;, \tag{3.2.a}$$

$$\text{Im}\,X_L^m = \Theta\sigma^m\bar\Theta \;. \tag{3.2.b}$$

It is well known that one can define a somewhat more economical object than the general scalar superfield, namely, the chiral superfield $\phi(X_L,\,\Theta_L)$. The latter unifies only $2^2 = 4$ complex fields. In fact, this superfield can be called the analytical one according to the notion of Grassmann analyticity [14]. Really, the chiral superfield is the general superfield (2.4) obeying the Cauchy-Riemann condition and it is independent of $\bar Z$

$$\Phi(X,\,\Theta,\,\bar\Theta) = \phi = \phi(Z) \;,\quad if \quad \bar D_{\dot\alpha}\,\Phi = 0 \;. \tag{3.3}$$

$\bar D_{\dot\alpha}$ is a spinor derivative.

In the presence of gravity we single out the (4, 4) real hypersurface by the equations (3.2.a) replacing Eq. (3.2.b) by the equation [3]

$$\text{Im}\,X_L^m = H^m(X,\Theta,\bar\Theta) \;, \tag{3.4}$$

where $H^m(X,\Theta,\bar\Theta)$ is some axial real superfield. The variation of the action principle will lead us to the equations of motion. The latter are geometrically the equations of a supersurface which single out a supersurface of a "minimal" supervolume. The general coordinate transformations are now analytic transformations, which can be written in a compact form

$$Z' = Z + \lambda(Z) \;, \tag{3.5}$$

where $\lambda(Z)$ are arbitrary functions independent of $\bar Z$. One can rewrite Eq. (3.5) at a greater length

$$X_L'^m = X_L' + \lambda^m(X_L,\,\Theta_L) \;,\quad \Theta_L'^\mu = \Theta_L^\mu + \lambda^\mu(X_L,\Theta_L) \;. \tag{3.6}$$

It turns out [3] that the supergroup (10) is inherent to the conformal or Weyl supergravity. To pass to the Einstein supergravity group one has to impose in addition the condition of the supervolume conservation [3,4]

$$Ber\ \frac{\partial Z'}{\partial Z} = 1 \quad \text{or, infinitesimally,} \quad \partial_m \lambda^m - \partial_\mu \lambda^\mu \tag{3.7}$$

where *Ber* means the Berezinian which is a superanalogue of the Jacobian of transformations. As the latter, the Berezinian has the multiplicative property

$$Ber\ \frac{\partial Z'}{\partial Z} \times Ber\ \frac{\partial Z''}{\partial Z'} = Ber\ \frac{\partial Z''}{\partial Z}$$

which ensures the group property of condition (4.1). Thus Einstein supergravity is a theory based on the supervolume conserving transformations (3.7) in the complex (4, 2) superspace. The axial gravitational superfield $H^m(X, \Theta, \bar{\Theta})$ singles out the (4, 4) real hypersurface in this superspace. $H^m(X, \Theta, \bar{\Theta})$ has 64 fields in its decomposition in powers of $\Theta, \bar{\Theta}$. An analysis shows that it describes the following physical and auxiliary fields off mass shell: the vierbein e_a^m (16 - 6 - 4 = 4 degrees of freedom) gravitino ψ_α^m (4×4 - 4 = 12 degrees of freedom) and auxiliary fields $A^m(X)$, $j(X)$, $p(X)$ (4 + 1 + 1 = 6 degrees of freedom). When counting the number of degrees of freedom we extract the gauge degrees connected with transformations (3.5) and (3.7).

Thus, the axial gravitational superfield describes 12 bosonic and 12 fermionic degrees of freedom off mass shell and, respectively, 2 + 2 on mass shell, free graviton and gravitino with helicities ± 2 and ± 3/2.

4.

The non-minimal Siegel-Gates formulation of $N = 1$ supergravity [10] was translated by Sokatchev into a somewhat different geometric language as follows. One has to supplement the complex superspace (4, 2) above with two additional spinorial coordinates $\bar{\phi}_L^{\dot{\mu}}$ to obtain the (4, 4) complex superspace with coordinates

$$Z = (X_L^m, \Theta_L^\mu, \bar{\phi}_L^{\dot{\mu}}) \ .$$

To preserve the concept of chirality we have to introduce a "projective structure" and shall consider the subgroup of coordinate transformations leaving invariant the old chiral complex (4, 2) superspace (X_L^m, Θ_L^μ) In other words we shall consider transformations

$$\delta X_L^m = \lambda^m(X_L, \Theta_L) \ , \qquad \delta \Theta_L^\mu = \lambda^\mu(X_L, \Theta_L) \ , \qquad \delta \bar{\phi}_L^{\dot{\mu}} = \bar{\rho}^{\dot{\mu}}(X_L, \Theta_L) \ . \tag{4.1}$$

The embedding of the old real (4, 4) superspace is achieved now by defining again a hypersurface

$$X^m = \text{Re } X_L{}^m \ , \qquad\qquad \Theta^\mu = \Theta_L{}^\mu \ , \qquad\qquad \bar\Theta^{\dot\mu} = \bar\Theta_R{}^{\dot\mu}$$

$$H^m(X,\Theta,\bar\Theta) = \text{Im } X_L{}^m \ , \qquad H^m(X,\Theta,\bar\Theta) = \phi_L{}^\mu - \Theta_L{}^\mu \ , \ \bar H^{\dot\mu}(X,\Theta,\bar\Theta) = \bar\phi_L{}^{\dot\mu} - \bar\Theta_R{}^{\dot\mu} \ .$$

The number of prepotentials becomes larger as the spinorial ones $H^{\dot\mu}(X,\Theta,\bar\Theta)$ and $H^{\dot\mu}(X,\Theta,\bar\Theta)$ are added. Due to the projective structure of the transformations (11), now two Berezinians have multiplicative property, the old one

$$Ber\left(\frac{X'_L,\Theta'_L}{X_L,\Theta_L}\right) \ , \text{ and the new one } Ber\left(\frac{X'_L,\Theta'_L,\bar\phi'_L}{X_L,\Theta_L,\phi_l}\right) \ .$$

We can now restrict the supergroup (4.1) by a group covariant restriction

$$\left[Ber\left\{\frac{X'_L,\Theta'_L,\bar\phi'_L}{X_L,\Theta_L,\bar\phi_L}\right\}\right]^{3n+1} = \left[Ber\left\{\frac{X'_L,\Theta'_L}{X_L,\Theta_L}\right\}\right]^{2n} \ , \tag{4.3}$$

where n is a parameter. In the cases $n = -\frac{1}{3}$ and in the case without restriction (4.2) the spinorial prepotentials can be gauged away completely. The first case corresponds to the minimal Einstein supergravity considered above while the second one corresponds to the Weyl or conformal supergravity also mentioned above in the minimal formulation. For general n we have the Einstein supergravity however with a larger set of fields, 20 + 20 instead of 12 + 12. There is an interesting peculiar value of the parameter n, $n = 0$, at which the whole left supervolume is conserved. This case is very degenerate and after nontrivial considerations one can see that it corresponds also to a (12 + 12) formulation but with different content of auxiliary fields. This version was considered recently in component form by Sohnius and West and earlier by Akulov, Volkov and Soroka (linearized form).

5.

The corresponding differential geometry in real superspace can be arranged as follows. First of all one can define the covariant spinor derivatives [5b] (see also Siegel and Gates' paper [10])

$$D_\alpha \Phi_B(X,\Theta,\bar\Theta) = E_\alpha{}^M \partial_M \Phi_B + \omega_{\alpha B}{}^C \Phi_C \tag{5.1}$$

where α, B, C are indices of the "tangent superspace", $M = (m, \mu, \dot\mu)$ are those of the "world superspace", $\partial_M = (\partial_m, \partial_\mu, \bar\partial_{\dot\mu})$. The superanalogues of tetrad $E_\alpha{}^m$ and of connection $\omega_{\alpha B}{}^C$ are expressed wholly in terms of derivatives of H^M. The explicit expressions in terms of H^M can be found in [5b]. The vector derivative is

$$D_a = \frac{i}{4} \, \bar{\sigma}_a{}^{\dot{\alpha}\alpha} \, \{\bar{D}_{\dot{\alpha}} \, , D_\alpha\} \, . \tag{5.2}$$

To define the components of torsion $T_{AB}{}^C$ and curvature tensor $R_{ABD}{}^E$ one has to consider either commutator or anti-commutator according to a grade of indices A, B containing in a condensed form both vector (grade 0) and spinor (grade 1) indices, e.g., $A = (a, \alpha, \dot{\alpha})$

$$[D_A \, , \ D_B \} \, \Phi_D = - \, T_{AB}{}^C D_C \Phi_D - R_{ABD}{}^E \, \Phi_E \, . \tag{5.3}$$

All these tensor components are combinations of basic superfields [15]

$$R \, , \ G_{\alpha\dot{\beta}} \, , \ W_{\alpha\beta\gamma} \tag{5.4}$$

which in turn had been expressed in terms of the prepotential H^m [6b].

In all the calculations the use of the normal gauge is very effective. The normal gauge was introduced in [6a]. It is a superanalogue of the Riemannian normal coordinate system where the metrics $g_{mn}(X)$ is determined completely by the Riemann tensor R_{mnkl} its covariant derivatives and their products.

$$g_{mn}(X) = \eta_{mn} - \frac{1}{3} \, R_{mknl}(X_0) \, (X^k - X_0^k) \, (X^l - X_0^l)$$
$$+ \frac{1}{6} \, R_{mknl,p}(X_0) \, (X^k - X_0^k) \, (X^l - X_0^l) \, (X^p - X_0^p) + \dots \, . \tag{5.5}$$

Analogously, the gravitational prepotential $H^m(X, \Theta, \bar{\Theta})$ is expressed in the normal gauge in terms of basic superfields (5.4), their covariant derivatives, and products. The effectiveness of the use of the normal gauge can be demonstrated by proving that all constraints are fulfilled automatically in the framework of the complex approach, etc. Let us show, e.g., that the constraint $T_{\alpha\beta}{}^\gamma = 0$ holds. The dimension of $T_{\alpha\beta}{}^\gamma$ is $cm^{-1/2}$, the dimensions of the basic superfields are $[R] = [G_{\alpha\dot{\beta}}] = cm^{-1}$ and $[W_{\alpha\beta\gamma}] = cm^{-3/2}$. Products of basic superfields or their covariant derivatives have even higher negative dimension. The quantity of dimension $cm^{-1/2}$ cannot be constructed and therefore the proof of vanishing $T_{\alpha\beta}{}^\gamma$ is completed.

6. ACTION AND EQUATIONS OF MOTION

The action for pure supergravity in the real superspace is simply the supervolume [8b] while in the presence of matter fields it is written as [6c]

$$\mathcal{I} = \int d^4X \, d^4\Theta \; Ber\|E_M{}^A\| \left\{ \frac{1}{\chi^2} + \mathcal{L}_{mat}(\Phi, D\Phi) \right\} \tag{6.1}$$

where $\mathcal{L}_{mat}$ is the Lagrange density which differs from the flat one by replacing all the derivatives in it by gravitational covariant ones (13), (14). In the case of pure supergravity the action in terms of H^m has the form

$$\mathcal{I} = \int d^4X \, d^4\Theta \; (\text{Det} \, \|\Delta \, \sigma_a \, \bar{\Delta} \, H^m\|)^{1/3} \, (\text{Det}\|\delta_c{}^d + \partial_c H^b \, \partial_b H^d\|)^{1/2}$$

$$where \quad \Delta_\alpha = \partial_\alpha + \Delta_\alpha H^n \, \partial_n \, . \tag{6.2}$$

The equations of motion are derived from eq. (19) by a straightforward variational procedure. For the gravitational superfield they have a form

$$G_{\alpha\dot\alpha} = \chi^2 \, V_{\alpha\dot\alpha} \tag{6.3}$$

where χ is the gravitational constant. The left hand side of this equation is the super-analogue of the Einstein tensor, $R_{mn} - \frac{1}{2} g_{mn} R$, while the right had side is the super-current, which is the superanalogue of the stress-energy tensor, T_{mn}. Just this form was suggested and proven in the linearized approximation [4], when we have postulated the Ferrara-Zumino supercurrent as the source of the gravitational superfield.

Now this postulate is completely confirmed. Note that the variational principle (6.2) gives only Eq. (6.3) and nothing more. In particular, in the absence of matter one can conclude that $R = constant$ and this constant can be identified with the cosmological term. Their vanishing is connected with an assumption that $H(X,\Theta,\bar\Theta)$ have to decrease rapidly with increasing X.

7. LEVY SUPERFORM

In fact, we deal with the complex geometry of the complex (4, 2) superspace. However, above we passed to the real (4, 4) superspace. It is not very natural. Schwarz has shown that one can derive the action principle in the framework of complex superspace. To this end the superanalogue of the Levy form can be used. We deal with the complex (4, 2) superspace

$$Z^M = \{X_L{}^m, \Theta_L{}^\mu\} \, . \tag{7.1}$$

The (4, 4) hypersurface is singled out by imposing 4 real conditions

$$f^k(Z, \bar{Z}) = 0 \, , \quad f^{k*} = f^k \, , \quad k = 0, 1, 2, 3 \, . \tag{7.2}$$

An example of this condition is Eq. (3.7)

$$\text{Im } X_L{}^m - H^m(\text{Re } X_L{}^m, \Theta_L, \bar{\Theta}_R) = 0 \; .$$

Evidently, this surface remains the same if one replaces f in Eq. (7.2) by

$$f'^k(Z, \bar{Z}) = \eta_l{}^k(Z, \bar{Z}) \, f^l(Z, \bar{Z}) \tag{7.3}$$

where $\eta_l{}^k$ is an arbitrary nonsingular matrix. The invariance is required under general analytic transformations (3.7), (4.1) and simultaneously under reparametrizations (7.3). The real tangent $(4, 4)$ plane to this hypersurface is defined by the real equation

$$\frac{\partial f^k}{\partial Z^M} \, dZ^M + \frac{\partial f^k}{\partial \bar{Z}^M} \, d\bar{Z}^M = 0 \tag{7.4}$$

i.e. dZ and $d\bar{Z}$ lie in this hyperplane if they satisfy (7.4). The maximal complex hyperplane in real hyperplane (7.4) has the equation

$$\frac{\partial f^k}{\partial Z^M} \, dZ^M = 0 \; . \tag{7.5}$$

Its complex dimension is $(0, 2)$. Indeed, using the more detailed form of (7.5)

$$\frac{\partial f^k}{\partial X^m} \, dX^M + \frac{\partial f^k}{\partial \Theta^\mu} \, d\Theta^\mu = 0 \tag{7.6}$$

one can express all the dX^m in terms of $d\Theta^\mu$. Now the 2- form is introduced

$$\omega^k = \frac{\partial^2 f^k}{\partial Z^M \, \partial \bar{Z}^N} \, dZ^M \, d\bar{Z}^N \, (-1)^{P(M)P(N)} \tag{7.7}$$

where $P(M)$ is the grade of the index M, i.e., 0 for $M = m$ and 1 for $M = \mu, \dot{\mu}$. The Levy form is the value of this 2-form at point Z of the hypersurface with dZ being in the complex tangent superplane

$$\omega^k = \Gamma_a{}^k \, \sigma^a_{\mu\dot{\nu}} \, d\Theta^\mu \, d\bar{\Theta}^{\dot{\nu}} \; . \tag{7.8}$$

Note that the matrix $\Gamma_a{}^k$ coincides up to a factor with the so-called central vierbein from [6c]. We denote its determinant as

$$\Gamma = \text{Det} \|\Gamma_a{}^k\| \; . \tag{7.9}$$

Now our action (6.2) can be represented as the integral over the whole $(4, 2)$ complex superspace

$$\mathcal{I} = \int dZ \, d\bar{Z} \left(\text{Det} \left\| \frac{\partial f^{k}}{\partial X^{m}} \right\| \right)^{4/3} \Gamma^{-1/3} \, \Pi\delta(f^{k}(Z, \bar{Z})) \ . \tag{7.10}$$

This definition emphasizes the equality of fields and coordinates. Indeed, the integration is performed under the whole complex (4, 2) superspace, and the hypersurface is singled out by δ-functions. Now there is no necessity to pass to the real superspace. It would be instructive to know how one can, with the help of the Levy form and other devices, obtain in the complex geometry framework the higher invariants needed to construct the multi-loop counterterms.

8. REMARKS

So, now we know how to formulate the $N = 1$ supergravity in terms of fields, we know its formulation in the (4, 4) real superspace with all constraints, and finally, we can formulate it in pure geometric terms in the (4, 2) complex superspace. We know today also the $N = 2$ supergravity in terms of fields [16,17] and a rather long list of constraints for its formulation in the real (4, 8) superspace [12,13]. Sokatchev [18] has developed an approach to the $N = 2$ supergravity using a complex (4, 8) superspace. However, some external constraints have to be imposed here though their number reduces considerably as compared to the real approach [12,13]. Unfortunately we know very little about the most interesting higher N extended supergravities. It is necessary to find a key to the construction of adequate geometric approaches. We would like to emphasize that such a key can be obtained from the knowledge of constraints and understanding of its meaning in the real superspace approach. We can illustrate it by an example of the $N = 1$ supergravity. In an interesting paper [11] Gates, Stelle and West elucidated the meaning of constraints in this case. What is important for us is that in particular they mean the preservation of the notion of chiral superfield in supergravity. This statement is nontrivial because the chiral superfield is singled out of the general scalar superfield by the spinor differentiation, recall Eq. (3.3)

$$\Phi(X, \Theta, \bar{\Theta}) = \phi = \phi(Z) \ .$$

In the curved case we have

$$\{\bar{D}_{\dot{\alpha}} \, , \, \bar{D}_{\dot{\beta}} \} \, \Phi = T_{\dot{\alpha}\dot{\beta}}{}^{A} D_{A} \, \Phi \tag{8.1}$$

and the definition (3.3) is consistent provided there are constraints on the torsion components.

$$T_{\dot{\alpha}\beta}{}^{\cdot\gamma} = T_{\dot{\alpha}\beta}{}^{\cdot a} = 0 \ . \tag{8.2}$$

On the other hand Eq. (3.3) is the Cauchy-Riemann condition for Grassmann analyticity. It means that the superfield Φ depends on $Z = (X_L \ , \ \Theta_L)$, and is independent of $\bar{Z}$. Thus, one can rediscover our approach [3-6] by solving the corresponding constraints. Searching an adequate complex geometry for the $N = 2$ supergravity one can apply to the solution of its constraints [12,13]. In this case they mean in particular the preservation in curved superspace both of chiral superfield and of the so-called Fayet-Sohnius multiplet. The latter is singled out of the general scalar superfield with one $SU(2)$ index by the differential conditions

$$D_{\alpha\{i} \ \Phi_{j\}} = 0 \ , \qquad\qquad\qquad \bar{D}_{\dot{\alpha}\{i} \ \Phi_{j\}} = 0 \tag{8.3}$$

where braces stand for symmetrization in indices i,j. It is desirable to find a solution of the corresponding constraints and to express this solution in terms of superfields defined on some complex superspaces. This problem is not solved yet. However, in the flat space equations (8.3) turns out to be the condition of Grassmann analyticity in two pairs of spinor variables [20]. Its solution singles out two complex (4, 2) spaces which merge in an interesting way under transformations of the internal symmetry $SU(2)$. However, even there the situation is far from being clear though we know the field content of the supergravity multiplet [16,17]. A further elucidation is needed to find the intrinsic complex geometry of the $N = 2$ supergravity. It is worthwhile to emphasize the great importance of the central charge already in the $N = 2$ case. So, e.g., the Fayet-Sohnius multiplet ((8.3)) has the central charge. One can expect that at $N = 2$ central charges will occupy an analogous or maybe even more central place.

Note that the hypercomplex analysis could become of importance in the higher extended supergravities along the complex one [14].

At present the supersymmetric way is becoming more probable in a search of the grand unified theory. It is very desirable that this way would be the most geometrized one. The spinor Grassmann variables then are necessary; however, their number has to be as small as possible. The more of them, the more constraints have to be imposed to exclude a huge number of superfluous variables. It would be very desirable to manage with the smallest possible number of them. A primary superspace can be rather vast; however, afterwards a set of their coordinates will be identified with superfields. The final picture will have to be clear and geometric.

REFERENCES

[1]. Gürsey F., Radicati L., Phys. Rev. Lett. 13 (1964) 173.

[2]. Gürsey F., Pais A., Radicati L., Phys. Rev. Lett. 13 (1964) 299.

[3]. Ogievetsky V., Sokatchev E., Phys. Lett. B79 (1978) 222.

[4]. Ogievetsky V., Sokatchev E. in Proc. of IV Intern. Conference on Nonlocal and Nonlinear Field Theory, Alushta, April, 1976, JINR D2-9788, Dubna, p. 183; see also Nucl. Phys. B124 (1977) 309.

[5]. Ogievetsky V., Sokatchev E., Yad. Phys. 31 (1980): a) 264; b) 821. English trans. JINR preprints a) E2-12469; b) E2-12511.

[6]. Ogievetsky V., Sokatchev E., Yad. Phys. 32 (1980): a) 862; b) 870; c) 1142. English trans. JINR preprints a) E2-80-127; b) E2-80-138; c) E2-80-139.

[7]. Schwarz A. S., Yad. Phys. 34 (1981) 1144, Nucl. Phys. B171 (1980) 154.

[8]. Wess J., Zumino B. a) Phys. Lett. B66 (1977) 361; b) Phys. Lett. B74 (1978) 51.

[9]. Wess J. in Proc. GIFT Seminar, Salamanca, Springer p. 81.

[10]. Siegel W., Gates S. J., Nucl. Phys. B147 (1979) 77.

[11]. Gates S. J., Stelle K. S., West P. C., Nucl. Phys. B169 (1980) 347.

[12]. Wess J. in Proc. of Seminar *Group-Theory Methods in Physics*, Zvenigorod v. 2 (1979) 235, Nauka, Moscow, 1980.

[13]. Breitenlohner E., Sohnius M., Nucl. Phys. B165 (1980) 483.

[14]. Galperin A. S., Ivanov E., Ogievetsky V., Pisma JETP 33 (1981) 176. English version, JINR preprint E2-80-790.

[15]. Grimm R., Wess J., Zumino B., Nucl. Phys. B152 (1979) 255.

ON THE PHYSICS OF
DIMENSIONAL REDUCTION

Peter G. O. Freund

The Enrico Fermi Institute
and the Department of Physics
The University of Chicago,
Chicago, Illinois 60637

Abstract

Certain abstract constructs in ordinary four-dimensional Physics are found to acquire simple geometrical interpretations in the context of Kaluza reduction from higher dimensions. Among such constructs we treat the operation of charge conjugation, the (abelian or non-abelian) Lorentz force, and the forces on point particles moving in the scalar fields of a $GL(N,R)/O(N,R)$ σ-model. Some cosmological and "experimental" aspects of dimensional reduction are discussed.

1. INTRODUCTION

In four space-time dimensions a unified field theory may involve many fields of spins up to 2. In general, such a theory also leads to some arbitrary choices: of gauge group, of matter fields and also some parameters. By dimensional reduction one can eliminate most of these arbitrary elements. One requires the complicated theory in four dimensions to be that limit of a very simple theory in a higher number of dimensions, in which the extra dimensions have "curled" up into a compact manifold of very small characteristic size (say of order the Planck length, $l_{\text{Planck}} \sim 10^{-33}$ *cm*). This procedure was first demonstrated by Kaluza and further developed by Klein, Jordan and others [1]. These authors all added but one space-like dimension to the four physical dimensions and thereby arrived at unified theories of gravity, an abelian

191

gauge field and neutral scalar matter starting from simple 5-dimensional gravity. Addition of further dimensions has resulted in nonabelian gauge fields and charged scalar matter fields [2,3]. This technique was extensively used in extended supergravity where an eleven-dimensional master-theory [4] then yields the supergravities in lower (in particular in four [5]) dimensions. Invariably the higher dimensional theory owes its simplicity to its geometrical structure. Yet in the lower-dimensional physics there are phenomena to which geometry does not seem to be relevant. I would like to discuss here a sample of such phenomena and show how their geometrical origins can be easily revealed via dimensional reduction.

I have in mind

  (i)the forces acting on charged point particles moving in abelian or non-abelian gauge fields and in the scalar fields of the $GL(N,R)/O(N)$ σ-model [3] and

  (ii)the operation of the charge conjugation which unlike the space and time inversion operations admits of no simple geometric interpretation in 4-dimensions.

I shall also discuss certain technical features of dimensional reduction and possible cosmological consequences thereof.

It is a great pleasure to dedicate this work to Feza Gürsey whose imaginative use of algebraic and geometric ideas in physics has had a deep and very fruitful impact on our field.

2. DIMENSIONAL REDUCTION OF GRAVITY THEORIES

Start from gravity theory in a D-dimensional space-time (1 time and $D-1$ space dimensions). If $D-d$ space-like dimensions are compactified then in a local direct product basis e_M of the tangent space of the D-dimensional space-time manifold, the metric can be written in the form

$$\gamma_{MN} = \begin{pmatrix} g_{\mu\nu} + 16\pi G\, g_{ab} A_\mu{}^a A_\nu{}^b & \{16\pi G\}^{1/2}\, g_{np} A_\mu{}^p \\ \{16\pi G\}^{1/2}\, g_{mp} A_\nu{}^p & g_{mn} \end{pmatrix} \tag{2.1}$$

where $M, N \ldots = 1, 2, \ldots, D$; $\mu, \nu = 1, \ldots, d$; $m, n, \ldots = d+1, d+2, \ldots, D$; and all fields $g_{\mu\nu}$, A_μ^p, g_{mn} depend only on $x^1, \ldots, x^d$. The basis e_M obeys the commutation relations

$$[e_\mu, e_\nu] = [e_\mu, e_n] = 0 \ , \quad [e_m, e_n] = \frac{e}{\{16\pi G\}^{1/2}} f_{mn}^p e_p \tag{2.2}$$

(f_{mn}^p the structure constants of the gauge group). This is not a coordinate basis as far as the compactified dimensions are concerned. Here and below, G is the Newton constant and e the gauge coupling constant. The scalar fields g_{mn} play an important role. The γ_{MN} or more properly, the corresponding D-bein e^A_M ($\gamma_{MN} = e^A_M e^B_N \eta_{AB}$ with η_{AB} the D-dimensional Minkowski metric) parametrize the coset space $GL(D,R)/O(D,R)$ so that gravity in D-dimensions has an $O(D)$ gauge invariance. Upon compactification of $(D-d)$-dimensions this is reduced to $O(d) \times O(D-d)$ gauge invariance with the $O(d)$ part implemented via the $g_{\mu\nu}$ (or the corresponding d-bein) and $O(D-d)$ via the scalar field g_{mn} (or the corresponding $(D-d)$-bein). So these scalar fields parametrize the coset space $GL(D-d)/O(D-d)$. The lagrangian for these scalar fields constructed in reference 3 is indeed that of a nonlinear or "σ-model" [6]. So the gauging of $O(D-d)$ is implemented without gauge fields, but rather via scalar fields. Of course the fields A^a_μ are gauge fields, but not of $O(D-d)$. Indeed, there are but $(D-d)$ such vector fields, while the dimensions of $O(D-d)$ is $\binom{D-d}{2} > D-d$ for $D-d > 3$. The A^a_μ gauge some $(D-d)$-dimensional Lie group.

We now consider the equations [7] of geodesic or self-parallel curves:

$$\nabla_u u = 0 \qquad\qquad (2.3)$$

where u is the tangent to the geodesic curve at the point corresponding to, say, parameter value s. Expand u in the basis e_M: $u = u^M e_M$. The general properties of covariant differential yield

$$\nabla_u u = \nabla_{u^M e_M} u^N e_N = \left(\dot{u}^M + \Gamma^M_{AB} u^A u^B \right) e_M$$

where $\dot{u} = \dfrac{du^M}{ds}$, so that the geodesic equation becomes

$$\dot{u}^M + \Gamma^M_{AB} u^A u^B = 0 . \qquad\qquad (2.4)$$

In a coordinate basis $u^M = \dot{x}^M$ so that the geodesic equation takes the familiar form $\ddot{x}^M + \Gamma^M_{AB} \dot{x}^A \dot{x}^B = 0$. The basis (2.2) is not a coordinate basis. If in the horizontal lift basis the components of u^N are $u^\mu = \dot{x}^\mu$, $u^n = 1/m \, I^n$, (m is, for the time being, an arbitrary parameter), then [3] the local direct product basis (2.2)

$$u^\nu = \dot{x}^\nu \quad , \qquad u^n = \frac{1}{m} \left(I^n - e A^n_\rho \, m\dot{x}^\rho \right) \qquad\qquad (2.5)$$

so that

$$\dot{u}^\nu = \ddot{x}^\nu \quad , \qquad \dot{u}^n = \frac{1}{m} \left(\dot{I}^n - e \, A^n_\rho \, m \, \ddot{x}^\rho - \frac{e}{m} \, \partial_\sigma A^n_\rho \, m\dot{x}^\rho \, m\dot{x}^\sigma \right) . \qquad\qquad (2.6)$$

Inserting the expressions (2.5) and (2.6) into the geodesic equations (2.4) and using the connection coefficients appropriate to the basis (2.2) we can work out the geodesics of the reduced geometry. This will be done in some detail in the next section.

3. GEOMETRICAL DERIVATION OF FORCES ACTING ON POINT PARTICLES

The law of motion of a point particle in Einstein gravity is geometrical: the particle moves along a geodesic of the world geometry. Yet no comparable geometrical picture exists for the motion of a point particle in abelian or non-abelian gauge fields or in scalar fields. Here, by requiring the point particle to follow geodesics of the higher-dimensional geometry, we rederive the well-known abelian and non-abelian Lorentz force expressions, as well as the expression of the force due to the $GL(N,R)/O(N)$ scalar fields. The derivation of the abelian Lorentz force along these lines has been known for a very long time [1]. The non-abelian force law [8] has also been rederived by dimensional reduction [9] (in a basis less convenient than (2.2)). The force for motion in the scalar fields of the $GL(N,R)/(O(N))$ σ-model seems not to have been obtained before.

To derive the equations of motion of a charged point particle in a nonabelian gauge field, a) set the scalar fields g_{mn} equal to the (x-independent!) Killing metric of the group gauged by the A_μ^a and b) let the d-dimensional space-time manifold be flat: $g_{\mu\nu} = \eta_{\mu\nu} \equiv$ Minkowski metric. For this special case the connection coefficients obtained from (2.1) are

$$\Gamma_{mn}^r = \tfrac{1}{2} f_{mn}^r \ , \quad \Gamma_{\mu n}^r = \Gamma_{n\mu}^r = \tfrac{e}{2} \left\{ f_{an}^r A_\mu^a + eA^{\alpha r} (F_n)_{\alpha\mu} \right\} ,$$

$$\Gamma_{\mu n}^\rho = \Gamma_{n\mu}^\rho = \tfrac{e}{2} (F_n)_\mu^{\ \rho} \ , \quad \Gamma_{mn}^\rho = 0 \ , \quad \Gamma_{\mu\nu}^\rho = 0 , \qquad (3.1)$$

$$\Gamma_{\mu\nu}^r = \tfrac{e}{2} \left\{ \partial_\mu A_\nu^r + \partial_\nu A_\mu^r + e^2 A_\alpha^r \left\{ (F_k)^\alpha_{\ \mu} A_\nu^k + (F_k)^\alpha_{\ \nu} A_\mu^k \right\} \right\} ,$$

where $F_{\rho\sigma}^n = \partial_\rho A_\sigma^n - \partial_\sigma A_\rho^n + e f_{pq}^n A_\rho^p A_\sigma^q$

and latin (greek) indices are raised and lowered with the Killing (Minkowski) metric. Inserting this and Eqs. (2.5) and (2.6) into the geodesic equations (2.4) we find the well-known (gauge-invariant) result [8]:

$$m \ddot{x}_\nu = e F_{\nu\alpha}^b \, I^b \, \dot{x}^\alpha , \qquad \dot{I}^n + e f_{ab}^n B_\rho^{\ a} \, \dot{x}^\rho \, I^b = 0 \qquad (3.2)$$

(the parameter m introduced in Eqs. 2.5 and 2.6 thus plays the role of mass of the particle).

For the special case of a one-parameter abelian group ($D = d + 1$) this gives the familiar Lorentz-force and the conservation of "electric" charge. To obtain the point particle equations in the g_{mn}-scalar fields, first set the coupling constant $e = 0$ to eliminate the A^a_μ fields and then set the Newton constant $G = 0$ and $g_{\mu\nu} = \eta_{\mu\nu}$. The connection coefficients are now

$$\Gamma^r_{\mu n} = \Gamma^r_{n\mu} = \tfrac{1}{2} g^{rs} \partial_\mu g_{sn} \quad , \quad \Gamma^\rho_{mn} = -\tfrac{1}{2} \partial^\rho g_{mn} \, ,$$

$$\Gamma^r_{mn} = \Gamma^r_{\mu\nu} = \Gamma^\rho_{\mu n} = \Gamma^\rho_{m\nu} = \Gamma^\rho_{\mu\nu} = 0 \tag{3.3}$$

and the desired equations of motion take the form

$$m\,\ddot{x}_\nu = -\,\partial_\nu g_{ab}\, J^a J^b = g_{ac} \left(A^c_{\ b}\right)_\nu J^a J^b \, ,$$

$$\dot{J}^n - g^{na}\left(\dot{x}^\rho \partial_\rho g_{ab}\right) J^b = \dot{J}^n + \left(A^n_{\ b}\right)_\rho \dot{x}^\rho J^b = 0 \, , \tag{3.4}$$

$$J^n \equiv (m)^{-1/2}\, I^n \quad , \quad \left(A^a_{\ b}\right)_\mu \equiv -\, g^{ac}\, \partial_\mu g_{cb} \, .$$

Notice that the equation for J^n involves a covariant derivative with respect to the "composite gauge field" $\left(A^a_{\ b}\right)_\mu$. They possess the expected $GL(D-d)$ symmetry (as can easily be seen) and this is at the root of the appearance of the terms quadratic in I^A. (Unlike the equations 3.2 which only involve $\dot{x}^\mu I^a$ type nonlinearities, here we also encounter these new $J^A J^B$-nonlinearities). In the "abelian" case $D = d + 1$, with the notation $g_{DD} = \phi$, $I_D \equiv I$ we have the equations

$$m\ddot{x}_\nu = -\,\partial_\nu \phi\, J^2$$

$$\dot{J} = J\, \dot{x}^\rho \partial_\rho \ln \phi \, .$$

Thus, in spite of their apparently nongeometric structure in d-space, the Lorentz force, its nonabelian generalization and the corresponding force for the $GL(D-d,R)/O(D-d)$ σ-model have all been derived from a unified geometrical principle: geodesic motion in D-space.

4. GEOMETRIZATION OF CHARGE CONJUGATION

Of the three discrete symmetries, space inversion (P), time reversal (T), and charge conjugation (C), the first two have very clear geometrical meaning while the last one is an abstract operation (first discovered in the context of the Dirac equation), devoid of any simple geometrical interpretation. Now, C can be defined in higher dimensional spaces as well, and in general it is still an abstract operation. There is an interesting case, however, in which C acquires a simple geometrical interpretation, which we now describe. In general, a field $\phi_{\alpha\ldots}$ need not be charge-self-conjugate, i.e. $\phi_{\alpha\ldots}^C \neq (-1)^\epsilon \phi_{\alpha\ldots}$ ($\epsilon \equiv C$-parity). Yet one can rewrite any field theory in terms of a set of charge-self-conjugate fields $\phi_{\alpha\ldots} \pm \phi_{\alpha\ldots}^C$. On the other hand, there exist field theories which cannot be written in terms of Lorentz (or generally) covariant fields such that at least one of these fields *not* be charge-self-conjugate (this, in particular, forbids the presence of fields carrying a conserved charge in any space-time-dimension, or of more than one Majorana field of a given spinor type in a space-time dimension in which Majorana fields exist but Weyl fields do not). We call such theories C-trivial. As an example, think of a gravitating polynomially self-interacting scalar field. One can even speak of strong C-triviality if all fields are naturally assigned even C-parity. Now it will be argued that while C-trivial theories exist, they are hardly realistic as they contain no Dirac fermions or charged particles. But assume that a certain 4-dimensional theory of interest can be obtained by dimensional reduction from a D-dimensional theory ($D > 4$). It can then happen that the D-dimensional theory is C-trivial (with respect to D-dimensional charge conjugation), without its 4-dimensional descendant being C-trivial (examples will be given shortly). In such a case, the non-trivial 4-dimensional C-operation must originate in some other, usually geometric, symmetry of the D-dimensional theory, and as such acquire a geometric meaning itself.

As a first example, consider pure D-dimensional gravity. It is obviously C-trivial in D-dimensions. Reduce it to any dimension $d < D$ and the resulting fields are (see Section 2) d-dimensional gravity $g_{\mu\nu}$, the gauge fields A_ν^n and the scalar fields g_{mn}. Consider for simplicity the case that the gauge group is $SU(N)$ (so that $D-d = N^2 - 1$). The d-dimensional charge conjugation then leaves unchanged d-gravity $g_{\mu\nu}$ and those A_ν^n that gauge the $SO(N)$ subgroup of $SU(N)$, and changes the sign of the remaining $N(N+1)/2$ A_μ^n's (neutral vector mesons have negative C-parity!). At this point it is convenient to simplify notation by labeling by $n=d+1, \ldots,$ $d + N(N-1)/2$ the gauge fields of $O(N)$ and by $n = d + N(N-1)/2 + 1, \ldots, D$ the other gauge fields. The effect of C on the scalar fields is then simply

$$g_{mn} \rightarrow -g_{mn} \quad \text{if } m \le d + \frac{N(N-1)}{2} \text{ and } n > d + \frac{N(N-1)}{2}$$

$$\text{or if } n \le d + \frac{N(N-1)}{2} \text{ and } m > d + \frac{N(N-1)}{2}$$

$$g_{mn} \rightarrow g_{mn} \quad \text{otherwise.}$$

What is the coordinate transformation in D dimensions that upon dimensional reduction induces these transformation properties of the reduced fields? It is a reflection of the last $\{N(N+1)-2\}/2$ coordinates. So what we perceive as charge conjugation in d-dimensions is but a "mirroring"--a reflection of $\{N(N+1)-2\}/2$ coordinates--in D-dimensions. Just as P corresponds to a reflection of ordinary space, so charge conjugation involves a reflection in the extra space dimensions. In C-trivial theories, then, antimatter is an "artifact" of dimensional reduction. In the original D-dimensional theory matter and antimatter are identical, and in the dimensionally reduced theory the concept antimatter is a device of keeping track of "lost reflections".

Now it is straightforward to generalize these ideas to the case when the gauge group is not $SU(N)$ when less familiar but mathematically equally well known involutive automorphisms of the gauge group's Lie algebra are to be used for the definition of C in d-dimensions.

We now come to the case of interest: supergravity in 11-dimensions. This theory involves (in 11-dimensions) the graviton, one Majorana gravitino and the antisymmetric rank three tensor fields, all of which are charge-self-conjugate and have different tensorial or spinorial ranks. Thus, 11-dimensional supergravity is C-trivial. If $N = 8$ supergravity turns out to be a) realistic and b) a dimensional reduction of 11-dimensional supergravity, then the concept of antimatter as used in 4-dimensional Physics is connected with a reflection of 4 of the 7 extra space dimensions. Matter/antimatter discrimination is only meaningful in 4-dimensions, not in the original 11-dimensional theory where matter and antimatter are identical.

5. ARE THE HIGHER DIMENSIONS PHYSICAL?

The discussion of the previous sections has revealed that certain physical constructs such as the Lorentz force law (and its nonabelian and σ-model generalizations) and the operation of charge conjugation acquire simple geometrical interpretations when viewed in the context of dimensional reduction. Without any further ado (based also on earlier considerations) [2,3] we shall therefore adopt here the view that the real theory does indeed involve $D > 4$ space-time dimensions and that our 4-dimensional Physics is to be viewed as the result of dimensional reduction. This presumes the existence of meaningful (e.g., finite non-renormalizable) quantum field theories in D-dimensions. Specifically, we shall invoke $D = 11$ as suggested by supergravity.

We now consider some features of such a view. First of all, at small distances comparable to the size of the compactified dimensions ($\sim l_{Planck}$) we expect the laws to be D-dimensional rather than 4-dimensional. In particular, in the "very" early stages of the universe its dimensionality should be D. Second, allowing a harmonic dependence of fields on the compactified coordinates we would find new excitations, most of mass of the order of the Planck mass. Could some of these excitations be light and could they correspond to bound states in the reduced $N = 8$ supergravity theory?

Finally, there is the problem of graviphotons (or "antigravity") [10] that can be most simply, though unfortunately not exclusively, understood in terms of dimensional reduction. [11] Finding a feasible experimental discrimination between $N = 8$ supergravity in 4-dimensions as a fundamental theory or as a dimensionally reduced form of a more basic 11-dimensional theory remains an important open problem. [12]

REFERENCES

[1]. Th. Kaluza Berl. Ber. 966 (1921). O. Klein, Z. Physik 37 (1926) 895; in *New Theories in Physics*, International Institute for Intellectual Cooperation, Paris, 1939, p. 77. P. Jordan, Z. Phys. 157 (1959) 112. W. Pauli, *Theory of Relativity*, Pergamon Press, N. Y. 1958.

[2]. B. de Witt in *Relativity, Groups and Topology*, edited by B. and C. de Witt (Gordon and Breach, N. Y. 1964). A. Trautman, Rep. Math Phys. 1 (1970) 29. R. Kerner, Ann. Inst. H. Poincare, 9 (1968) 143.

[3]. Y. M. Cho and P. G. O. Freund, Phys. Rev. 12 (1975) 1711. Y. M. Cho, J. Math. Phys. 16 (1975) 2029.

[4]. E. Cremmer, B. Julia and J. Scherk, Phys. Lett. 76B (1978) 409. W. Nahm, Nucl. Phys. B135 (1978) 149.

[5]. E. Cremmer and B. Julia, Phys. Lett. 80B (1978) 48.

[6]. This point, implicit in reference 3, has become clear to me in the course of interesting discussions with T. L. Curtright which I gratefully acknowledge.

[7]. R. L. Bishop and R. J. Crittenden, *Geometry of Manifolds*, Academic Press, N. Y. 1964, p. 113.

[8]. S. K. Wong, Nuovo Cim. 65A (1970) 689.

[9]. N. K. Nielsen, Nucl. Phys. B167 (1980) 249.

[10]. J. Scherk, in *Supergravity*, P. van Nieuwenhuizen and D. Z. Freedman, editors, North-Holland, Amsterdam, 1979, p. 43. C. K. Zachos, Phys. Lett. 76B (1978) 1329.

[11]. I wish to thank A. Salam for emphasizing this point to me.

[12]. This work was supported in part by the NSF.

BACKLUND TRANSFORMATIONS AND DEFORMATIONS OF LINEAR DIFFERENTIAL EQUATIONS WITH APPLICATIONS TO DIOPHANTINE APPROXIMATIONS

G. Chudnovsky

Department of Mathematics
Columbia University
New York, New York 10027

1. INTRODUCTION

We examine here some new methods borrowed from mathematical physics that are used in analytic number theory and especially in the study of diophantine approximations. Our final product is summarized in a few new results on measures of irrationality of the constants of classical analysis.

The main instrument is the method of Backlund transformations (*BTs*), now actively employed in field theory. We refer the reader to [14] where that particular kind of *BTs* we speak about is presented together with physical applications. We employ *BTs* in their particular role as an effective method of constructing generalized Padé approximations which are linear combinations with polynomial coefficients of a given set of functions having zeros of high orders at fixed points. *BTs* furnish another important instrument. The relationship between different *BTs* gives recurrence relations between consecutive Padé approximants known as "contiguous relations" [1, 5, 10]. These recurrences (contiguous relations) are applied directly to the examinations of diophantine approximations to values of functions satisfying linear ordinary differential equations (l.o.d.e). The relationship between *BTs* and problems of diophantine approximations can be extended much further since recurrence relations between *BTs* generating good rational approximations to numbers, generate at the same time lattice (quantum Hamiltonian) completely integrable systems. See examples given in paper [14] of the

Toda lattice, *XXX*-model and the nonlinear Schrödinger equation. We are planning to continue the study of this relationship with particular attention to the properties of correlation functions. It should be noted that the reverse relationship between diophantine approximations and field theory may hopefully be nontrivial as well. We want to remind the reader of the wonderful arguments of Lüscher [15] on the absence of particle production for the $O(N)$-σ-model using the transcendence of linear forms of logarithms of algebraic numbers.

This work was supported in part by the John Simon Guggenheim Memorial Foundation and by the United States Air Force under contract AFOSR-81-0190.

2. BACKLUND TRANSFORMATIONS AND CONTIGUOUS RELATIONS FOR SOLUTIONS OF LINEAR DIFFERENTIAL EQUATIONS

According to our approach [16-19] we treat *BTs* as isomonodromy deformations. We present the treatment of *BTs* in the case of system of functions satisfying l.o.d.e. We first consider equations with rational function coefficients. We start with the n×n matrix case:

$$\frac{dY}{dx} = \sum_{j=1}^{m} \sum_{r=1}^{s} \frac{Y U_j^{(r)}}{(x - a_j)^r} \tag{2.1}$$

and with the fundamental solution $Y_b(x)$ of Eq. (2.1) such that $Y_b(b) = I$. Locally the solutions $Y_b(x)$ can be represented as a formal power series in the neighborhood of $x = a_j$ as

$$Y_b(x) = (x - a_j)^{W_j^{(1)}} \tilde{G}_b^{(j)}(x) \, \bar{Y}_b^{(j)}(x) \tag{2.2}$$

where $\bar{Y}_b^{(j)}(x)$ and $\bar{Y}_b^{(j)}(x)^{-1}$ are holomorphic at $x = a_j$ and $\tilde{G}_b^{(j)}(x)$ is an entire function in a local parameter $(x - a_j)^{-1}$. The precise form of $\tilde{G}_b^{(j)}(x)$ is determined by a sector in which x is lying and by the corresponding Stokes multipliers.

Under additional assumptions on Eq. (2.1) it is possible to represent $Y_b(x)$ as

$$Y_b(x) = V_j(x - a_j)^{W_j^{(1)}} \cdot \exp\left\{ \sum_{i=1}^{q_j} W_j^{(i+1)} (x - a_j)^{-1} \right\} \bar{Y}_b^{(j)}(x) \, C_j$$

where $\bar{Y}_b^{(j)}(x) = I + \mathbb{O}(x - a_j)$ and the $W_j^{(i+1)}$ are diagonal matrices for $i = 0, \ldots, q_j$.

By a system of equations contiguous to Eq. (2.1) (*BT* of Eq. (2.1)) we understand a similar system of l.o.d.e.

$$\frac{dY'}{dx} = \sum_{j=1}^{m} \sum_{r=1}^{s} \frac{Y' U_j^{(r)'}}{(x - a_j)^{r'}} \tag{2.3}$$

whose monodromy group and Stokes multipliers are the same as for Eq. (2.1). Two solutions $Y_b(x)$ of Eq. (2.1) and $Y'_b(x)$ of Eq. (2.3) are related by a linear relation with rational function coefficients

$$Y'_b(x) = Y_b(x) R(x) \tag{2.4}$$

where $R(x)$ is an $n \times n$ matrix with rational function entries and $R(b) = I$. The relation Eq. (2.4) is called a *BT* or a contiguous relation. The *BT* means a change in the gauge matrices C_j and the exponential matrices $W_j^{(1)}$ to C'_j and $W'^{(1)}_j$ leaving fixed monodromy invariants: $\exp(2\pi i W_j^{(1)})$, $W_j^{(i+1)}$, V_j for $i = 1, \ldots q_j$, $j = 1, \ldots m$.

By an "elementary" *BT* [16], we understand a *BT* changing the exponential matrix only at one point, say $x = a_{m+1} = \infty$ and changing the eigenvalues of $W_\infty^{(1)}$ by a unit. We present the explicit expression for $R(x)$ and $U_j^{(r)'}$ for the "elementary" *BTs* following [16].

Lemma 2.5: Let $U_\infty^{(1)} = \sum_{k=1}^{m} U_k^{(1)}$ have a normal form, $U_\infty^{(1)} = S^{-1} diag(l_1, \ldots, l_n) S$ and for

$$U_k = a_k U_k^{(1)} + U_k^{(2)} \quad \text{let} \quad \sum_{k=1}^{m} (S U_k S^{-1})_{ij} \neq 0$$

for $i, j = 1, \ldots n$. Then there exists an elementary *BT* Eq. (2.4) with $n \times n$ matrices $R(x)$ and $R(x)^{-1}$ being linear in x with coefficients expressed in terms of the following matrix A:

$$A = (l_i - l_j + 1) \left\{ \sum_{k=1}^{m} (S U_k S^{-1})_{ij} \right\}^{-1} S^{-1} E(i, j) S$$

for $E(i,j)_{kl} = \delta_{ik} \delta_{jl}$. The formulae for $R(x)$ and $U_j^{(r)'}$ are the following

$$R(x) = I + A(x - b), \qquad R(x)^{-1} = I - A(x - b), \text{ and}$$

$$U_j'^{(r)} = [I - (a_j - b)A] U_j^{(r)} [I + (a_j - b)A].$$

Expressions for *BTs* similar to that given by Lemma 2.5 can be given in "nonelementary" cases as well. For example, as we know (cf. [14,19]) *BTs* depend not only on integers (added to local multiplicities) but also on continuous parameters: matrices C_j. These continuous parameters appear in the following expression of *BT* with two different points $x_1, x_2 \neq \infty$ where multiplicities are changed by a unit.

Lemma 2.6: Starting from a fundamental solution $Y(x)$ we construct its *BT*

$$Y'(x) = Y(x) R(x)$$

corresponding to two distinct finite points x_1, x_2 and two arbitrary vectors $\vec{C}_1$ and $\vec{C}_2$. Namely, $R(x)$ and $R(x)^{-1}$ both have only a single simple pole. We have

$$R(x) = I + \frac{x_1 - x_2}{x - x_1} P \, , \qquad R(x)^{-1} = I - \frac{x_1 - x_2}{x - x_2} P$$

for a projection operator P, $P^2 = P$. The one dimensional projector P is defined by two vector solutions of Eq. (2.1):

$$\vec{F}_2 = \vec{C}_2 \, Y(x_2)^t \, , \qquad \vec{F}_1^{\,t} = \{Y(x_1)^t\}^{-1} \vec{C}_1^{\,t} \, .$$

Then the projector P is defined as

$$P = \vec{F}_1^{\,t} \cdot \vec{F}_2 \, \frac{1}{(\vec{F}_1, \vec{F}_2)} \, .$$

The *BTs* described by Lemma 2.6 iterated successively at different points x_i generate arbitrary multi-point Padé approximations to systems of functions satisfying l.o.d.e. From this point of view the effective construction of Padé approximations is performed. However, a simple and general expression of Lemma 2.6 does not really give sufficient information of arithmetic structure of Padé approximations. The reason for this is the necessity to solve actually Eq. (2.1) in order to construct P. This is in a sharp contrast with Lemma 2.5, where only coefficients of equations Eq. (2.1) were used. Only analytic asymptotics of Padé approximations can be established using analytic methods and iterations of *BTs*. We come here to nonlinear completely integrable lattice equations considered in [18, 19, 14]. The asymptotics are, of course, determined by the free energy of the corresponding lattice model. However, arithmetic properties (convergence in the p-adic domain) are to be analyzed separately in each case. Recurrences provided by elementary *BTs* are the main tool in this arithmetic analysis.

3. SEQUENCES OF RATIONAL APPROXIMATIONS

We study diophantine approximations to irrational numbers that are values of the logarithmic (or inverse trigonometric) function using linear recurrences supplied by formulas of *BTs* of section 2, that define numerators and denominators of rational approximations. The rational approximations under consideration are not the best possible rational approximations, and so we cannot completely describe continued fraction expansions [12] of even the simplest numbers like ln 2, π, $\pi/\sqrt{3}$ . . ., etc. Moreover, as numerical experiments show, continued fraction expansions to these numbers are random, and linear recurrences relating numerators and denominators of the best rational approximations cannot be incorporated into any consistent analytic scheme. It is our purpose, however, to find a "dense" family of rational approximations p_n/q_n to a number α which allows us to estimate the measure of the diophantine approximation of α. The word "dense" is used here as an equivalent of the following properties of q_n and p_n:

$$p_n,\, q_n \in Z; \qquad |q_n| \sim e^{an}; \qquad |q_n \alpha - p_n| \sim e^{bn}$$

with $a > 0$, $b < 0$ as $n \to \infty$

Obviously, these properties guarantee that α is irrational, and the measure of irrationality of α depends on the "density constant" $1 - a/b$. As we see later, under the conditions above, $|\alpha - p/q| > |q|^{-1 + a/b + \epsilon}$ for all $\epsilon > 0$ and $p,\, q \in Z$, $|q| \geq q_0(\epsilon)$.

The existence of "dense" families of rational approximations is established using appropriate explicit linear recurrences (see section 2) with coefficients expressed as functions of n. Asymptotics of solutions to these recurrences then define the "density constants" and hence, the measure of the diophantine approximation. Our main examples include studies of Padé approximations to powers of the logarithmic function [13] and the improved Padé approximations with more rapid convergence. The improved Padé approximations provide new, better bounds for the measure of diophantine approximations to constants of classical analysis. Most attention is devoted to two particular numbers, ln 2 and $\pi/\sqrt{3}$, for which new "dense" families of rational approximations are presented. New measures of diophantine approximations for these two numbers improve considerably upon those established in [2,3].

Proofs of irrationality of the number θ follow basically the same pattern: one establishes the existence of approximations p/q to θ such that

$$0 < |\theta - p/q| < \frac{1}{\psi(N)} \quad \textit{for } p,\, q \in Z; \quad |p|,\, |q| \leq N$$

and $\psi(N)/N \to \infty$ for $N \to \infty$. We present a general version of this pattern for finding the measures of diophantine approximations based on simultaneous approximations of powers of θ:

Lemma 3.1: Let θ be a complex number and let us suppose we have the system of linear forms

$$R_N = \sum_{i=0}^{m-1} \theta^i \cdot P_{i,N}$$

with rational integer coefficients $P_{i,N}$ satisfying the following properties:

(i) For $N \to \infty$ and $|\alpha| < 1$, $|R_N| \sim \alpha^N$ or $1/N \log |R_N| \to \log \alpha$ as $N \to \infty$.

(ii) For any $i = 0, 1, \ldots m-1$, $1/N \log |P_{i,N}| \leq \log \beta$ as $N \to \infty$.

Then for any rational integers p, q we have

$$|\theta - p/q| < |q|^{-\chi-\epsilon} \text{ for } |q| \geq q_0(\epsilon)$$

and for any $\epsilon > 0$. Here

$$\chi = (m-1)\left\{1 - \frac{\log \beta}{\log \alpha}\right\}$$

<u>Proof</u>: Let $|\theta - p/q| < |q|^{-\chi-3\epsilon}$, $|q| \geq q_0(\epsilon)$. We define

$$N = -\left[(1 + \epsilon)(m - 1)\frac{\log |q|}{\log \alpha}\right].$$

Let $\theta = \delta + p/q$, so

$$R_N = \sum_{i=0}^{m-1} \frac{p_i}{q_i} P_{i,N} + \delta \cdot C_m \cdot \max(|P_{i,N}|) + \mathbb{O}(\delta^2 q^{m-1}\max|P_{i,N}|)$$

Now

$$q^{m-1} R_N = A_N + \delta C' \cdot q^{m-1}\max(|P_{i,N}|) + \mathbb{O}(\delta^2 q^{m-1}\max(|P_{i,N}|))$$

for any integer $A_N \in \mathbb{Z}$. By a choice of N

$$|q^{m-1} R_N| < |q|^{-\epsilon/2}.$$

If $A_N = 0$, then $|\delta| \geq |R_N| \cdot |\beta|^{-N(1+\epsilon)}$ or $|\delta| \geq |q|^{-2\epsilon}$. If however $|A_N| \geq 1$, then $|\delta| \geq q^{-m+1} \beta^{-N(1+\epsilon)} \geq |q|^{-(\chi+3\epsilon)}$ QED.

In particular, if we approximate θ alone by a "dense" sequence of rational numbers, then we obtain a very useful corollary referred to above:

Corollary 3.2: Let us assume that there exists a sequence of rational integers P_n, Q_n such that

$$\log |P_n| \sim \bar{a} \cdot n, \qquad \log |Q_n| \sim \bar{a} \cdot n \tag{3.3}$$

as $n \rightarrow \infty$ and

$$\log |Q_n \theta - P_n| \sim \bar{b} \, n \tag{3.4}$$

as $n \rightarrow \infty$, where $\bar{b} < 0$. Then the number θ is irrational and, moreover,

$$|\theta - p/q| > |q|^{\bar{a}/\bar{b} - 1 - \epsilon}$$

for all rational integers p, q with $|q| \geq q_o(\epsilon)$.

If some family of rational approximations P_n/Q_n to θ is found, then the determination of sizes of P_n and Q_n as in Eq. (3.3) is called an "arithmetic" asymptotic, while the error of approximation, as in Eq. (3.4) is called an "analytic" asymptotic of rational approximation.

The most essential thing in the correct determination of "analytic" and "arithmetic" asymptotics of rational approximations is to establish the (linear) recurrence formulas connecting successive P_n and Q_n.

Knowing them, it is rather easy to determine "analytic" asymptotics of P_n and Q_n using the following Poincaré lemma:

Lemma 3.5: Let

$$\sum_{i=1}^{m} a_i(n) X_{n+i} = 0 \tag{3.6}$$

be a linear recurrence with coefficients depending on n such that

$$a_i(n) \rightarrow a_i \quad as \quad n \rightarrow \infty.$$

Let the roots of the "limit" characteristic polynomial

$$\sum_{i=0}^{m} a_i \lambda^i = 0$$

have distinct absolute values:

$$|\lambda_1| > \ldots > |\lambda_m|.$$

Then there are m solutions $X_n^{(j)}$: $j = 1, \ldots, m$ of Eq. (3.6) such that

$$\log |X_n^{(j)}| \sim n \log |\lambda_1|: \quad j = 1, \ldots m \text{ as } n \to \infty$$

and there is only one (up to a scalar multiplication) solution $\bar{X}_n$ of Eq. (3.6) such that

$$\log |\bar{X}_n| \sim n \log |\lambda_m|$$

Though the recurrences Eq. (3.6) are extremely important for the construction of diophantine approximations, they do not solve the problem of "arithmetic" asymptotics, since of $a_m(n)$ being polynomial in n, the expression for X_n tend to have complicated denominators. Only in very special cases these expressions have the denominator growing not faster than geometric progression.

4. PADE APPROXIMATIONS AND CORRESPONDING CONTIGUOUS RELATIONS

One of the situations, in which we know both the "arithmetic" and "analytic" a-symptotics, is the case when the numbers under consideration are values of (generalized) hypergeometric functions. The corresponding family of rational approximations is a specialization of a system of Padé approximations to (generalized) hyper-geometric functions [5, 6]. The "analytic" asymptotic is determined using the Riemann boundary value problem that is associated with the monodromy group of the corresponding differential equation. From the point of view of applications, we present the corresponding results for systems of functions with only the simplest singularities. For completeness we remind the readers of the definition of the Padé approximation (cf. [9]).

Definition 4.1: Let $f_1(x), \ldots, f_n(x)$ be functions analytic at $x = 0$ and let $m_1, \ldots m_n$ be non-negative integers (called weights). Then the polynomials $P_1(x), \ldots, P_n(x)$ of degree at most $m_1, \ldots, m_n$ are called Padé approximants to $f_1(x), \ldots, f_n(x)$ if

$$R(x) = \sum_{i=1}^{n} P_i(x) f_i(x)$$

has a zero at $x = 0$ of order $\geq \sum_{i=1}^{n} (m_i + 1) - 1$. The function $R(x)$ is called the remainder function.

The asymptotic of Padé approximations for logarithmic and similar functions [13, 5] is given in the following:

Theorem 4.2: Let $w_1, \ldots, w_n$ be distinct (mod Z) complex numbers. Let $f_1(x), \ldots, f_n(x)$ be one of the following systems of functions:

(i) $f_i(x) = (1 - x)^{w_i}$ for $i = 1, \ldots, n$;
(ii) $f_i(x) = {}_2F_1(1; w_i; c; x)$ for $i = 1, \ldots, n$;
(iii) $f_i(x) = \log (1 - x)^{i-1}$ for $i = 1, \ldots, n$.

Furthermore, let

$$R(x) = \sum_{i=1}^{n} P_i(x) f_i(x)$$

be the remainder function in the Padé approximation to $f_1(x), \ldots, f_n(x)$ with weights $m_1, \ldots, m_n$ at $x = 0$. Let

$$m_i = M + m_i^o \quad and \quad m_i^o/M \to 0 \quad as \quad M \to \infty \ for \ i = 1, \ldots, n.$$

Then, the asymptotics of $|R(x)|$ and $|P_i(x)|$ is determined everywhere in $\mathcal{C}$ using the following notations:

$$r_n^-(x) = \min\left\{ |1 - \zeta_n^j \sqrt[n]{1 - x}| : j = 0, \ldots, n - 1 \right\}$$

$$r_n^+(x) = \max\left\{ |1 - \zeta_n^j \sqrt[n]{1 - x}| : j = 0, \ldots, n - 1 \right\}$$

Then for any $x \neq 0$, 1, ∞ where $r_n^-(x) < r_n^+(x)$ we have

$$|R(x)| \cong r_n^-(x)^{nM} \left\{1 + \mathcal{O}(1/M)\right\}, \quad |P_i(x)| \cong r_n^+(x)^{nM} \left\{1 + \mathcal{O}(1/M)\right\}$$

for $i = 1, \ldots, n$ as $M \to \infty$. Here $\zeta_n^j = \exp\left\{ (2\pi i)j/n \right\}$.

One should note that for $n = 2$ and $f(x) = f_1(x)/f_2(x)$ being the Hilbert transform of a positive continuous measure with support on the finite interval:

$$f(x) = \int_a^b \frac{d\mu(x)}{x - x'}$$

similar results hold (the Szego theorem [7]).

Unlike "analytic" asymptotic, denominators of coefficients of Padé approximants $P_i(x)$ cannot be easily determined. To find denominators of Padé approximants and to determine "arithmetic" asymptotic, the recurrences relating Padé approximants with contiguous weights should be analyzed. These recurrences for systems of functions satisfying Fuchsian linear differential equations are called contiguous relations following Riemann [1]. We concentrate here only on the case of logarithmic functions as a particular case of Gauss's hypergeometric function.

The recurrences that are consequences of Gauss's contiguous relations between ${}_2F_1$ functions can be presented in the form:

$$F(m+1, l, k; z) = F(m, l, k-1; z) + z\,F(m, l, k; z)$$

$$F(m, l+1, k; z) = F(m, l, k-1; z) + (z - 1)\,F(m, l, k; z)$$

Specifications of initial conditions $F(1, 1, k; z)$ gives us $P_n(z)$, $Q_n(z)$, $R_n(z)$ in the Padé approximation problem for $\ln (1 - 1/z)$:

$$P_n(z) \ln (1 - 1/z) + Q_n(z) = R_n(z)$$

where $R_n(z) = \mathcal{O}(z^{-(n+1)})$ as $|z| \to \infty$. $P_n(z)$ and $Q_n(z)$ are polynomials of degree n and $n-1$ respectively.

(i) If

$$F_1(1, 1, k; z) = \frac{(-z)^{2-k} - (1 - z)^{2-k}}{k - 2} \quad \textit{for } k \neq 2$$

and $F_1(1, 1, 2; z) = \ln (1 - 1/z)$, then

$$R_n(z) \equiv F_1(n+1, n+1, n+2; z) \ ;$$

(ii) If $F_2(1, 1, k; z) = \delta_{k2}$, then

$$P_n(z) \equiv F_2(n+1, n+1, n+2; z) \ ;$$

(iii) If

$$F_3(1, 1, k; z) = \frac{(-z)^{2-k} - (1 - z)^{2-k}}{k - 2} \quad \textit{for } k \neq 2$$

and $F_3(1, 1, 2; z) = 0$, then

$$Q_n(z) \equiv F_3(n+1, n+1, n+2; z) \ .$$

These recurrences are usually substituted by a single three term recurrence:

$$(n + 1)\, X_{n+1} - (2n + 1)\, (z - 2)\, X_n + n\, z^2\, X_{n-1} = 0 \qquad (4.2)$$

satisfied by $X_n(z) = P_n(z)$, $Q_n(z)$, or $R_n(z)$. Nevertheless, the recurrence Eq. (4.2) conceals the most remarkable arithmetical properties of $P_n(z)$ and $Q_n(z)$:

(iv) Coefficients of $P_n(z)$ are rational integers, and

(v) Coefficients of $Q_n(z)$ are rational numbers with the common denominator dividing the least common multiplier of $1, \ldots, n$ which we denote by $\mathrm{lcm}(1, \ldots, n)$ (i. e. growing not faster than $\exp\{(1+\mathcal{O}(1))n\}$ as $n \to \infty$).

Statements iv) and v) do not follow immediately from Eq. (4.2), since one may suspect the denominator grows as $n!$ due to the division by $n + 1$ in order to find X_{n+1}. However, previous matrix recurrences together with Eq. (4.2) immediately imply iv) and v). Hence one needs *all* the recurrences simultaneously.

Comparing Eq. (4.2) with the classical recurrence relations for the Legendre polynomials [7] an immediate relation is established

$$P_n(z) \, z^{-n} = \bar{P}_n(x); \qquad x = 1 - 2/z$$

where $\bar{P}_n(x)$ is the Legendre polynomial of degree n:

$$\bar{P}_n(x) = \frac{1}{2^n \, n!} \frac{d^n}{dx^n} \left\{ (x^2 - 1)^n \right\}$$

Similarly

$$R_n(z) \, z^{-n} = \bar{Q}_n(x); \qquad x = 1 - 2/z$$

where $\bar{Q}_n(x)$ is the Legendre function of the second kind and consequently

$$R_n(z) = \frac{1}{2} P_n(z) \ln (1 - 1/z) + Q_n(z) \; .$$

Because of the connection with the classical theory of orthogonal polynomials one can write

$$Q_n(z) = \int_0^1 dx_1 \frac{P_n(z) - P_n(x_1)}{z - x_1} \; .$$

This shows that coefficients of $Q_n(z)$ do have denominators but they are in the form $1/m$ where $m = 1, \ldots .n$ which proves the properties of P_n and Q_n. The asymptotics of P_n and Q_n follows from Eq. (4.2) according to Poincaré's lemma.

The properties i) - v) of Padé approximants and remainder function in the Padé approximation problem for the logarithmic function enables us to estimate the measure of diophantine approximation of logarithms of rational numbers. For this we simply use Corollary 3.2. The "arithmetic" and "analytic" asymptotics are given by the properties iv) and v) and the recurrence Eq. (4.2) together with Lemma 3.5. We start with the number ln 2, which corresponds in the above mentioned scheme to $z = -1$. Corollary 3.2 on "dense" sequences of approximations immediately gives us as in [2]:

$$|q \ln 2 - p| > q^{-(3.660137409...+\epsilon)}$$

for the rational integers p, q provided that $|q| \geq q_1(\epsilon)$ for any $\epsilon > 0$.

Similarly, Padé approximations to the logarithmic function at points of a Gaussian field $\mathbb{Q}(i)$ give us a measure of the diophantine approximation to $\pi/\sqrt{3}$ [2]:

$$|q \, \pi/\sqrt{3} - p| > |q|^{-(7.30998634...+\epsilon)}$$

for the rational integers p, q provided that $|q| \geq q_2(\epsilon)$ for any $\epsilon > 0$. Moreover results of computer experiments show that [10]:

$$|\pi/\sqrt{3} - p/q| > |q|^{-8.31}$$

for rational integers p, q with $|q| \geq 2$. Historically the first explicit published result on the measure of irrationality of $\pi/\sqrt{3}$ belongs to Danilov [11]. Later similar bounds with different exponents were obtained independently by several researchers (including Wirsing and Beukers) [2, 8, 20]. All these results were based on the same system of Padé approximations to logarithmic function and, hence, on Hermite technique. The difference in exponent is explained by different accuracy in the computation of the asymptotics. We present now the recurrence leading to the exact asymptotics together with explicit solutions to the recurrences: the exponent 7.309. . . in the measure of irrationality of $\pi/\sqrt{3}$ is connected with the following nice three-term linear recurrence:

$$n(2n+1)(4n-3) X_{n+1} + (7 \times 16 \, n^3 - 7 \times 12 \, n^2 - 6 \, n + 5) X_n$$

$$+ (4n+1)(2n-1)(n-1) X_{n+1} = 0,$$

$X_n \sim x^n$ as $n \to \infty$ where $x^2 + 7x + 1 = 0$.

There are two solutions p_n and q_n of this recurrence such that

$$q_n \in Z \quad \text{for all } n$$

and

$$p_n \, \text{lcm}(1, \ldots, 2n) \in Z \text{ for all } n,$$

by Poincaré theorem

$$|p_n| \sim (2 + \sqrt{3})^{2n}, \quad |q_n| \sim (2 + \sqrt{3})^{2n}.$$

Then

$$|q_n \, \pi/\sqrt{3} - p_n| \sim (2 - \sqrt{3})^{2n} .$$

The expression for q_n ($\in \mathbb{Z}$) is rather simple:

$$q_n = \frac{1}{2^{2n-1}} \sum_{m=0}^{n-1} \binom{2n-1}{m} \binom{4n-2m-2}{2n-1} 3^{n-m} .$$

This number is indeed an integer and it can be represented in a number of different ways since it is connected with the values of Legendre polynomials. Namely

$$q_n = \sum_{m=0}^{2n-1} \binom{2n-1}{m} \binom{2n+m-1}{m} \rho^m, \quad \rho = \tfrac{1}{2} (i\sqrt{3} - 1) .$$

Similarly an expression exists for the "near integer" solution p_n:

$$p_n = \sum_{m=0}^{2n-1} \frac{(2n+m-1)!}{(m!)^2 (2n-1-m)!} \, \sigma(m) \, \rho^m - \sigma(2n-1) \, q_n$$

for $\sigma(m) = 1 + 1/2 + \dots 1/m$. This explains why the denominator of p_n divides lcm $(1, \dots 2n)$.

Asymptotics of p_n and q_n give a weak measure of irrationality of $\pi/\sqrt{3}$:

$$|q \, \pi/\sqrt{3} - p| > |q|^{\chi - \epsilon} \text{ for } |q| > q_3(\epsilon) \text{ and}$$

$$\chi = \frac{\ln (2 + 3^{1/2}) + 1}{\ln (2 - 3^{1/2}) + 1} = -7.3099863 .$$

Though the possibilities of Padé approximation to the logarithmic function are completely exhausted by the bounds above, the improved rational approximations to the logarithmic function give new measures of irrationality. In Section 5 we present new recurrences and new bounds of the measure of diophantine approximation to $\pi/\sqrt{3}$ and ln 2. These new measures are remarkable because for the first time they do not make use of Hermite's technique of 1873 [4].

Nevertheless Hermite's technique of Padé approximations to powers of the logarithmic function provide a nontrivial bound for the measure of irrationality of π. For this we use Padé approximations to the system of functions in iii) of Theorem 4.2 for $n = 6$, cf. [3]. The bound obtained so far is the following:

$$|q \, \pi - p| > |q|^{-18.88999444\dots}$$

for the rational integers p, q provided that $|q| \geq q_4(\epsilon)$ for any $\epsilon > 0$.

Similarly Padé approximants to binomial functions of i) in Theorem 4.2 provide the effectivization of Thue-Siegel theorem on diophantine approximations to certain classes of algebraic numbers. From the results of [3] it follows that for all algebraic numbers in the following fields, for example, an effectivization of Thue-Siegal theorem is valid:

$$\mathbb{Q}(\sqrt[3]{2}), \quad \mathbb{Q}(\sqrt[3]{3}), \quad \mathbb{Q}(\sqrt[3]{6}), \quad \mathbb{Q}(\sqrt[5]{2}), \quad \mathbb{Q}(\sqrt[5]{5})$$

For example, for any irrational number $\alpha \in \mathbb{Q}(\sqrt[3]{2})$ we have

$$|\alpha - p/q| > c(\alpha) \, |q|^{-2.429\ldots}$$

for arbitrary rational integers p, q. Here the constant $c(\alpha) > 0$ depends effectively on $H(\alpha)$ only.

The construction of "dense" sequences of approximations to algebraic numbers leads to linear recurrences satisfied by Padé-type polynomials. These polynomials are called Thue polynomials and are defined as follows. Let α be an algebraic number of degree m. Then for every integer $n \geq 0$ there exists (Thue) polynomials $A_n(x)$, $B_n(x)$ with rational integer coefficients of degree of at most $nm+m/2$ such that

$$\alpha A_n(x) - B_n(x) = (x-\alpha)^{2n+1} \, C_n(x)$$

where $C_n(x)$ is a polynomial. These polynomials provide a "dense" sequence of rational approximations p_n/q_n to α, if one starts with an additional good rational approximation $x=p/q$ to α and put $p_n/q_n = B_n(p/q)/A_n(p/q)$.

We consider cubic irrationalities α that are solutions of the algebraic equation $f(\alpha,1)=0$ for the binary cubic form $f(x,y)=ax^3 + bx^2y+cxy^2+dy^3$ with rational integers a, b, c, d.   With this form we associated an invariant $\mathcal{D}$ (discriminant): $\mathcal{D} = -27a^2d^2+18abcd+b^2c^2-4ac^3-4b^3d$, and two (quadratic and cubic) covariants

$$H(x,y) = -.25 \begin{vmatrix} \dfrac{\partial^2 f}{\partial x^2} & \dfrac{\partial^2 f}{\partial x \partial y} \\[2mm] \dfrac{\partial^2 f}{\partial x \partial y} & \dfrac{\partial^2 f}{\partial y^2} \end{vmatrix}, \quad G(x,y) = \begin{vmatrix} \dfrac{\partial f}{\partial x} & \dfrac{\partial f}{\partial y} \\[2mm] \dfrac{\partial H}{\partial x} & \dfrac{\partial H}{\partial y} \end{vmatrix}.$$

We present recurrences and expressions for the Thue polynomials associated with α in the homogeneous form, when an initial approximation to α is x/y (see $y=1$ above). The approximants $A_n(x,y)$ and $B_n(x,y)$ are homogeneous polynomials in x, y of degree $3n+1$ with rational coefficients and satisfy $\alpha A_n(x,y)-B_n(x,y)=(x-\alpha y)^{2n+1}C_n(x,y)$ for polynomial $C_n(x,y)$. The following three-term linear recurrence

$$(n+\tfrac{2}{3})Z_{n+1}(x,y)+(2n+1)G(x,y)Z_n(x,y)-9(3n+1)\mathcal{D}f(x,y)^2 z_{n-1}(x,y) = 0 \qquad (4.3)$$

has both $A_n(x,y)$ and $B_n(x,y)$ as its solutions. This enables us to express $A_r(x,y)$ and $B_r(x,y)$ in terms of Jacobi polynomials. Namely, two linearly independent polynomial solutions of the recurrence (4.2) are

$$X_n^+(x,y) = P_n^{(\nu,-\nu)}(-27\mathscr{D})^{-1/2}(\frac{G(x,y)}{f(x,y)})(\sqrt{-27\mathscr{D}}\, f(x,y))^n\binom{n-\nu}{n}^{-1}, \text{ and}$$

$$X_n^-(x,y) = P_n^{(-\nu,\nu)}(-27\mathscr{D})^{-1/2}(\frac{G(x,y)}{f(x,y)})(\sqrt{-27\mathscr{D}}\,)f(x,y))^n\binom{n-\nu}{n}^{-1}$$

for $\nu=1/3$, where $P_n^{(\alpha,\beta)}(x)$ are Jacobi polynomials.

Thue polynomials $A_n(x)=A_n(x,1)$, $B_n(x)=B_n(x,1)$ satisfy differential (and difference-differential) equations in addition to the three-term recurrence (4.2). If we put $f(x)=f(x,1)$, then $y(x)=A_n(x)$, $B_n(x)$ are two linearly independent solutions of the following Fuchsian l. o. d. e. of the second order:

$$f(x)y''(x)-2nf'(x)y'(x)-\frac{n(3n+1)}{2}f''(x)y(x) = 0 .$$

The sequence of dense approximations to α given by B_n/A_n gives the same exponent in the measure of irrationality for any irrational number from the cubic field $\mathbb{Q}(\alpha)$ of the form:

$$|\beta-p/q| > cH(\beta)^{-\text{æ}}|q|^{-\text{æ}}$$

for all irrational $\beta\in\mathscr{K}=\mathbb{Q}(\alpha)$ and an effective constant $c>0$ depending only on $\mathscr{K}$. For $\alpha=\sqrt[3]{3}$ we have æ$=2.4297...$(above), for $\alpha=\sqrt[3]{2}$ we have æ$=2.692661...$,etc.

The second class of numbers is given by the values of modular functions and is closely connected to Heegner-Stark studies of one-class discriminants. These numbers are very interesting because they admit few unusually large partial fractions in the initial part of their continued fraction expansions [21]. The modular function is

$$f(z)=q^{-1/48}\prod_{n=1}^{\infty} (1+q^{n-1/2}), \quad q=e^{2\pi iz} \text{ for Im } z>0.$$

The six appropriate numbers here are $f(\sqrt{d})$ for which $\mathbb{Q}(\sqrt{d})$ $(d<0)$, the imaginary quadratic field, has a class number $h=1$ and $|d|\equiv 3(mod 8)$: $d=-3, -11, -11, -43, -67, -163$. In all these cases we obtain the corresponding (effective) exponent æ<3 of the measure of irrationality of algebraic numbers α from the field $\mathscr{K}$, $\mathscr{K}=\mathbb{Q}(f)$, $f=f(\sqrt{d})$.

E.g. for $d=-11$, $f^3-2f^2+2f-2=0$ and æ$=2.326120...$; for $d=-19$, $f^3-2f-2=0$ and æ$=2.535262...$; ...; for $d=-163$, $f^3-6f^2+4f-2=0$ and æ$=2.882945...$.

G. Chudnovsky

5. NEW DIOPHANTINE APPROXIMATIONS

Here we present recurrences and their solutions that provide "dense" sequences of rational approximations to ln 2, $\pi/\sqrt{3}$ and π with "density constants" better than rational approximations given by the Padé approximation to the logarithmic function. Following Lemma 3.5, one can determine the "analytic" and "arithmetic" asymptotics of solutions of these recurrences. We follow the general scheme of Section 4 but with the matrix and scalar recurrences of more complicated form reflecting a different monodromy structure.

The new, better measure of irrationality of ln 2 is based on a new set of contiguous relations, reflecting the presence of apparent singularities. These recurrences are the following:

$$G(m+1, n, k; z) = G(m, n, k-2; z)+(2z-1)G(m, n, k-1; z)+(z^2-z)\,G(m, n, k; z)$$

$$G(m, n+1,k; z) = G(m, n, k-2; z) + (z - z^2)\,G(m, n, k; z)\,. \tag{5.1}$$

Solutions to these recurrences are completely determined by initial conditions $G(1, 1, k; z)$.

There are two kinds of initial conditions that determine sequences P_n and Q_n:

(i) If $G_1(1, 1, k; z) = \delta_{k2}$, then

$$P_n \equiv G_1(N_1, N_2, N_3; -1)$$

$$N_1 = [0.88\,n], \quad N_2 = [0.12\,n], \quad N_3 = n$$

(ii) If

$$G_2(1, 1, k; z) = \frac{(1 - z)^{2-k} - (-z)^{2-k}}{k - 2} \quad for\ k \neq 2\,,$$

and $G_2(1, 1, 2; z) = 0$. Then

$$Q_n \equiv G_2(N_1, N_2, N_3; -1)\,;$$

$$N_1 = [0.88\,n], \quad N_2 = [0.12\,n]\,, \quad N_3 = n$$

The specialization $z = -1$, above, corresponds to specialization of ln 2 as a value of the logarithmic function ln $(1 - 1/z)$. Two solutions P_n and Q_n of Eq. (5.1) with initial conditions i) and ii) satisfy the following familiar properties:

(i) P_n are rational integers.

(ii) Q_n are rational numbers whose denominators divide lcm $(1, \ldots .n)$.

Now the remainder $R_n = P_n \ln 2 - Q_n$ again arises from the recurrences Eq. (5.1) and is very small, when $n \to \infty$ so that Q_n/P_n determines a very good rational approximation to ln 2.

Numbers Q_n, P_n and R_n satisfy a scalar recurrence with coefficients depending on n which is not three-term recurrence but a four-term linear recurrence. The asymptotics of its solutions is determined according to Lemma 3.5 by roots of quartic polynomial. Numerically one has

$$\log |P_n| \sim 1.5373478. \ldots \times n , \quad \log |Q_n| \sim 1.5373478. \ldots \times n$$

and

$$\log |P_n \ln 2 - Q_n| \sim - 1.77602924\ldots \times n$$

as $n \to \infty$.

Hence one has the following measure of irrationality of ln 2:

$$|q \ln 2 - p| > |q|^{-3.2696549} .$$

The best exponent in the measure of diophantine approximation to ln 2 we can achieve this way requires more complicated recurrences than Eq. (5.1) corresponding to more apparent singularities. The properties iii) and iv) hold, but the asymptotics of P_n, Q_n and R_n are substituted by the following ones:

$$\log |P_n| \sim 1.93902189. \ldots \times n ,$$

$$\log |Q_n| \sim 1.93902189. \ldots \times n ,$$

and

$$\log |P_n \ln 2 - Q_n| \sim - 1.93766649. \ldots \times n$$

as $n \to \infty$. Hence the measure of diophantine approximation of ln 2 is:

$$|q \ln 2 - p| > |q|^{-3.134400029\ldots}$$

for rational integers p and q with $|q| \geq q'$.

The new better measure of irrationality is connected with more sophisticated recurrence relations.

The three-term recurrence relation determining the system of rational approximations to $\pi/\sqrt{3}$ is the following:

$$Y_{n+1}\, a\, b(3,\,0)\, b(3,\,1)\, b(3,\,-1)\, d^{-1} + Y_{n-1}\, a'\, b(1,\,-2)\, b(1,\,-3)b(1,\,-1)\, d'^{-1}$$

$$+ Y_n \Big\{\, b(1,\,0)\, b(3,\,-1)\, \{cb(2,\,1) - e\, b(3,\,1)\}\, d^{-1}$$

$$+ b(3,\,-2)\, b(1,\,-1)\{\, c'\, b(2,\,-3) - e'\, b(1,-3)\}\, d'^{-1} - b(2,\,-1)\Big\} = 0$$

where

$a = 2\,(4n + 3)\,(12n + 13)\,\big\{\,(12n + 13)\,z - 6\,(n + 1)\big\}\,,$
$a' = z\,(12n - 7) - (6n - 4)\,,$

$b(i,\,j) = b(i;\,3n+j,\,4n+j)\,,$
$b(1;n,\,m) = 4\,(n + 1)\,(2n + 3)\,(4n + 9)\,,$
$b(2;n,\,m) = (4n - 4m + 5)\,(4n + 7) + (1 - 2z)\,(4n + 5)\,(4n + 7)\,(4n + 9)\,,$
$b(3;n,\,m) = 4\,(m + 1)\,(4n - 2m + 7)\,(4n + 5)\,,$

$c = (12n + 13)^2\,(4n + 4) - 12\,(12n + 13)\,(n + 1)$
$\qquad - 6\,(n + 1)\,(6n + 7)\,(4n + 9)\,z^{-1}(1 - z)^{-1}\,,$

$c' = z\,(1 - z)\,(12n - 7) - \dfrac{(6n - 4)\,(6n - 3)\,(4n + 1)}{(12n - 7)\,(4n - 1) - 4\,(3n - 2)}\,,$

$d = c\, b(2,\,1)\, b(2,\,0) - c\, b(1,\,1)\, b(3,\,0) - e\, b(2,\,0)\, b(3,\,1)\,,$
$d' = c'\, b(2,\,-2)\, b(2,\,-3) - c'\, b(3,\,-3)\, b(1,-2) - e'\, b(2,\,-2)\, b(1,\,-3)\,,$

$e = 6\,(n + 1)\,(6n + 7)\,\big\{\,(12n + 13)\,z - 6\,(n + 1)\big\}\,z^{-1}\,(1 - z)^{-1}\,,$

$e' = \dfrac{(8n + 2)\,(4n - 3)\,\big\{\,(12n - 7)z - 2\,(3n - 2)\big\}}{(12n - 7)\,(4n - 1) - 4\,(3n - 2)}\,.$

Though complex looking, this recurrence can be combined from simpler matrix recurrence relations corresponding to contiguous relations.

Again, there are two linearly independent solutions of this recurrence: one "integer" solution $Y_n^{(in)}$ and another "almost integer" $Y_n^{(nonin)}$.

We present here, for example, the expression for the integer solution of the recurrence, expressed in the familiar form as a sum of products of binomial coefficients:

$$Y_n^{in} = \sum_{i_1,i_2=0;\ i_1+i_2<4n}^{3n} \binom{3n}{i_1}\binom{3n}{i_2}\binom{2(4n-i_1-i_2-1)}{4n-i_1-i_2-1}$$

$$\times \Big\{(2(4n - i_1 - i_2)-3)^{-1}\, z^{-n+i_2}(z - 1)^{3n-i_2}(-4)^{(i_1+i_2-4n)}\Big\}\,.$$

For the approximation to $\pi/\sqrt{3}$ one takes $z = -3$ and puts $X_n = 12^{[n/4]}\, Y_{[n/4]}$.

The approximations $X_n^{(in)}$ and $X_n^{(nonin)}$ possess the following properties:

(i) $X_n^{(in)} \in Z$; The denominators of $X_n^{(nonin)}$ divide the lcm $(1, \ldots, n)$.

(ii) As $n \twoheadrightarrow \infty$, one has

$$\log |X_n^{(in)}| \twoheadrightarrow -1.66439185\ldots \times n$$

$$\log |X_n^{(nonin)}| \twoheadrightarrow -1.66439185\ldots \times n$$

and

$$\log | \pi/\sqrt{3} \, X_n^{(in)} + X_n^{(nonin)}| \twoheadrightarrow 2.2006689\ldots \times n \; .$$

Hence the measure of irrationality of $\pi/\sqrt{3}$ is considerably better than it was before

$$| q \, \pi/\sqrt{3} - p| > |q|^{-4.817441679\ldots}$$

as $|q| \geq q_0$.

The best measure of irrationality of $\pi/\sqrt{3}$ does not differ considerably from the one given above, but still this is the best measure of irrationality of a number connected with π. In fact one establishes the existence of a sequence with the following properties.

(i) Numbers P_n and Q_n are rational numbers, where the P_n are integers and the denominators of Q_n divide lcm$(1, \ldots, n)$.

(ii) We have

$$\log |P_n| \sim 2.191056949\ldots \times n \, , \quad \log |Q_n| \sim 2.191056949\ldots \times n$$

as $n \twoheadrightarrow \infty$.

(iii) The numbers Q_n/P_n approximate $\pi/\sqrt{3}$:

$$\log |P_n \, \pi/\sqrt{3} - Q_n| \sim -1.6658281013\ldots \times n$$

as $n \twoheadrightarrow \infty$.

This gives us the following measure of irrationality of $\pi/\sqrt{3}$.

$$|q \, \pi/\sqrt{3} - p| > |q|^{-4.792613804}$$

for all integers p and q with $|q| \geq q_0$. The expression for P_n is somewhat similar to the one given above.

One can generalize expressions of P_n to the sums of products of a large number of binomial coefficients, with the substitution of three-term linear recurrence by multi-term linear recurrences. This way we improve the measure of irrationality of π^2 (or of $\zeta(2)$ belonging to Apery) to the following:

$$|\pi^2 q - p| > |q|^{-7.5} ,$$

so that $|\pi q - p| > |q|^{-16}$ for rational integers p, q with $|q| \geq q_1$.

REFERENCES

[1]. B. Riemann, *Oeuvres Mathematiques*, Blanchard, Paris, 1968, pp. 353 - 363.

[2]. G. V. Chudnovsky, C. R. Acad. Sci. Paris, 288 (1979) A-607--A-609.

[3]. G. V. Chudnovsky, C. R. Acad. Sci. Paris, 288 (1979) A-965--A-967.

[4]. Ch. Hermite, C. R. Acad. Sci. Paris, 77 (1873) 18-24, 74-79, 226.

[5]. G. V. Chudnovsky, Cargese Lectures in *Bifurcation Phenomena in Mathematical Physics and Related Topics*, D. Reidel, Boston, 1980, 448-510.

[6]. G. V. Chudnovsky, J. Math. Pure Appl. 58 (1979) 445-476

[7]. G. Szegö, *Orthogonal Polynomials*, AMS, Providence, 1938.

[8]. K. Alladi, M. L. Robinson, Lecture Notes in Math., Vol. 751, Springer, 1979, 1-9.

[9]. G. A. Baker, Jr. *Essentials of Padé Approximants*, Academic Press, 1975.

[10]. G. V. Chudnovsky, Lecture Notes in Physics, Vol. 120, Springer 1980, 103-150.

[11]. V. Danilov, Math. Zametki, 24 (1978) No. 4.

[12]. A. Y. Khintchine, *Continued Fractions*, University of Chicago Press, 1964.

[13]. K. Mahler, Phil. Trans. Roy. Soc. London, 245A (1953) 371-398.

[14]. D. V. Chudnovsky, see this volume.

[15]. M. Lüscher, Nuclear Physics B135 (1978) 1.

[16]. D. V. Chudnovsky and G. V. Chudnovsky, Lett. Math. Phys. 4 (1980) 373.

[17]. D. V. Chudnovsky and G. V. Chudnovsky, J. Math. Pure Appl. (1982), No 1, 1.

[18]. D. V. Chudnovsky and G. V. Chudnovsky, Phys. Lett. 82A (1981) 271.

[19]. D. V. Chudnovsky and G. V. Chudnovsky, Phys. Lett. 87A (1981) 325.

[20]. F. Beukers, Lecture Notes in Math., Vol 888, Springer, 1981, 90.

[21]. H. M. Stark, in *Applications of Computers in Algebra*, Academic Press, 1970, 21.

BACKLUND TRANSFORMATIONS AND GEOMETRIC AND COMPLEX-ANALYTIC BACKGROUND FOR CONSTRUCTION OF COMPLETELY INTEGRABLE LATTICE SYSTEMS

D. V. Chudnovsky

Department of Mathematics
Columbia University
New York, NY 10027

Abstract

Topological and complex analytic background is given for the introduction of the concept of Backlund transformation (*BT*). Backlund
transformation is introduced from the point of view of isomonodromy
deformation. The axiom of permutability of *BTs* is equivalent to the
Baxter-Zamolodchikov introduction of completely integrable quantum
systems with factorized *S*-matrices. Many important concrete examples of classical and quantum completely integrable systems of field
theory and statistical mechanics are considered.

PART 1
COMPLETELY INTEGRABLE SYSTEMS

1. INTRODUCTION

The purpose of this paper is to examine the general structure of completely integrable classical and quantum systems in dimensions one and two, associated with iso-

spectral deformation. Our purpose is to examine the underlying structure of these systems, symmetries and group transformations that bring about complete integrability. We look on completely integrable systems with arbitrary operator variables (following our early suggestion [1]), so that quantum and classical systems are considered simultaneously. Completely integrable systems under investigation have symmetries associated with infinite dimensional extensions $g \otimes \mathscr{C}[\lambda, \lambda^{-1}]$ of finite dimensional Lie algebras g. The group of transformations (that are canonical transformations of a given system) is a group of Backlund transformations. We define the whole concept of Backlund transformations (*BTs*) from the point of view of complex analysis. *BTs* are considered as linear transformations with coefficients holomorphic in (a spectral) variable λ determined by positions and multiplicities of zeros and poles. These *BTs* are typically applied to the solutions of the Riemann boundary value problem [2, 3] equivalent to the original spectral problem.

In Section 2 we introduce the necessary abstract topological and categorial background for the introduction of completely integrable systems and the rules of composition of *BTs*. These laws of composition are formulated as laws of associativity in a category of vector spaces. In particular, axioms of symmetric monoidal categories provide geometric interpretation of axioms of factorization Baxter-Zamolodchikov for *S*-matrix elements. These axioms are later in the paper reformulated in the form of Baxter's lemma, which is a general statement describing completely integrable systems using the language of transfer matrices, depending on (the spectral parameter) λ. Using this lemma we propose the method of construction of completely integrable lattice systems associated with finite dimensional Lie algebras. For example, quantum mechanical systems generalizing *XXX* and *XXZ* models of statistical mechanics are presented; both of them correspond to the group $SO(3)$. Relationship with Onsager-Baxter models of statistical mechanics is presented.

In Section 3 *BTs* for solutions of the Riemann boundary value problem are introduced and interpreted as isomonodromy deformations that add integers to local multiplicities of matrix solutions. As a particular example we have chosen *BTs* of solutions of matrix Dirac equation. Iterative applications of *BTs* are used as the method of decomposition of two dimensional completely integrable systems into a sequence of lattice completely integrable systems. One can substitute isospectral deformation system of partial differential equations (of Korteweg-deVries (KdV) or non-linear Schrödinger type) by equivalent lattice completely integrable systems which also approximate the original system of p.d.e. Example of nonlinear operator Schrödinger equation is described in detail. The Toda lattice system (non-Abelian Toda lattice) occurs in a particular case of a lattice system associated with the non-linear Schrödinger equation. It should be noted that some classes of completely integrable difference-differential equations can be represented as a result of applications of *BTs* to two dimensional isospectral deformation equations. The present paper is a preliminary report in the area of Backlund transformations in their relation with quantum systems through geometric generalization of *S*-matrices.

2. CATEGORIAL FORMULATION OF FACTORIZATION AXIOMS FOR S-MATRICES

The concept of a "factorized" *S*-matrix and similar notions in the theory of completely integrable classical and quantal Hamiltonian systems [8, 11] can be explained in the most general way from the abstract point of view of category theory.

The key abstract notion is the notion of monoidal Abelian category [7], i.e., a category possessing associativity and commutativity. All considerations below are the result of the joint work of L. Breen, G. Chudnovsky and D. Chudnovsky. We present here only part of the axiomatic description leaving the detailed exposition to separate comprehensive publications.

We refer readers to [7] for general definitions of category theory. Categories corresponding to factorized *S*-matrices are monoidal symmetric categories (without identity). These categories $B = \langle B, \otimes, \alpha, \gamma \rangle$ are characterized by the existence of a bifunctor $\otimes: B \times B \to B$ and two natural isomorphisms α, β. Explicitly,

$$\alpha = \alpha_{x,y,z}: x \otimes (y \otimes z) \cong (x \otimes y) \otimes z$$

and

$$\gamma_{x,y}: x \otimes y \cong y \otimes z$$

are natural for all x, y, $z \in B$. Isomorphisms α and γ express "associativity and commutativity up to isomorphism" and one assumes commutativity of two diagrams [see 7, chap. 7] describing compatibility of α and γ. The first diagram is called pentagon:

$$
\begin{array}{ccccc}
x \otimes \big(y \otimes (z \otimes v)\big) & \xrightarrow{\ \alpha\ } & (x \otimes y) \otimes (z \otimes v) & \xrightarrow{\ \alpha\ } & \big((x \otimes y) \otimes\big) z \otimes v \\
\Big\downarrow {\scriptstyle 1 \otimes \alpha} & & & & \Big\uparrow {\scriptstyle \alpha \otimes 1} \\
x \otimes \big((y \otimes z) \otimes v\big) & & \xrightarrow{\hspace{4cm}} & & (x \otimes (y \otimes z)) \otimes v
\end{array}
\tag{2.1}
$$

which is to be commutative for all x, y, z, v, $\in B$. The second diagram called hexagon assures commutativity of the diagram

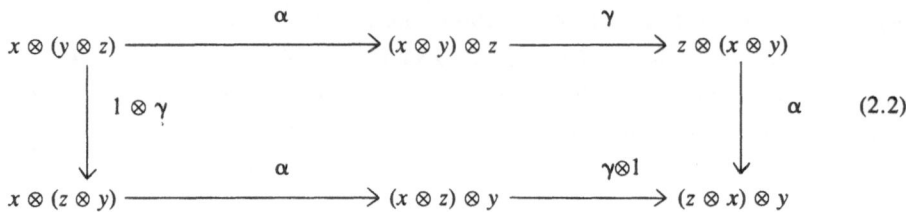

for all $x, y, z \in B$. The last axiom of symmetry is

$$\gamma_{x,y} \circ \gamma_{y,x} = 1 \qquad\qquad (2.3)$$

for all $x, y \in B$. The axiom 2.3 is responsible for the unitarity of the S-matrix.

This general definition of monoidal symmetric category is applied in the case when B is a category of vector spaces and the bifunctor $\otimes: B \times B \to B$ is a natural operation of tensor products of vector spaces. Unlike the case in topological applications, our category B is strictly associative, which means that the isomorphism $\alpha = \alpha_{x,y,z}: x \otimes (y \otimes z) \cong (x \otimes y) \otimes z$ is fixed as an identity isomorphism for arbitrary vector spaces $x, y, z \in B$. The isomorphism $\gamma = \gamma_{x,y}: x \otimes y \cong y \otimes x$ is the one that determines the S-matrix. From this point of view we can call vector spaces $x, y, v, z, \in B$ (objects of B) states of particles. Typically B is a category of finite dimensional vector spaces, and if v is an n-dimensional vector space from B, then generators of v: $A_1, \ldots, A_n$ are describing n-plets of particles. If v' is another vector space from B with generators $A'_1, \ldots, A'_m$, then the isomorphism

$$\gamma_{v, v'}: v \otimes v' \cong v' \otimes v$$

can be called an abstract S-(scattering) matrix, describing the scattering of an n-plet $(A_1, \ldots, A_n)$ on an m-plet $(A'_1, \ldots, A'_m)$. In the chosen bases $(A_1, \ldots, A_n)$ and $(A'_1, \ldots, A'_m)$ the S-matrix $\gamma_{v,v'}$ is represented as an $nm \times nm$ matrix with elements $S_{ij,kl}$, corresponding to bases $A_i \otimes A'_j$ and $A'_k \otimes A_l$ in $v \otimes v'$ and $v' \otimes v$, respectively:

$$\gamma_{v,v'} (A_i \otimes A'_j) = \sum_{k,l} S_{ij,kl} A'_k \otimes A . \qquad (2.4)$$

Because of strict associativity the pentagon diagram 2.1 is trivial and the hexagon axiom 2.2 turns into a triangle (the notion and definition belongs to L. Breen):

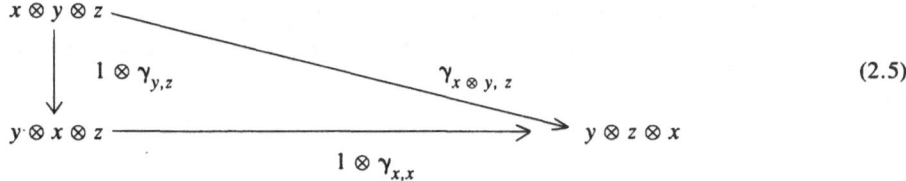

$$(2.5)$$

The diagram (2.5) is denoted by $[x; y, x]$.

In the general case of monoidal symmetric categories the coherence theorem of MacLane [7, Chap. 7] states "every diagram commutes" provided pentagon, hexagon and unitarity axioms (2.1), (2.2) and (2.3) are satisfied. In particular, in the case of strictly associative category, the axiom of triangle 2.5 and the unitarity axiom (2.3) imply all the other associativity-commutativity axioms. This allows us to call the triangle axiom (2.3) "factorization axiom" for an abstract S-matrix $\gamma_{x, y}$. Apparently, readers familiar with Zamolodchikov's interpretation of factorization [8, 9, 10] will find triangular axiom (2.5) different from the traditional "hexagon" (not the hexagon of (2.2)) form of factorization equations. This traditional "hexagon" is based on two evaluations of a natural isomorphism $x \otimes y \otimes z \to z \otimes y \otimes x$ (or a law described by J. L. Verdier as $ABC = CBA$ [13]).

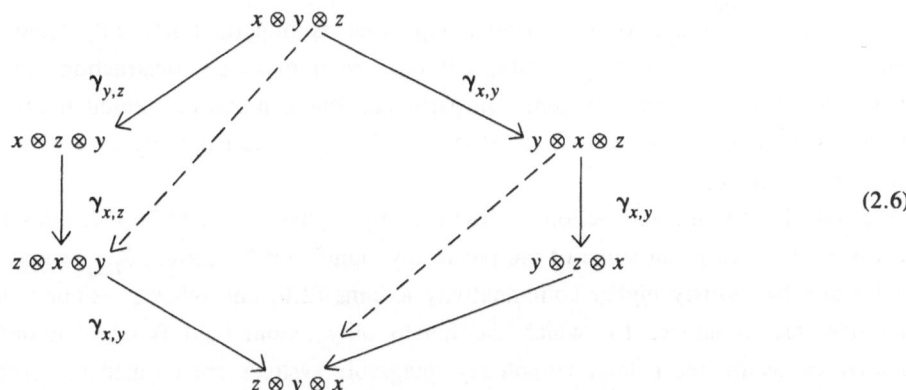

$$(2.6)$$

The assumption of commutativity of (2.6) is the traditional way of formulation of factorization axiom of S-matrix. However, the commutativity of the diagram (2.6) is a consequence of the triangle axiom (2.5) (as a coherence theorem of MacLane states). For this deduction one should use functoriality of $\gamma_{x,y}$. In fact, dividing (2.6) into two triangles and one rectangle, one sees that the commutativity of triangles follow from (2.5) and the commutativity of rectangle is the consequence of functoriality of γ. The functoriality of γ itself, even without the axiom (2.5) is a very powerful axiom, which we are expressing later under the name of the Baxter Lemma.

Remark 2.7: Instead of assuming functoriality, one may equivalently deduce factorization axioms (2.6) from triangle axiom (2.5) and the following new commutative diagram.

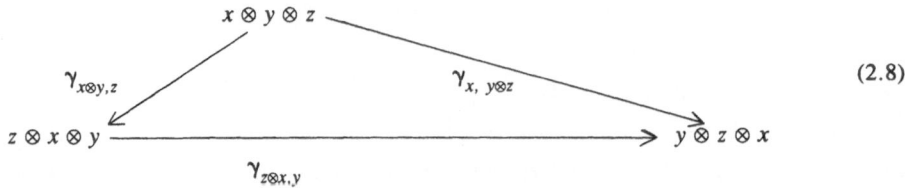

$$(2.8)$$

The abstract definition of S-matrix requires explanation on its connection with the concept of S-matrices associated with quantal completely integrable systems that had been studied before [8, 9, 12, 10, 15]. The definition (2.4) of S-matrix elements shows exact correspondence with "Zamolodchikov algebra" [8, 10, 54]. The only difference with the earlier definitions of factorization equations and the abstract definition (2.6) is that in [8,10] only generating vector spaces $V(\Theta)$ corresponds to the multiplet of particles $A_i = A_i(\Theta)$: $i = 1, \ldots n$ moving with speed $\mathrm{tg}\Theta$ at a straight line. The S-matrix $S_{ij,kl} = S_{ij,kl}(\Theta, \Theta')$ then corresponds to a matrix of all two-body scattering of a multiplet $A_i(\Theta)$: $i = 1, \ldots, n$ at $A'_j(\Theta)$: $i = 1, \ldots, m$. The natural assumption is, additionally, $S_{ij,kl} = S_{ij,kl}(\Theta - \Theta')$.

The abstract definition of factorization equations is important not only from the point of view of mathematical clarity, but because it expresses obstructions to the satisfaction of factorization equations. In particular, one expresses Zamolodchikov tetrahedron factorization axioms [14] for strings as the higher associativity axiom instead of triangular axiom (2.5).

Not only three-string interaction is described by higher commutativity axioms like (2.8) instead of (2.5), but some of the physically significant S-matrices $S_{ij,kl}$ which are not factorizable, satisfy higher commutativity axioms (2.8) and others. Among such S-matrices are S-matrices for which the functoriality axiom (and Baxter lemma) is satisfied so, as we see below, completely integrable systems are defined but factorization equations are not satisfied. For example, S-matrices with Abelian symmetries corresponding to Abelian varieties of higher dimension [16] seem to be not necessarily factorized but satisfy higher commutativity axioms reflecting the existence of higher obstructions.

3. DEFINITION OF BACKLUND TRANSFORMATIONS

The definition of BT can be roughly presented as a linear transformation with coefficients rational in λ, acting on multivalued $n \times n$ matrix solutions of a certain

Riemann boundary value problem. This general definition of *BT* as isomonodromy deformation is given for an arbitrary multi-valued $n \times n$ solution $\phi(\lambda)$ of the Riemann boundary value problem in the λ-plane. This definition is usually formulated in terms of local properties (local multiplicities) of global solution $\phi(\lambda)$ and consists of adding apparent singularities and changing the local multiplicities of different branches of $\phi(\lambda)$ by integers. This procedure was described in classical language in [19, 20, 21] as contiguous relations in [22, 23], as "addition of integers" [24, 25] or "dressing of the vacuum" [2], or for equations of the Painleve type, as "Schlesinger transformations" [26, 27]. We describe the space of data for the *BT*, which turns out to have much more complex structure, then just change in the local multiplicities.

The $n \times n$ matrix function $\phi(\lambda)$ henceforth is a solution of the Riemann boundary value problem in the λ-plane. In applications $\phi(\lambda)$ as function of parameters $x, t, \ldots$ satisfies an ordinary differential equation (o.d.e) with coefficients being rational in λ. At a point $\lambda = a$, where multivalued matrix $\phi(\lambda)$ is singular, the behavior of $\phi(\lambda)$ is determined following [19-21], by the exponential matrices describing the character of regular and irregular singularities:

$$\phi(\lambda) = \tilde{\phi}(\lambda) \cdot (\lambda - a)^{W_a} \cdot H_a(\lambda) , \tag{3.1}$$

Here $\tilde{\phi}_a(\lambda)$ is holomorphic and invertible at $\lambda = a$, the term $(\lambda - a)^{W_a}$ corresponds to a regular singularity of $\phi(\lambda)$ at $\lambda = a$ and W_a is called a regular *exponential* matrix, while $H_a(\lambda)$ corresponds to irregular singular behavior at $\lambda = a$. Typically, $H_a(\lambda)$ is a combination of exponents in $(\lambda - a)^{-1}$ and is not unique, but differs by right matrix multipliers (called Stokes multipliers) in different sectors in the neighborhood of a. Under additional assumption (satisfied for physical models below), W_a and H_a can be simultaneously reduced to diagonal form. Then locally at $\lambda = a$ we have:

$$\phi(\lambda) = \tilde{\phi}(\lambda) \cdot (\lambda - a)^{W_a^{(0)}} \cdot \exp\left\{\sum_{j=1}^{q_a} W_a^{(j)} (\lambda - a)^{-j}\right\} \cdot C_a \tag{3.2}$$

where $\tilde{\phi}_a(\lambda)$ is holomorphic and invertible at a; $W_a^{(0)}$ and $W_a^{(j)}$ $(j = 1, \ldots q_a)$ are diagonal matrices and C_a is so called connection matrix. Moreover norming for $\phi(\lambda)$ at infinity is usually chosen in the form $C_\infty = I$.

The *BT* is in its general form a linear transformation of $\phi(\lambda)$ with rational function coefficients in λ:

$$\phi'(\lambda) = L(\lambda) \phi(\lambda) \tag{3.3}$$

for $n \times n$ matrix $L(\lambda)$ rational in λ.

In general $\phi(\lambda)$ satisfies the Riemann boundary value problem in λ on a Riemann surface g_A, matrix $L(\lambda)$ is rational on a Riemann surface g_A. *BTs* Eq. (3.3) are defined according to their action on monodromy data of $\phi(\lambda)$. Any linear transformation Eq. (3.3) may change the regular exponential matrices W_a into W'_a:

$$\phi'(\lambda) = \tilde{\phi}'(\lambda) (\lambda - a)^{W'_a}$$

for at most regular singularity at $\lambda = a$. In general, W'_a is not a diagonal matrix. However, as local monodromy of $\phi'(\lambda)$ is the same as $\phi(\lambda)$, $\exp(2\pi i\, W'_a) = V_a$, where V_a is a local monodromy matrix of $\phi(\lambda)$ at $\lambda = a$. Hence the most general expression for W'_a is: $W'_a = \log V_a/(2\pi i)$.

E. g., let $\lambda = a$ be an apparent singularity of $\phi'(\lambda)$, i.e., a regular singularity of $\phi'(\lambda)$, but a regular point of $\phi(\lambda)$. Then $V_a = I$ and general expression for W'_a turns out to be

$$W'_a = S_a^{-1} \cdot \mathrm{diag}(m_1, \ldots, m_n) \cdot S_a \,,$$

where S_a is an arbitrary nonsingular $n \times n$ matrix. In this expression S_a describes continuous parameters (moduli) of *BT* data and $\tilde{m} = (M_1, \ldots, m_n)$ is a sequence of rational integers. *These integers are added to local multiplicities of the old matrix* $\phi(\lambda)$ *at* $\lambda = a$ [22, 24, 25, 31].

Hence, according to [37] the space of *BT* data is determined by the isomonodromy deformation conditions:

$$\exp\{2\pi i\, W_a\} = \mathrm{const.}$$

for all a (being a singular or regular point of $\phi(\lambda)$). According to [32], this condition can be expressed as follows. Let W_a be reduced to its Jordan normal form

$$W_a = S_a^{-1} \cdot (J_{\rho_1}, \ldots, J_{\rho_l}) \cdot S_a \,, \tag{3.4}$$

where J_{ρ_i} is a Jordan block of the size k_i corresponding to the eigenvalue of ρ_i, and $k_1 + \ldots + k_l = n$. While Jordan normal form is unique, the choice of S_a is not. E.g., if W_a have a nontrivial centralizer G_{W_a} and S_a reduces W_a to a normal form, then for an invertible $S \in G_{W_a}$, $S_a \cdot S$ again reduces W_a to a normal form.

The space of all possible *BT* data of $\phi(\lambda)$ is described now as a space of all possible regular exponential matrices W'_a such that

$$W'_a = S_a^{-1} \cdot (J_{\rho_1} + m_1^a I_{k_1}, \ldots, J_{\rho_l} + m_l^a I_{k_l}) \cdot S_a$$

where

(i) $m_1^a, \ldots, m_l^a$ are integers and I_{k_i} are $k_i \times k_i$ unit matrices, and

(ii) S_a are arbitrary matrices satisfying Eq. (3.4).

Remark 3.5: It was explained in [37] that for a a regular singularity of $\phi(\lambda)$, BT data consists of

(i)"moduli", that is arbitrary invertible matrices S_a, and

(ii)integers $m_1^a, \ldots , m_n^a$ interpreted as orders of poles of branches of $\phi'(\lambda)$ at an apparent singularity of $\lambda = a$ of $\phi'(\lambda)$.

We mentioned that BT $\phi'(\lambda)$ with the given data $(W'_a) \in \text{Ln } (W_a)$ exists for a generic $\phi(\lambda)$.

The necessary condition for the existence of BT with the data i) and ii) is that

$$\sum_a \sum_{i=1}^{n} m_i^a = 0$$

Following this description, the BT $B_{\mathcal{B}}(\phi(\lambda)) \equiv \phi'(\lambda)$ is defined as a linear transformation

$$\phi'(\lambda) = L(\lambda) \, \phi(\lambda)$$

such that

$$\phi'(\lambda) = \bar{\phi}'(\lambda) \, (\lambda - a)^{W'_a} = \bar{\phi}' \, (\lambda - a)^{\text{diag } \vec{m}^a} \cdot S_a \tag{3.6}$$

for an apparent singularity a, and for an essential singularity $\lambda = a$, the local expansion for $\phi'(\lambda)$ is the same as in Eq. (3.2): $S_a = C_a$, but local multiplicities of $W_a^{(0)}$ are changed to $W_a^{(0)} + diag \, \vec{m}^a$. The "group" of BT is an Abelian group (infinite free Abelian group), if one *fixes* continuous part of BT data ii) -- connection matrices S_a. However, the continuous part of BT data turns the "group" of BT into a highly non-Abelian object. We present in sections 5 and 6 the simplest examples of these "groups".

Lemma 3.7: Let $\phi(\lambda)$ satisfy o.d.e.

$$\frac{\partial \, \phi(\lambda)}{\partial \, x} = U(\lambda) \, \phi(\lambda) \, ,$$

and let $\phi'(\lambda)$ be a BT 3.1 of $\phi(\lambda)$ for which BT data are x-independent. Then $\phi'(\lambda)$ satisfies l.o.d.e.

$$\frac{\partial \, \phi'(\lambda)}{\partial \, x} = U'(\lambda) \, \phi'(\lambda) \, .$$

Here $U'(\lambda)$ is regular at those points in the λ-plane where $U(\lambda)$ is regular. In particular, if $U(\lambda) = A \cdot \lambda + U$ and A is diagonal, then $U'(\lambda) = A \cdot \lambda + U'$.

The proof of the first part of the lemma is based on the expression of

$$\frac{\partial}{\partial x} \phi'(\lambda) \cdot \phi'(\lambda)^{-1} = U'(\lambda)$$

in the neighborhood of a regular point $\lambda = a$ of $U(\lambda)$. One may assume that at this point $\lambda = a$, $\phi(\lambda)$ is itself regular. The second part of the lemma follows from an explicit expression of $U'(\lambda)$

$$U'(\lambda) = \frac{\partial L(\lambda)}{\partial x} \cdot L(\lambda)^{-1} + L(\lambda) U(\lambda) L(\lambda)^{-1} .$$

4. AN EXAMPLE OF BACKLUND TRANSFORMATIONS

Let us give a simple expression for the Backlund transformation of $\phi(\lambda)$, assuming that we apply this transformation to regular points μ_1, μ_2 of $\phi(\lambda)$, where $\mu_1 \neq \mu_2$. In other words, we add a zero at one regular point and a pole at another regular point of $\phi(\lambda)$, cf. [2].

We consider the Backlund transformation of $\phi(\lambda)$ applied to μ_1 and μ_2. The transformed $\phi'(\lambda) = B_{\mathscr{B}}(\phi(\lambda))$ has the form:

$$\phi'(\lambda) = \mathscr{L}(\lambda) \cdot \phi(\lambda) \tag{4.1}$$

and

$$\mathscr{L}(\lambda) = I + \frac{\mu_1 - \mu_2}{\lambda - \mu_1} P \quad , \quad \mathscr{L}(\lambda)^{-1} = I - \frac{\mu_1 - \mu_2}{\lambda - \mu_1} P , \tag{4.2}$$

and P is a projection operator, $P^2 = P$. Let $\phi(\lambda)$ satisfy the equation in x

$$\frac{\partial \phi(\lambda)}{\partial x} = U(\lambda) \phi(\lambda) \tag{4.3}$$

where $U(\lambda)$, as a function of λ, doesn't have singularities at $\lambda = \mu_1$ or $\lambda = \mu_2$. Then $\phi'(\lambda)$ satisfies the similar differential equation

$$\frac{\partial \phi'(\lambda)}{\partial x} = U'(\lambda) \phi'(\lambda) \tag{4.4}$$

with

$$U'(\lambda) = \frac{\partial \mathscr{L}(\lambda)}{\partial x} \cdot \mathscr{L}(\lambda)^{-1} + \mathscr{L}(\lambda) U(\lambda) \mathscr{L}(\lambda)^{-1} . \tag{4.5}$$

The proper choice of P is made by the choice of *Ker P* and *Im P*. For this we take two complementary subspaces L^+ and L^- of $\mathscr{C}^n$ and define

$$Ker P = \phi(\mu_1) L^- , \quad Im P = \phi(\mu_2) L^+ \tag{4.6}$$

A similar approach, based on the Riemann boundary value problem with Zeros, was used by Zakharov-Mikhailov-Shabat [2].

The case of the one-dimensional projector is especially simple.

For this we take the solution of the direct and adjoint linear problems:

$$\frac{\partial \vec{\phi}_{\mu_2}{}^t}{\partial x} = U(\mu_2)\, \vec{\phi}_{\mu_2}{}^t \quad , \quad \frac{\partial \vec{\psi}_{\mu_1}{}^t}{\partial x} = -\, U(\mu_1)\, \vec{\psi}_{\mu_1}{}^t \ . \tag{4.7}$$

The eigenvectors $\vec{\phi}_{\mu_1}$ and $\vec{\psi}_{\mu_2}$ can be obtained from the matrix $\phi(\lambda)$:

$$\vec{\phi}_{\mu_2}{}^t = \phi(\mu_2)\, C_2{}^t \quad , \quad \vec{\psi}_{\mu_1}{}^t = C_1{}^t\, \phi(\mu_1)^1 \ . \tag{4.8}$$

The projector P is nothing but

$$P = \vec{\phi}_{\mu_2}{}^t \cdot \vec{\psi}_{\mu_1} \frac{1}{(\vec{\phi}_{\mu_2}, \vec{\psi}_{\mu_1})} \tag{4.9}$$

where $(\vec{\phi}_{\mu_2}, \vec{\psi}_{\mu_1})$ is a scalar product.

5. SCATTERING INTERPRETATION OF BACKLUND TRANSFOR-MATIONS

In this section we briefly discuss the class of two dimensional equations solvable via the Inverse Scattering Method (*ISM*) and describe the action of *BT* in terms of scattering data. We follow the exposition of [28, 29, 33, 35]. One starts in *ISM* with a spectral problem in x:

$$\frac{\partial \phi}{\partial x} = (\lambda \cdot A + U)\, \phi \ , \tag{5.1}$$

where A is a diagonal $n \times n$ matrix. With Eq. (5.1) is connected the group of gauge transformations. One chooses a subalgebra g_A of $gl(n, \mathscr{C})$ of matrices commuting with A: $g_A = \{g \in gl(n, \mathscr{C}) \mid [A, g] = 0\}$ and the corresponding decomposition of an algebra $gl(n, \mathscr{C}) = g_A \otimes g_F$. The linear problem Eq. (5.1) is invariant under the gauge transformations with the gauge group G_A corresponding to g_A:

$$\phi \rightarrow \phi' = G \cdot \phi \ , \quad U' = G \cdot U \cdot G^{-1} - \frac{\partial G}{\partial x}\, G^{-1}$$

for λ-independent matrix $G \in G_A$. In particular, one can assume that $U \in g_F$. Let us consider rapidly decreasing potentials $U = U(x)$ for which $U \rightarrow 0$ as $|x| \rightarrow \infty$. Then one takes two solutions ϕ^+ and ϕ^- of Eq. (5.1) such that $\phi^{\pm}(x, \lambda) \rightarrow e^{\lambda x A}$ as $x \rightarrow \pm \infty$. The transition (or scattering) matrix $S(\lambda)$ is defined as $S(\lambda) = \phi^-(x, \lambda)^{-1} \phi^+(x, \lambda)$.

To classify two-dimensional systems as "solvable" via *ISM* [28, 29, 33, 35] we introduce an infinite dimensional Lie algebra $g_{\tilde{A}} = g_A \otimes \mathscr{C}\ [[\lambda, \lambda^{-1}]]$ (say, Fourier power series with coefficients in g_A). Then a two-dimensional evolution equation on potential $U(x,t)$ is called "solvable" via *ISM* if the evolution in t of its scattering coefficient $S(\lambda,t)$ is linear:

$$\frac{\partial S(\lambda,t)}{\partial t} = [\ h(\lambda,t)\ ,\ S(\lambda,t)] \tag{5.2}$$

with some $h(\lambda,t) \in g_{\tilde{A}}$. If $h(\lambda,t)$ is entire in λ, AKNS and Calogero-Degasperis [29] showed that Eq. (5.2) is equivalent to a certain nonlinear integro-differential equation on $U(x,t)$. We concentrate on particular examples of these equations considered in [33, 35]. Corresponding systems are connected with matrix (operator) *NLS*. This corresponds to the case of block matrices:

$$A = \begin{pmatrix} I_{n_1} & 0 \\ 0 & -I_{n_2} \end{pmatrix}, \quad n_1 + n_2 = n, \quad \text{and} \quad U = \begin{pmatrix} 0 & P \\ Q & 0 \end{pmatrix}.$$

Eq. (5.2) in the simplest case $h(\lambda) = A \cdot \lambda^2 \in g_{\tilde{A}}$ is the matrix (coupled) *NLS* equation

$$iP_t = P_{xx} + 2\,P\,Q\,P\ , \qquad -\,iQ_t = Q_{xx} + 2\,Q\,P\,Q$$

or

$$iP_t = P_{xx} + 2\,P\,P^\dagger\,P$$

(where $P_t = \frac{\partial P}{\partial t}$, $P_x = \frac{\partial P}{\partial x}$, etc. ,). Similarly for $h(\lambda,t) = \lambda^3 \cdot A$ one gets matrix modified Korteweg deVries equation

$$P_t = P_{xxx} + 3\,P_x P P + 3\,P P P_x \qquad (P = Q)\ .$$

These equations and other equations of type 5.2 (e.g. different chiral field models and nonlinear σ-models) possess infinitely many first integrals (conservation laws), some of which are local. However, unlike in the scalar case, when the systems of local first integrals is complete, in the matrix case there is no simple complete system of commuting first integrals. In fact, in cases when the gauge group G_A is non-Abelian one has a system of local and nonlocal conservation laws, generating an infinite dimensional non-Abelian Lie algebra isomorphic, essentially, to $g_{\tilde{A}}$.

The best way to describe this algebra and to understand for what internal symmetries of the system Eq. (5.2) noncommuting conservation laws are responsible, is to use the Noether theorem and study canonical transformations associated with *ISM*. These transformations are *BT* of $\phi(\lambda)$ studied in section 3.

First of all we look at what effect *BT* is going to make on the scattering matrix $S(\lambda)$ of a linear problem 5.1. We call a new potential $U'(x)$ ($U' \in g_F$) a *BT* of a potential $U(x)$ if the new scattering matrix $S'(\lambda)$ of the linear problem

$$\left\{\tfrac{\partial}{\partial x} - \lambda A - U\right\} \phi'(\lambda, x) = 0$$

satisfies

$$S'(\lambda) = B(\lambda)^{-1} \cdot S(\lambda) \cdot B(\lambda) \tag{5.3}$$

for a matrix $B(\lambda) \in G_A$ for every λ. Hence, we identify here as the set of *BT* data, the infinite dimensional group $G_{\tilde{A}}$ associated with $g_{\tilde{A}} = g_A \otimes \mathscr{C}\,[[\lambda, \lambda^{-1}]]$. Through its action on scattering data 5.3, the Lie group $G_{\tilde{A}}$ acts as a group of *BT* on a space of potentials $U(x)$. E.g., since $G_A \subset G_{\tilde{A}}$, gauge transformations $U \to U^1 = B^{-1} \cdot U \cdot B$ are examples of *BTs*.

Remark 5.4: From the discussion of section 1 it is clear that the group of *BT* of section 3 applied to matrix eigenfunctions $\phi(\lambda)$ of Eq. (5.1) is larger than $G_{\tilde{A}}$. One of the reasons for this is the fact that *BT* of $\phi(\lambda)$, as defined in section 3, generates a new potential $U^1(x)$ (cf. remark 3.7) which does not necessarily exponentially decay at infinity. The condition of decay of $U'(x)$ at infinity imposes additional restrictions on gauge matrices S_μ in the data i).

The *BTs* of Eq. (5.3) are responsible for a large ($\cong g_{\tilde{A}}$) symmetry structure of two-dimensional systems "solvable" by *ISM*. In fact, infinitesimal action of *BTs* 5.3 generate all evolution equations 5.2. We put

$$B(\lambda) = I_n + \delta h(\lambda) , \qquad U' = U + \delta U_t$$

as $\delta \to 0$, so the equation 5.3 determines the evolution equations 5.2.

6. COMPLETELY INTEGRABLE EQUATIONS GENERATED BY COMPOSITIONS OF BT'S

BT 3.3 can be considered as a generalized Lie transformation. The whole algebra of *BT* is still under investigation, since in the most general case it is not obvious, how the

composition law is formulated for arbitrary *BT* data. Successive applications of different *BTs* and their commutation laws give rise to difference (lattice) equations expressed
as algebraic equations on elements of $L(\lambda)$. We denote *BT* $\phi'(\lambda)$ of $\phi(\lambda)$ corresponding to *BT* data $\mathcal{B}$ as

$$B_{\mathcal{B}}(\phi(\lambda)) \equiv \phi'(\lambda) = L(\lambda) \, \phi(\lambda) \,, \tag{6.1}$$

and, in order to avoid ambiguity, $L(\lambda)$ is also denoted as $L_{\mathcal{B}}(\lambda)$. The law of composition of two *BTs* can be expressed as follows

$$B_{\mathcal{B}_1 \otimes \mathcal{B}_2}(\phi(\lambda)) = B_{\mathcal{B}_2}\left(B_{\mathcal{B}_1}(\phi(\lambda))\right) \tag{6.2}$$

or

$$L_{\mathcal{B}_1 \otimes \mathcal{B}_2}(\phi(\lambda)) = L_{\mathcal{B}_2}(\lambda) \cdot L_{\mathcal{B}_1}(\lambda) \,.$$

In order to understand what lattice models arise from conditions of commutativity of
BTs, we restrict ourself first to 2-dimensional difference equations. In this case one
takes an Abelian subgroup of a group of *BTs*, generated by two independent *BTs*.
Then the space of *BT* data under consideration is a free Abelian group $A = \{n\mathcal{B}_1 \oplus n\mathcal{B}_2\}_{n,\,m \in Z}$ with two generators $\mathcal{B}_1$ and $\mathcal{B}_2$. This example is indeed the
simplest one and a lot of different Abelian groups A can be constructed. Starting from
an arbitrary $\phi(\lambda)$ we consider its *BT* corresponding to an arbitrary element of A.
$\phi_{n,m}(\lambda) \equiv B_{n\mathcal{B}_1 \oplus m\mathcal{B}_2}(\phi(\lambda))$. $\phi_{n,m}$ can be inductively defined in terms of successive
one step applications of *BTs*:

$$\phi_{n+1,m}(\lambda) = L^1_{n,m}(\lambda) \, \phi_{n,m} \quad \left(= B_{\mathcal{B}_1}(\phi_{n,m}(\lambda)) \right) \,;$$
$$\phi_{n,m+1}(\lambda) = L^2_{n,m}(\lambda) \, \phi_{n,m} \quad \left(= B_{\mathcal{B}_2}(\phi_{n,m}(\lambda)) \right). \tag{6.3}$$

The commutativity conditions between *BTs* are expressed in the form of Eq. (6.2).
Since A has rank 2 and is Abelian, the only equation of the form 6.2 which is to be
checked is $B_{\mathcal{B}_1 \oplus \mathcal{B}_2} = B_{\mathcal{B}_2 \oplus \mathcal{B}_1}$. In terms of notations 6.3 this operator relation amounts
to the following system of quadratic difference equations on coefficients of $L^1_{n,m}(\lambda)$ and
$L^2_{n,m}(\lambda)$.

$$L^1_{n,m+1}(\lambda) \, L^2_{n,m}(\lambda) = L^2_{n+1,m}(\lambda) \, L^1_{n,m}(\lambda) \tag{6.4}$$

which is equivalent to $\phi_{n+1,m+1}(\lambda) = B_{\mathcal{B}_1 \oplus \mathcal{B}_2}(\phi_{n,m}(\lambda)) = B_{\mathcal{B}_2 \oplus \mathcal{B}_1}(\phi_{n,m}(\lambda))$.

Equation 6.4 can be compared with the Hirota's and [4, 6] method of introduction of difference-difference analogs of two dimensional isospectral deformation equations. In fact, Eq. (6.4) in limit cases turn into Lax [44] representation of two dimensional isospectral deformation equations such as the Korteweg de Vries equations (or difference-differential equations such as the Toda lattice).

First of all we consider continuous limit in variable m. We assume that

$$L^1_{n,m} = L_n(\lambda, x) \quad , \quad L^1_{n,m+1} = L_n(\lambda, x) + \epsilon \frac{\partial L_n(\lambda, x)}{\partial x} + \mathcal{O}(\epsilon^2)$$
$$L^2_{n,m} = I + \epsilon A_n(\lambda, x) + \mathcal{O}(\epsilon^2)$$

(6.5)

Substituting Eq. (6.5) as $\epsilon \to 0$ into Eq. (6.4) one gets a familiar operator identity equivalent to a system of difference-differential equations associated with the inverse scattering method:

$$A_{n+1}(\lambda) L_n(\lambda) - L_n(\lambda) A_n(\lambda) = \frac{\partial L_n(\lambda)}{\partial x}$$

(6.6)

Further, making continuous limiting process in variable n operator equation 6.6 turns into Lax-Zakharov-Shabat pair representation [40, 41]:

$$\frac{\partial L(\lambda)}{\partial x} - \frac{\partial A(\lambda)}{\partial t} = [A(\lambda), L(\lambda)]$$

(6.7)

The significance of passing from Eq. (6.4) to Eq. (6.6) and Eq. (6.7) lies in the realization of the fact that *BTs* themselves produce one and two dimensional nonlinear p.d.e. equations of completely integrable type.

We present now a brief discussion of scattering interpretation of *BT* data for an arbitrary two dimensional system of equation 6.7.

We define formal scattering matrix for solutions of l.o.d.e. with coefficients depending on λ:

$$\frac{\partial \phi(\lambda, x)}{\partial x} = U(\lambda, x) \phi(\lambda, x) \quad ,$$

(6.8)

where $U(\lambda, x)$ can be assumed to be rational in λ:

$$U = \sum_{j=1}^{m} \sum_{r=1}^{q_j} U_{i,r} (\lambda - a_i)^{-r} + \sum_{r=0}^{q_\infty} U_{i,\infty} \lambda^{-r} .$$

The main assumption here is simple behavior of $U(\lambda, x)$ at x-infinity: $U(\lambda, x) \to U_\infty$ as $|x| \to \infty$, where U_∞ is x-independent. Hence we can define scattering matrix $S(\lambda)$ of linear problem 2.8 as

$$S(\lambda) = \phi_-(x, \lambda)^{-1} \cdot \phi_+(x, \lambda) .$$

Here, as usual, ϕ_+ and ϕ_- denote two fundamental solutions of Eq. (2.8) for which $\phi_\pm(x, \lambda) \rightarrow e^{U_\infty(\lambda)x}$ as $x \pm \infty$. We leave outside of the scope of this paper nontrivial problem concerning the possibilities of reconstruction of $U(\lambda)$ from $U_o(\lambda)$ and $S(\lambda)$. (This inverse scattering problem is absolutely unnecessary in our approach via lattice models.)

The *BT* of a (solution of a) linear differential operator, $L_\lambda = \frac{\partial}{\partial x} - U(\lambda,x)$, is a new linear differential operator $L'_\lambda = \frac{\partial}{\partial x} - U'(\lambda,x)$, and is given by

$$L'_\lambda = L(\lambda) \cdot L_\lambda \cdot L(\lambda)^{-1} \,. \tag{6.9}$$

It is clear that Eq. (6.9) is equivalent to the *BT* expression Eq. (3.3) with $\phi(\lambda)$ and $\phi'(\lambda)$ being fundamental solutions of, respectively, $L_\lambda \phi = 0$ and $L'_\lambda \phi' = 0$. This definition of *BT* allows us to express *BT* data as elements of $\check{G}$. The scattering matrices $S'(\lambda)$ of L'_λ and $S(\lambda)$ of L_λ are related by a linear fractional transformation

$$S'(\lambda) = B(\lambda)^{-1} \cdot S(\lambda) \cdot B(\lambda) \tag{6.10}$$

for an element $B(\lambda)$ of $\check{G}$. The space of *BT* data is, hence, parameterized by x-independent invertible $B(\lambda)$ and is only a subspace of a more general space of *BT* data described in section 3, since one must additionally assume in Eq. (6.9) that $U'(\lambda,x) \rightarrow U_\infty(\lambda)$, if $|x| \rightarrow \infty$. This new representation of the group of *BTs* from section 3 shows that lattice difference-differential and p.d.e. Eq. (6.4), Eq. (6.6) and Eq. (6.7) all possess infinite dimensional Lie group of canonical (*BT*) transformations.

We present now one of the descriptions of (two dimensional) nonlocal conservation laws of nonlinear p.d.e. [27] existing according to the Noether theorem. The first similar description of nonlocal conservation laws for nonlinear σ-model belong to Luscher and Pohlmeyer [57]. The system Eq. (6.7) is equivalent to the consistency of two linear problems of the form Eq. (6.8):

$$\begin{aligned} \frac{\partial \phi(\lambda,x,t)}{\partial x} &= A(\lambda,x,t) \, \phi(\lambda,x,t) \,; \\ \frac{\partial \phi(\lambda,x,t)}{\partial t} &= L(\lambda,x,t) \, \phi(\lambda,x,t) \,. \end{aligned} \tag{6.11}$$

Following our description of the scattering matrix, we assume $A(\lambda)$ and $L(\lambda)$ to have a simple behavior as $|x| \rightarrow \infty$: $A(\lambda) \rightarrow A_\infty(\lambda)$, $L(\lambda) \rightarrow L_\infty(\lambda)$, where $A_\infty(\lambda)$ and $L_\infty(\lambda)$ are both x (and t) -independent. This assumption makes us consider Eq. (6.7) as an equation of evolution in t. We choose an eigenfunction $\phi(\lambda,x)$ of Eq. (6.11) as, e.g., $\phi(\lambda,x) \rightarrow \exp\{A_\infty(\lambda)x+L_\infty(\lambda)t\}$ as $x \rightarrow -\infty$ (cf. $\phi_-(\lambda,x)$ above). We remark that $\phi(\lambda,x,t)$ has a singularity at $\lambda = a$ only if $A(\lambda)$ (or $L(\lambda)$) has a singularity at $\lambda = a$. In the case considered in [57], $A(\lambda) = A/(\lambda - 1)$ (and $L(\lambda) = L/(\lambda - 1)$ in different notations), $\phi(\lambda)$ can be represented as

$$\sum_{n=0}^{\infty} \phi_n \lambda^{-n} \quad , \quad \phi_o = I$$

because $A(\infty) = L(\infty) = 0$. In the expansion of Eq. (3.2)

$$\phi(\lambda,x,t) = \tilde{\phi}_a(\lambda) \cdot H_a(\lambda) \ , \quad \tilde{\phi}_a(\lambda) = \sum_{n=0}^{\infty} \phi_{a,n} (\lambda - a)^n$$

the elements of $\phi_{a,n}(x)$ can be determined by at most n-fold integration involving elements of $A(\lambda)$ with initial condition containing elements of $L(\lambda)$. The nonlocal quantities $\phi_{a,n}(\infty)$ are in reality nonlocal conservation laws for the time evolution equation (6.7):

Lemma 6.12: Let $\phi(\lambda,x,t)$ be a solution of Eq. (6.11) with boundary conditions $\phi(\lambda,x) \rightarrow \exp\{A_\infty(\lambda)x + L_\infty(\lambda)t\}$ as $x \rightarrow -\infty$. Let the expansion of $\phi(\lambda,x,t)$ in the form of Eq. (3.2) be

$$\phi(\lambda,x,t) = \left\{ \sum_{n=0}^{\infty} \phi_{a,n} (\lambda - a)^n \right\} H_a(\lambda)$$

where $H_a(\lambda)$ is a singular part and is present only if $\lambda = a$ is a singularity of $A(\lambda)$ or $L(\lambda)$. Then for every $n = 0, 1, 2, \ldots \phi_{a,n}(+\infty,t)$ is a nonlocal conservation law of the system Eq. (6.7). In general, for arbitrary constants $C_1, \ldots, C_n, \ldots$ we have

$$\frac{d}{dt} \left\{ \sum_{n=0}^{\infty} C_n \phi_{a,n}(\infty) \right\} = 0$$

according to equation of evolution Eq. (6.7).

Not all the integrals $\phi_{a,n}$ are independent, and it is possible to show that the number of independent nonlocal first integrals is at most $\infty^{n^2 d}$, where $d = \max\{$ ord $A(\lambda)$, ord $L(\lambda)\}$, and ord stands for the order of $A(\lambda)$ in λ, i.e., the sum of multiplicities of poles of $A(\lambda)$ in the λ-plane. Though the number of integrals seems sufficient for the complete integrability of Eq. (6.7), it should be noted that a) nonlocal integrals are not in involution, in general; and b) under the additional gauge symmetry some integrals cease to exist or become dependent.

7. COMPLETELY INTEGRABLE LATTICE SYSTEMS ASSOCIATED WITH BT'S

According to the results of section 6, in order to construct difference-differential systems of equations (6.4), (6.6) and (6.7) it is enough to construct *BTs*, (or infinitesimal operators, corresponding to these *BTs*), (3.1) with $L(\lambda)$ rational in λ. One should bear in mind Lemma 3.7 according to which with *BT* data, which are x and t-independent, define a transformation preserving equation (6.7). We start with presentation of such *BT*, corresponding to addition of one pole and one zero to $\phi(\lambda)$ presented above in section 4. If μ_1 and μ_2 are two distinct finite points, then *BT* of $\phi(\lambda)$ is defined as

$$\phi'(\lambda) = \left\{ I + \frac{\mu_1 - \mu_2}{\lambda - \mu_1} P \right\} \phi(\lambda),\tag{7.1}$$

where P is a projector of rank one, defined in terms of $\phi(\mu_1)$ and $\phi(\mu_2)$:

$$\vec{\phi}_2{}' = \phi(\mu_2) \cdot \vec{C}_1{}', \quad \vec{\psi}_1 = \vec{C}_1 \cdot \phi(\mu_1)^{-1}$$

for arbitrary vectors $\vec{C}_1$, $\vec{C}_2$ from $\mathscr{C}^n$ and

$$P = \vec{\phi}_2{}' \cdot \vec{\phi}_1 \cdot \frac{1}{(\vec{\phi}_2, \vec{\phi}_1)}.$$

Lattice models generated by *BT* Eq. (7.1) were, in fact previously considered in [37] as quantum systems for $n = 2$. We use this example in order to show how a sequence of *BTs* (of the form of Eq. (7.1)) generates lattice models 6.6 even when auxiliary differential operators (infinitesimal operators of *BT*) are not presented explicitly. Let us take a sequence of *BTs* with *BT* data $\mathscr{B}_o$, $\mathscr{B}_1$, . . . , $\mathscr{B}_N$, . . . (e.g., one can take $\mathscr{B}_N = \mathscr{B}_o$) and consider step-by-step applications of *BTs* to a given function $\phi(\lambda) = \phi_o(\lambda)$:

$$\phi_{N+1}(\lambda) = L_N(\lambda) \phi^N(\lambda)\tag{7.2}$$

with $L_N(\lambda) = L_{\mathscr{B}_N}(\lambda)$ and $\phi_{N+1}(\lambda) = B_{\mathscr{B}_N}(\phi_N(\lambda))$.

In Eq. (7.2) one considers elements of $L_N(\lambda)$ as lattice variables labeled by an integer N. In the "homogeneous" case $\mathscr{B}_N = \mathscr{B}_o$, when one successively applies the same *BT* with the same *BT* data $\mathscr{B}_N = \mathscr{B}_o$, $L_N(\lambda)$ can be considered as an $n \times n$ matrix with elements rationally depending on λ and labeled by N, where singularities in λ and their orders do not depend on N. System of difference equations 7.2 generated by *BTs* gives rise to a sequence of commuting Hamiltonians describing difference-differential equations. These Hamiltonians are given as coefficients of an equation $P(\lambda, \mu) = 0$ describing the curve

$$\det \left(L_M(\lambda). . . L_N(\lambda). . . L_0(\lambda) - \mu I \right) = 0\tag{7.3}$$

(where the curve is algebraic or transcendental, respectively, whenever $M < \infty$ or $M = \infty$). Eq. (7.3) is nothing but a characteristic polynomial $\chi(\mu)$ with coefficients depending on λ, of a matrix $L(\lambda) = \{L_M(\lambda) . . .L_N(\lambda). . .L_0(\lambda)\}$ which determines the BT

$$\phi_{M+1} = B_{\mathcal{B}_M} (. . . (B_{\mathcal{B}_N} (. . .(\phi_0(\lambda)). . .). . .). . .)= L(\lambda) \, \phi_0(\lambda).$$

Coefficients of Eq. (7.3) are, typically, involutive Hamiltonians defining completely integrable dynamical systems. The definition 7.3 lacks, however, a symplectic structure which turns the coefficients of Eq. (7.3) into involutive quantities. The definition of symplectic structure is supplied by so called Baxter lemma (see sections 2 and 11) defining symplectic structure, and simultaneously, quantization of the corresponding Hamiltonian. According to this lemma [10], commutativity relations between elements of $L_N(\lambda)$ are defined as

$$R(\lambda,\mu) \, (L_N(\lambda) \otimes L_N(\mu)) = (L_N(\mu) \otimes L_N(\lambda)) \, R(\lambda-\mu) \tag{7.4}$$

and also by assumption elements of $L_N(\lambda)$ and $L_M(\mu)$ commute for $N \ne M$.[1]

The form of (7.4) the commutativity condition determines in the semiclassical limit the Poisson brackets between elements of $L_N(\lambda)$. The statement that coefficients of Eq. (7.3) are commuting Hamiltonians is still valid in the quantum case under certain additional conditions and with a proper definition of $\det(L(\lambda) - \mu I)$ for operator valued $L(\lambda)$.

As a very special case of system (3.2) we take a sequence of BTs generated by successive iterations of Eq. (7.1):

$$\phi_{N+1}(\lambda) = L_N(\lambda) \, \phi_N(\lambda) ,$$
$$L(\lambda) = \left\{ \delta_{i,j} + \frac{U_{i,j}^{(N)}}{\lambda - \mu_N} \right\}_{i,j=1}^n , \tag{7.5}$$

where $[U_{i,j}^{(N)} , U_{i',j'}^{(M)}] = 0$ if $N \ne M$ and for a fixed N operators $U_{i,j}^{(N)}$ obey rules of commutation of the algebra $sl(n)$. Commuting Hamiltonians defined as coefficients of the corresponding equation (7.3) can be called generalized "six-vertex" or XXX-models. Indeed, models of statistical mechanics correspond to the choice of $U_{i,j}^{(N)}$ in Eq. (7.5) in terms of n-dimensional representations of $sl(n)$ by matrices $u_{i,j}$ and $U_{i,j}^{(N)}$ acting in the Fock space $\otimes_{N=0}^{M} (\mathcal{C}^N)$ as $U_{i,j}^{(N)} = I \otimes . . .\otimes u_{i,j} \otimes . . . \otimes I$ (that is, non-trivial only on the Nth place). Such systems are completely integrable in the clas-

---

[1]The $n^2 \times n^2$ matrix $R(\lambda,\mu)$ is called an S-matrix of system 7.2 [10]. One should note, however, that $R(\lambda,\mu)$ may not satisfy factorization equations [8, 10].

sical case and in the quantum case belong to systems solvable via the Bethe Ansatz (in particular, free energy can be expressed in closed form).   Particular attention is devoted below to a special case $n = 2$, where one of the Hamiltonians generated by Eq. (7.3) is a genuine generalization of the XXX-model (Heisenberg ferromagnet) and is of the form:

$$H''_N = \sum_{n=1}^{N} \left\{ \sigma_n^1 \sigma_{n+1}^1 + \sigma_n^2 \sigma_{n+1}^2 + \sigma_n^3 \sigma_{n+1}^3 \right\}, \tag{7.6}$$

with $[\sigma_n^i, \sigma_m^j] = \delta_{nm} \sigma_n^k$ for a permutation (i, j, k) of $(1, 2, 3)$ and σ_n^i corresponding to elements of the universal enveloping algebra of $sl(2)$.

Other lattice models are examined in the next chapter where particular attention is devoted to lattice models generated by elementary BTs applied to solutions of the Dirac equation (see also section 8).   This way two dimensional operator nonlinear Schrodinger equations are substituted by one dimensional Hamiltonian lattice models equivalent to and approximating original system.   It should be noted that in view of the same geometric structure generated by the same S-matrix, lattice nonlinear Schrodinger Hamiltonians (given below) are equivalent to the generalized XXX-models such as Eq. (7.6).

8. LATTICE NONLINEAR SCHRODINGER EQUATION AND RELATED MODELS

This section is devoted to the investigation of "generator" of the group of BT for solutions of the Dirac equation and study of the lattice systems, appearing as a result of successive applications of BT. From the point of view of examples of NLS of section 5, we write matrices in block form

$$\begin{pmatrix} M_{11} & M_{12} \\ M_{21} & M_{22} \end{pmatrix}$$

with M_{11} an $n_1 \times n_1$ matrix, M_{22} an $n_2 \times n_2$ matrix, etc.   In the ISM language the action of "generators" of BT B_μ^+ consists in changing the scattering matrix $S(\lambda)$ (for potential U) into a new scattering matrix $S'(\lambda)$ (for transformed potential $U' = B_\mu^+(U)$) satisfying

$$S'(\lambda) = \begin{pmatrix} (\lambda - \mu)I & 0 \\ 0 & I \end{pmatrix}^{-1} \cdot S(\lambda) \cdot \begin{pmatrix} (\lambda - \mu)I & 0 \\ 0 & I \end{pmatrix}. \tag{8.1}$$

From the point of view of BT of section 1 as changing regular exponential matrices, the BT data are

$$\begin{pmatrix} -I & 0 \\ 0 & I \end{pmatrix} \text{ at } \lambda = \infty \text{ and } S_\mu^{-1} \begin{pmatrix} -I & 0 \\ 0 & I \end{pmatrix} S_\mu \text{ at } \lambda = \mu , \qquad (8.2)$$

where I is the $n_1 \times n_1$ unit matrix.

The construction of BT $\phi'(\lambda)$ of $\phi(\lambda)$ proceeds as follows. At $\lambda = \infty$, let $\phi(\lambda) = \{I + \phi^{(1)}\lambda^{-1} + \ldots\} \cdot Sing_\infty(\lambda)$, where $Sing_\infty(\lambda)$ is a diagonal matrix. BT $\phi'(\lambda)$ with the data of Eq. (8.2) has an expansion

$$\phi'(\lambda) = \{I + \phi'_\infty + \ldots\} \begin{pmatrix} \lambda & 0 \\ 0 & I \end{pmatrix} Sing_\infty(\lambda) \quad \text{at } \lambda = \infty$$

and

$$\phi'(\lambda) = \tilde{\phi}'(\lambda) \cdot (\lambda - \mu)^{S_\mu^{-1}} \cdot \begin{pmatrix} -I & 0 \\ 0 & I \end{pmatrix} S_\mu \text{ at } \lambda = \mu .$$

Linear transformation from $\phi(\lambda)$ to $\phi'(\lambda)$ corresponding to the data of Eq. (8.2) is given by the formula

$$\phi'(\lambda) = \begin{pmatrix} \lambda - \mu + A B & A \\ B & I \end{pmatrix} \qquad (8.3)$$

or, more compactly

$$\phi'(\lambda) = L_\mu^+ \cdot \phi(\lambda)$$

where $A = - (\phi_\infty^{(1)})_{12}$ and, for $\phi(\mu)$

$$S_\mu = \begin{pmatrix} S_{11} & S_{12} \\ S_{21} & S_{22} \end{pmatrix}$$

with $BS_{12} + S_{22} = 0$.

The existence of B is necessary and sufficient condition for the existence of BT with data 8.2. Let us apply BT with data 8.2 to solution of matrix Dirac equation (see Section 5):

$$\frac{\partial \phi(\lambda)}{\partial x} = \begin{pmatrix} \lambda I & P \\ Q & -\lambda I \end{pmatrix} \phi(\lambda) . \qquad (8.4)$$

In this case asymptotic expansion of $\phi(\lambda)$ is irregular at infinity and $\phi(\lambda)$ can be taken in the form [9, 23]:

$$\phi(\lambda) = \{I + \phi_\infty + \ldots\} e^{\lambda \sigma_3 x} , \quad \sigma_3 = \begin{pmatrix} I & 0 \\ 0 & -I \end{pmatrix}.$$

Then the BT $\phi'(\lambda)$ with the data (8.2) $\phi'(\lambda) = L_\mu^+ \cdot \phi(\lambda)$ again satisfies a Dirac equation:

$$\frac{\partial \phi'(\lambda)}{\partial x} = \begin{pmatrix} \lambda I & P' \\ Q' & -\lambda I \end{pmatrix} \phi'(\lambda)$$

The relationship between P and Q and P' and Q' and coefficients of transformation $L_\mu^+(\lambda)$ is the following: $A = -\frac{1}{2}P$, $P' = -\mu P + \frac{1}{2}P_x - \frac{1}{2}PBP$, $Q' = -2B$.

Here B, satisfying $B S_{12} + S_{22} = 0$, is the ratio of two components of a solution $\phi(\mu)S_\mu$ of the Dirac equation, Eq. (8.4), and hence satisfies a Ricatti equation: $B_x = -2\mu B + Q - BQB$. This brings the relationship between P, Q and P', Q' into the form:

$$P_x = 2 P' + 2 \mu P - \frac{1}{2} P Q' P$$

$$Q_x' = -2Q - 2 \mu Q' + \frac{1}{2} Q' P Q' \qquad (8.5)$$

Finally, we can consider BT Eq. (8.3) from $\phi(\lambda)$ to $\phi'(\lambda)$ as only a step in the infinite sequence of BTs generated by data (8.2). One can iterate data starting from

$$\mathscr{B}_{\mu_N} = \begin{pmatrix} a(\lambda - \mu_N)I & 0 \\ 0 & a^{-1}I \end{pmatrix} \text{ for } N \geq 0$$

for a sequence of c-numbers $\ldots$, μ_N, $\ldots$, μ_0. Then, starting from $\phi_0(\lambda) \equiv \phi(\lambda)$ we generate $\phi_{N+1}(\lambda) \equiv B_{\mathscr{B}_{\mu_N}}(\phi_N(\lambda)) = B_{\mathscr{B}_{\mu_N}} \ldots \cdot B_{\mathscr{B}_{\mu_0}} \phi(\lambda)$.

The sequence $\phi_N(\lambda)$ can be extended to negative N if one chooses

$$\phi_0(\lambda) = B_{\mathscr{B}_{\mu_{-1}}}(\phi_{-1}(\lambda)) \text{ or } \phi_{-1}(\lambda) = \left\{B_{\mathscr{B}_{\mu_{-1}}}\right\}^{-1}(\phi_0(\lambda))$$

where $\left\{B_{\mathscr{B}_\mu}\right\}^{-1}$ is itself a BT with data

$$\begin{pmatrix} a^{-1}(\lambda - \mu)^{-1} \cdot I & 0 \\ 0 & aI \end{pmatrix}.$$

If the initial $\phi(\lambda)$ satisfies the Dirac equation (8.4), then all $\phi_N(\lambda)$ will satisfy similar equations

$$\frac{\partial \phi(\lambda)_N}{\partial z} = \begin{pmatrix} \lambda I & P^{(N)} \\ Q^{(N)} & -\lambda I \end{pmatrix} \phi(\lambda)$$

also. Introducing new notations $Q_N = Q^{(N+1)}$, $P_N = P^{(N)}$ we can rewrite the system of differential equations on on Q_N , P_N in the form:

$$P_{n,x} = 2a^{-1} P_{n+1} + \mu_n P_n - \tfrac{1}{2} a P_n Q_n P_n \, ,$$

$$Q_{n,x} = - 2a^{-1} Q_{n-1} - \mu_n Q_n + \tfrac{1}{2} a P_n Q_n P_n \, . \tag{8.6}$$

This system of equations is a Hamiltonian one. First of all, if P_n and Q_n are classical matrix variables, $P_n = (P_{n,ij})$, $Q_n = (Q_{n,ij})$ with Poisson brackets defined as usual by $\{P_{n,ij} Q_{m,kl}\} = \delta_{il}\delta_{jk}$, then the corresponding Hamiltonian is

$$\mathcal{H} = \sum_n \text{Tr}\left\{ \tfrac{1}{4} a P_n Q_n P_n Q_n - 2 \mu_n P_n Q_n - 2 a^{-1} P_n Q_{n-1}\right\} .$$

Similarly, one can consider the case when P_n , Q_n are Heisenberg variables:

$$[P_n , Q_m] = \delta_{nm}$$

and the corresponding quantum Hamiltonian is

$$\mathcal{H}_Q = \sum_n \text{Tr}\left\{ \tfrac{1}{4} a Q_n^2 P_n^2 + \tfrac{1}{2} P_n Q_n - 2 \mu_n P_n Q_n - 2 a^{-1} P_n Q_{n-1}\right\} .$$

The Hamiltonians $\mathcal{H}$ and $\mathcal{H}_Q$ are completely integrable in the sense that they possess infinitely many local commuting Hamiltonians. Moreover, if one imposes periodic boundary conditions, $\{ P_{n+N} = P_n, Q_{n+N} = Q_n \}$, then the corresponding Hamiltonians are indeed completely integrable in the sense that they possess the involutive first integrals in the number equal to half the dimension.

The system Eq. (8.6) and higher flows commuting with it possess remarkable properties making it a *true* lattice equivalent to *NLS*. First of all, one considers the evolution according to the second flow, commuting with Hamiltonians $\mathcal{H}$ or $\mathcal{H}_Q$ (the nth flow is such, where P_i and Q_i are of degrees $n+1$). If the time is denoted by t, then the time evolution equations are the following:

$$P_{n,t} = \left\{ 4a^{-2} P_{n+2} + 4a^{-1} (\mu_n + \mu_{n+1})P_{n+1} + 4 \mu_n^2 P_n \right.$$

$$- P_{n+1} Q_{n+1} Q_{n+1} - P_{n+1} Q_n P_n + P_n Q_{n-1} P_n + P_n Q_n P_{n+1}$$

$$\left. - 2a \mu_n P_n Q_n P_n + \tfrac{1}{4} a^2 P_n Q_n P_n Q_n P_n + 2 P_n Q_{n-1} P_n\right\}$$

$$\tag{8.7}$$

$$- Q_{n,t} = \left\{ 4a^{-2} P_{n-2} + 4a_{-1} (\mu_n + \mu_{n-1})P_{n-1} + 4 \mu_n^2 P_n \right.$$

$$- P_{n-1} Q_{n-1} Q_{n-1} - P_{n-1} Q_n P_n + P_n Q_{n+1} P_n + P_n Q_n P_{n-1}$$

$$\left. - 2a \mu_n P_n Q_n P_n + \tfrac{1}{4} a^2 P_n Q_n P_n Q_n P_n + 2 P_n Q_{n+1} P_n\right\} \, .$$

In the natural continuous n-limit $P_n = P(x)$, $Q_n = Q(x)$, $P_{n\pm1} = P(x) \pm \delta P_x(x)$ and $Q_{n\pm1} = Q(x) \pm \delta Q_x(x)$. As $\delta \to 0$, $\mu_n \to -a^{-1}$ and $a \to 0$, the system (3.7) assumes its form as (coupled) *NLS* equation $P_t = P_{xx} + 2PQP$, $- Q_t = Q_{xx} + 2 QPQ$.

There is, however, a simpler method to use Eq. (8.6) directly to approximate solutions of *NLS* without solving the additional system of nonlinear o.d.e. Eq. (8.7). Indeed, one can suggest a more sophisticated transition to the continuous n limit in Eq. (8.6):

$$P_{n,x} = 2a^{-1} P_{n+1} + \mu_n P_n - \chi \frac{1}{2} a P_n Q_n P_n ,$$

$$Q_{n,x} = - 2a^{-1} Q_{n-1} - \mu_n Q_n + \chi \frac{1}{2} a P_n Q_n P_n .$$

Now we introduce a continuous variable y: $P_n(x) = P(x,y)$, $Q_n(x) = Q(x,y)$ and $P_{n\pm1} = P \pm \delta P_y + \delta^2/2 \, P_{yy} + \mathcal{O}(\delta^3)$ and $Q_{n\pm1} = Q \pm \delta Q_y + \delta^2/2 \, Q_{yy} + \mathcal{O}(\delta^3)$ as $\delta \to 0$ and $\mu_n \to a^{-1}$, $a = \delta^2$, $\chi = -4/\delta^2$, so that

$$P_x = 2 \delta^{-1} P_y + P_{yy} + 2PQP ,$$

$$Q_x = 2 \delta^{-1} Q_y - Q_{yy} - 2PQP$$

up to order $\mathcal{O}(\delta^3)$. After the change of variables $t = x$, $x_1 = y + 2\delta^{-1}x$, we get the *NLS* equations $P_t = P_{x_1 x_1} + 2PQP$ and $- Q_t = Q_{x_1 x_1} + 2QPQ$.

Hence, continuous limit of systems (3.6) describing a sequence of "elementary" *BTs* applied to Dirac equation yields a solution of *NLS*. In particular, we obtain a very simple way of solving *NLS* when it is enough to determine *all* successive *BTs* P_n , Q_n as functions of n and then to perform transition to continuos limit in n. In order to determine P_n and Q_n there is no need to solve on each step new nonlinear o.d.e. Eq. (8.6). It is enough to find eigenfunctions of Dirac equation (8.5) with initial potential and eigenvalues μ_n only. Since P_n and Q_n are ratios of eigenfunctions, they all are algebraic expressions in eigenfunctions of an original Dirac equation. In this sense to solve an equation like *NLS* there is no need to solve an inverse scattering problem, which determines the evolution of potential $U(x,t)$ through the scattering matrix $S(\lambda,t)$. From our point of view it is enough to solve completely, *direct* spectral problem (not a scattering one) and find eigenfunctions with their explicit dependence on eigenvalues μ.

Remark 8.8: In our paper [34] we proposed another decomposition theorem for *NLS* equations based on the reduction of matrix *NLS* equations to the common action of the first and second non-Abelian Toda lattice flows. As in the present paper, we used iterations of *BT*, but with different "generator" data:

$$\mathscr{B} = \begin{pmatrix} \lambda I & 0 \\ 0 & \lambda^{-1} I \end{pmatrix}$$

(addition of poles and zeros only at infinity). The decomposition theorem looked simpler, but did not change rapidly decreasing potentials into rapidly decreasing potentials.

Remark 8.9: One of the important factors in the application of *BT* is, from our point of view, the possibility to replace one, two or three dimensional systems of nonlinear p.d.e. by their lattice counterparts. These lattice models are a) equivalent and b) are approximations to initial systems of p.d.e. They are "equivalent" in a sense that introduction of a new discrete variable n generates canonical transformations $n \rightarrow n + 1$ of an original system. Moreover, this transformation, taken by itself, generates conserved currents. On the other hand, lattice systems turn out to be approximations to the original system of p.d.e. and regain their explicit form after passing to the continuous limit. The simplest example is sine-Gordon equation and its *BT*. The equation in light-cone coordinates has the form $\phi_{xt} = \sin \phi$, and its *BT* is known to be

$$\frac{1}{2} \left\{ \phi_n + \phi_{n+1} \right\}_x = \zeta^{-1} \sin \left\{ \frac{\phi_{n+1} - \phi_n}{2} \right\} ,$$

$$\frac{1}{2} \left\{ \phi_n - \phi_{n+1} \right\}_t = \zeta \sin \left\{ \frac{\phi_{n+1} - \phi_n}{2} \right\} .$$

Here ϕ_{n+1} is the BT of ϕ_n. In the continuum limit $x = n\delta$ as $\delta \rightarrow 0$ this *BT* regains the initial shape of the sine-Gordon equation. The lattice systems like *BT* for sine-Gordon equation given above, provide exceptionally good approximations to p.d.e. We present here a finite difference equivalent of sine-Gordon equation known as a discrete sine-Gordon equation (in Hirota form [35])

$$\sin \left\{ \frac{1}{4} \left(\phi(x+\delta,t+\epsilon) - \phi(x+\delta,t) + \phi(x,t) - \phi(x,t+\epsilon) \right) \right\} =$$

$$\frac{\delta\epsilon}{4} \sin \left\{ \frac{1}{4} \left(\phi(x+\delta,t+\epsilon) + \phi(x+\delta,t) + \phi(x,t) + \phi(x,t+\epsilon) \right) \right\} .$$

In the limit $\delta \rightarrow 0$ (or $\epsilon \rightarrow 0$) one recovers *BT* for sine-Gordon and in the limit ϵ, $\delta \rightarrow 0$ one recovers the sine-Gordon equation.

PART 2
QUANTUM COMPLETELY INTEGRABLE
LATTICE MODELS

9. SYMPLECTIC STRUCTURES ASSOCIATED WITH CLASSICAL ALGEBRAS

In many cases quantized mechanical systems and field theories are associated with representations of classical groups and do not arise directly from Heisenberg commutation relations. Perhaps the most famous example is the *XYZ*-model of Statistical Mechanics with Hamiltonian

$$H_{XYZ} = \sum_{n=1}^{N} \left\{ \mathcal{T}_1\, \sigma_n^1\, \sigma_{n+1}^1 + \mathcal{T}_2\, \sigma_n^2\, \sigma_{n+1}^2 + \mathcal{T}_3\, \sigma_n^3\, \sigma_{n+1}^3 \right\}$$

where σ_n^i arises from a given representation of $SU(2)$ or $SO(3)$ and obey the corresponding commutation relations $[\sigma_n^i, \sigma_n^j] = 2i\sigma_n^k$ where i, j and, k is a cyclic permutation of 1, 2 and 3, $(\sigma_n^k)^2 = I$, and $[\sigma_n^i, \sigma_m^j] = 0$ for $n \neq m$.

In general, when we consider quantization of a classical Hamiltonian H given in symplectic coordinates different from Darboux coordinates, one cannot a priori choose H as a differential operator. We see later that this is actually possible when H is written in symplectic coordinates arising from representation of a classical group. An ambiguity that remains is typical to Heisenberg relations, when quantization of H is defined up to the order of noncommuting operators. In general, let g be an arbitrary Lie algebra with generators $e_1, \ldots, e_m$ and commutation relations

$$[e_i, e_j] = \sum_{k=1}^{m} c_{ij}^k\, e_k$$

for structure constants c_{ij}^k. With this algebra g one can naturally associate a symplectic structure with Poisson brackets between variables $v_i : i = 1, \ldots, m$ as

$$\{ v_i, v_j \} = \sum_{k=1}^{m} c_{ij}^k\, v_k . \tag{9.1}$$

In this new canonical variable v_i with Poisson brackets defined by Eq. (9.1) one can write the Hamiltonian H. One wonders in what sense the symplectic structure is defined, because there is no reason to claim that the manifold is even dimensional (m can be arbitrary). However, we do have an even dimensional manifold with the same definition of Poisson brackets, if to take into account that variables v_i are not independent! Indeed, let A be an arbitrary Casimir operator for g:

$$A = \sum_{i_1 \ldots i_k} c_{i_1, \ldots, i_k} e_{i_1}, \ldots, e_{i_k} .$$

Then since A belongs to the center of the universal enveloping algebra of g, A commutes with all e_i. Hence the following function of v_i

$$a = \sum_{i_1 \ldots i_k} c_{i_1, \ldots, i_k} v_{i_1}, \ldots, v_{i_k}$$

has zero Poisson brackets with all functions of v_i and hence can be taken as a constant (though an arbitrary one). In other words, m-dimensional manifold with coordinates v_i is foliated into a family of "orbits" corresponding to fixed constant values of all Casimir operators of g written in coordinates v_i. Each of the orbits is apparently an even dimensional submanifold, where one can find by Darboux theorem [45] an equivalent Darboux symplectic structure. What is really important, is the existence of a global transformation allowing us to find this equivalent Darboux structure not on each orbit separately, but on all orbits simultaneously. This is done below for classical groups using infinitesimal operators of regular representations of g.

In the examples below the two dimensional symplectic manifold T is determined by the value of a single Casimir operator [46]. Below are listed commutation relations between generators of the corresponding groups and parametrizations of these generators in terms of operators p and q.

Examples:

1. $M(2)$:

$$[e_1, e_2] = 0 , \qquad [e_2, e_3] = e_1 , \qquad [e_3, e_1] = e_2 ;$$
$$\text{and } v_1 = R \cos q , \qquad v_2 = R \sin q , \qquad v_3 = -p .$$

2. $MH(2)$:

$$[e_+, e_-] = 0 , \qquad [e_3, e_+] = e_+ , \qquad [e_3, e_-] = -e_- ;$$
$$v_3 = p , \qquad v_+ = R\, e^q , \qquad v_- = R\, e^{-q} .$$

3. $SO(3)$:

$$[e_1, e_2] = e_3 , \qquad [e_2, e_3] = e_1 , \qquad [e_3, e_1] = e_2 ;$$
$$v_1 = i\left\{ vq + \tfrac{1}{2}(1 - q^2)p \right\} , \qquad v_2 = -vq + \tfrac{1}{2}(1 + q^2)p ,$$
$$v_3 = i\left\{ -v + qp \right\} .$$

4. $QU(2)$:

$$[e_1, e_2] = -e_3, \qquad [e_2, e_3] = e_1, \qquad [e_3, e_1] = e_2;$$

$$v_1 = \tfrac{1}{2}\left\{(\nu + \epsilon)\, e^{iq} + (\nu - \epsilon)\, e^{-iq} - 2\,\sin qp\right\},$$

$$v_2 = \tfrac{i}{2}\left\{(\nu + \epsilon)\, e^{iq} - (\nu - \epsilon)\, e^{-iq} + 2i\,\cos qp\right\},$$

$$v_3 = -i\epsilon + p.$$

5. $SL(2,R)$:

$$[e_1, e_2] = -2e_3, \qquad [e_2, e_3] = -2e_1, \qquad [e_3, e_1] = 2e_2;$$

$$v_1 = 2\,\nu q + (1 - q^2)p, \qquad\qquad v_2 = -2\,\nu q + (1 + q^2)p,$$

$$v_3 = 2\,\nu - 2qp.$$

Formulas for the parametrization of an arbitrary finite dimensional semi-simple Lie algebra in terms of differential operators can be presented similarly [46]. The Toda lattice system corresponding to $MH(2)$ can be written in variables $v_{3,n}, v_{+,n}, v_{-,n}$ as the following Hamiltonian

$$H_T = \tfrac{1}{2}\sum_n \left\{v_{3,n}^2 + v_{+,n+1}\,v_{-,n}\right\}$$

with Poisson brackets defined following the general formula Eq. (9.1):

$$\{v_{+,n}, v_{-,m}\} = 0, \qquad\qquad \{v_{3,n}, v_{\pm,m}\} = \pm\, v_{\pm,n}\,\delta_{n,m}.$$

After the substitution proposed above we obtain from H_T the ordinary Toda lattice Hamiltonian because $v_{3,n} = p_n$, $v_{+,n} = R\, e^{q_n}$, $v_{-,n} = R\, e^{-q_n}$. Here we choose a constant R to be the same for all values of n as we assume Casimir operators to have the same eigenvalues: this normalization condition can be removed.

An example of Toda lattice arises from vector bundles of rank 2 over $\mathscr{C}P^1$ equipped with a group structure induced by $MH(2)$. In terms of matrix notations we choose a "spectral" parameter λ in a base space $\mathscr{C}P^1$ and vector bundles are given by the following matrices

$$\mathscr{L}_n = \begin{pmatrix} \lambda + v_{3,n} & -v_{+,n} \\ v_{-,n} & 0 \end{pmatrix}.$$

Then completely integrable Hamiltonian H_T arises from this system of matrices as one of many commuting Hamiltonians given by an expansion of

$$\mathrm{Tr}\{\mathscr{L}_N(\lambda)\ldots\mathscr{L}_1(\lambda)\}$$

in powers of λ.

This construction can be widely generalized for an arbitrary classical group. In this sense there exists a completely integrable lattice system associated with an algebra g and written in Darboux coordinates prescribed by a parametrization of fixed orbits of g. This lattice system can be written in terms of variables $v_{i,n}$, where $v_{i,n}$ obey the rules of Eq. (9.1) for a fixed n: $i = 1, \ldots , m$ (and are associated with generators e_i of g), and have zero Poisson bracket $\{v_{i,n} \ v_{j,n'}\}$ for distinct n and n'. An example of such systems is provided by a Hamiltonian H_{XYZ} supra being a restriction of a canonical completely integrable statistical mechanics model associated with the algebra of $so(3)$.

10. INFINITESIMAL OPERATORS

In order to find in general case a parametrization of an arbitrary finite dimensional Lie algebra g one needs an irreducible representation of g realized on a space of functions and infinitesimal generators of these representations.

According to our definition of parametrization, the parametrization depends on the choice $e_1, \ldots , e_m$ of g. For consistency, generators $e_1, \ldots , e_m$ can be chosen as linearly independent tangent vectors constituting a basis of a tangent space to a Lie group of g at a unit element of a Lie group of g. In order to define these tangent vectors $e_1, \ldots , e_m$ one usually takes a one-parameter subgroup $\Omega_1(t), \ldots , \Omega_m(t)$ of the Lie group G passing through a unit matrix of the Lie group of g at $t = 0$. Then the generators of a Lie group G can be defined as tangent vectors to $\Omega_1(t), \ldots , \Omega_m(t)$:

$$e_i = \frac{d\Omega_i(t)}{dt}\Big|_{t=0} \quad \text{for } i = 1, \ldots , m.$$

This definition of the generators $e_1, \ldots , e_m$ shows that two locally isomorphic Lie algebras must have the same parametrization.

Now parametrization can be constructed if one has an irreducible representation of a Lie group of g on a space H of functions $\phi(x)$. We take m one-parameter subgroups $\Omega_1(t), \ldots , \Omega_m(t)$ and consider infinitesimal operators $E_1, \ldots , E_m$ on the space H of a representation $T(g)$ corresponding to the one parameter subgroups $\Omega_1(t), \ldots , \Omega_m(t)$. These operators are defined as follows:

$$E_i = \lim_{t \to 0} \frac{T(\Omega_i(t)) - I}{t} = \frac{d\Omega_i(t)}{dt}\Big|_{t=0}$$

for $i = 1, \ldots , m$. Since a representation of a Lie group G defines an infinitesimal representation of the Lie algebra g, operators $E_1, \ldots , E_m$ define a representation of generators $e_1, \ldots , e_m$. Hence $E_1, \ldots , E_m$ obey the same commutation rules as e_1, $\ldots , e_m$. Moreover, because the representation is irreducible, the Schur lemma implies that the Casimir operators for $e_1, \ldots, e_m$ and $E_1, \ldots, E_m$ are the same. Hence, one can consider expressions for $E_1, \ldots, E_m$ as parametrization of G. In case of classical groups G, the E_i are linear differential operators on a space H and, consequently, are defined in terms of representations of the Weyl group. In this case the parametrization of e_i by E_i is nothing but a parametrization of g in terms of elements of the Weyl group. This is exactly the case, when representations of a classical group are given in terms of conformal maps as in the case of $SU(2)$, when H is a space of polynomials of degree $2l$ and the representation $T_l(g)$ of a matrix g

$$g = \begin{pmatrix} \alpha & \beta \\ \gamma & \delta \end{pmatrix}$$

is given by the formula:

$$T_l(g) \cdot \phi(z) = (\beta z + \delta)^{2l} \phi\left(\frac{\alpha z + \gamma}{\beta z + \delta}\right).$$

The value of l (which can be half-integer or an arbitrary number depending on whether the representation is finite or infinite dimensional) is connected with the value of the Casimir operator corresponding to a given choice of generator. Such representations for subgroups of the Lorentz group define infinitesimal operators that are differential operators on a space H with polynomial coefficients. Other classical groups (e. g. subgroups of upper triangular 3×3 matrices) lead to more complicated differential operators E_i.

11. BAXTER METHOD FOR QUANTUM LATTICE MODELS

We describe lattice completely integrable models in their quantized form using S matrix techniques (see geometric interpretation of section 1). Following the pattern of statistical mechanics we construct our lattice models using the method of local transfer matrices as proposed by Baxter [48, 49] (Onsager-Baxter [48-51]) but with internal variables of transfer matrices corresponding to an arbitrary representation of g (element of the universal enveloping algebra $U(g)$).

The initial object is an $e^2 \times e^2$ matrix $R(\Theta) = R_{ij,kl}(\Theta)$ called an S-matrix and interpreted according to Section 1 as defining an isomorphism $R(\Theta)$: $V(\Theta_1) \otimes V(\Theta_2) \to V(\Theta_2) \otimes V(\Theta_1)$, $\Theta = \Theta_1 - \Theta_2$]. We assume the nonsingularity of $R(\Theta)$ for Θ close to zero. As usual, lattice system is defined in terms of *local* transfer $e \times e$ matrices $\mathcal{L}_n(\lambda)$ having operator entries from a Lie algebra g, depending on λ and satisfying the following commutation relations:

$$R(\Theta_1 - \Theta_2)\, (\mathcal{L}_n(\Theta_1) \otimes \mathcal{L}_n(\Theta_2)) = (\mathcal{L}_n(\Theta_2) \otimes \mathcal{L}_n(\Theta_1))\, R(\Theta_1 - \Theta_2) \tag{11.1}$$

$$R_{n,m}\, (\mathcal{L}_n(\Theta_1) \otimes \mathcal{L}_m(\Theta_2)) = (\mathcal{L}_m(\Theta_2) \otimes \mathcal{L}_n(\Theta_1))\, R_{n,m} \,, \; n \neq m \tag{11.2}$$

where the $e^2 \times e^2$ matrix $R_{n,m}$ is $(R_{n,m})_{ij,kl} = \delta_{il}\delta_{jk}$ for $n \neq m$. The second condition in (3.3) means that elements of matrices $\mathcal{L}_n(\Theta_1)$ and $\mathcal{L}_m(\Theta_2)$ commute in g if $n \neq m$ (which is a locality condition). Under the conditions (11.1) and (11.2) we have the system of Hamiltonians commuting according to relations in g. They are given by the expansion of

$$\mathrm{Tr}\{\mathcal{L}_n(\Theta). \, . \, .\mathcal{L}_1(\Theta)\}$$

in powers of Θ. Moreover we have "inhomogeneous" system of commuting Hamiltonians as well. The existence of the family of commuting quantum Hamiltonians is based on the most important statement called the Baxter lemma, see Section 2.

Lemma 11.1: (Baxter Lemma). Let $\Theta_1, \ldots, \Theta_N$ be arbitrary complex parameters. Then the Hamiltonians $\mathrm{Tr}\{\mathcal{L}_N(\Theta_N). \, . \, .\mathcal{L}_1(\Theta_1)\}$ and $\mathrm{Tr}\{\mathcal{L}_N(\Theta_N + c). \, . \, .\mathcal{L}_1(\Theta_1 + c)\}$ commute in $U(g)$ for all c.

Here and everywhere else in similar situations, for the $e \times e$ matrix $B = (b_{ij})$ with elements b_{ij} from $U(g)$ we denote by $\mathrm{Tr}B$ the sum of the diagonal elements,

$$\mathrm{Tr}\, B = \sum_{i=1}^{e} b_{ii}$$

where the B_{ii} are elements of $U(g)$.

The proof [10] follows directly from Eq. (11.1) and Eq. (11.2). In similar setting of Kostant-Kirillov coadjoint representation method, one can find a similar statement, known as the Kostant-Symes-Adler lemma.

In general the complete family of commuting Hamiltonians is larger than that given by $\mathrm{Tr}\{\mathcal{L}_N(\Theta_N + c). \, . \, .\mathcal{L}_1(\Theta_1 + c)\}$ for different c. In the classical case the complete family of first integrals is given by coefficients of the equation for the curve $\det\{\mathcal{L}_N(\Theta_N + c). \, . \, .\mathcal{L}_1(\Theta_1 + c) - d \cdot I\}\} = P(c,d) = 0$.

The definition of the corresponding curve can be generalized for the quantum case as well.

We describe $\mathcal{L}_n(\lambda)$ associated with the algebra $gl(n)$, in the case when $\mathcal{L}_n(\lambda)$ is a rational function of λ with a simple pole of the first order in λ. In this case the (factorized) S-matrix $R(\Theta)$ is of the following form:

$$R_{ij,kl}(\Theta) = \delta_{ik}\delta_{jl}\,\frac{\eta}{\Theta + \eta} + \delta_{il}\delta_{jk}\,\frac{\Theta}{\Theta + \eta} \tag{11.3}$$

for a constant η.

Lemma 11.2: Let

$$\mathcal{L}_n(\lambda) = I + \frac{U^{(n)}}{\lambda - \mu}$$

for $U^{(n)} = (U^{(n)}_{i,j})\,{}^e_{i,j=1}$ for some μ. Then relations (11.1) and (11.2) for the S-matrix (11.3) are equivalent to the following system of commutation relations

$$[U^{(n)}_{ab}, U^{(n)}_{cd}] = \eta(\delta_{ab}U^{(n)}_{cb} - \delta_{bc}U^{(n)}_{ad})\,,$$

$$[U^{(n)}_{ab}, U^{(m)}_{cd}] = 0 \quad \text{if } n \neq m\,.$$

(11.4)

We examine the relations (11.4) in detail only for the case $e = 2$, i.e., when we are working with 2×2 matrices. These relations give rise to the following form of the Casimir operator

$$A^{(n)} \equiv \{U^{(n)}_{11}\}^2 + U^{(n)}_{12}\,U^{(n)}_{21} + \eta\,U^{(n)}_{11}\,.$$

(11.5)

Naturally we assume that $\eta \neq 0$ since otherwise g is a commutative algebra. As we see relations (11.5) are equivalent to relation (9.1) for generators of the algebra $QU(2)$ (or even $SU(2)$). Indeed, one can put

$$2e_{1,n} = \eta^{-1}\{U^{(n)}_{12} + U^{(n)}_{21}\}\,, \quad e_{2,n} = \eta^{-1}\,U^{(n)}_{11}\,, \quad 2e_{3,n} = \eta^{-1}\{U^{(n)}_{21} - U^{(n)}_{12}\}\,.$$

(11.6)

In this notation the relations of Eq. (11.5) are equivalent to $QU(2)$ relations between the $e_{i,n}$:

$$[e_{1,n}, e_{2,n}] = -e_{3,n}\,, \quad [e_{2,n}, e_{3,n}] = e_{1,n}\,, \quad [e_{3,n}, e_{1,n}] = e_{2,n}\,.$$

(11.7)

Now the Hamiltonians from Lemma 11.3 can be represented in terms of the operators $e_{i,n}$. Hence we can use our parametrization to write them down in canonical Heisenberg variables p_i, q_i arising from the representation of the Weyl algebra. The corresponding system is a generalization of the Toda lattice in view of the remark that the algebra $MH(2)$ gives rise to the Toda lattice as a limiting case of the $QU(2)$ algebra.

Let us describe some of the Hamiltonians belonging to the families of commuting Hamiltonians described in Lemma 11.3 and associated with the group $QU(2)$. We consider the homogeneous case first, where

$$\mathcal{L}(\lambda) = \begin{pmatrix} I + \{\eta/(\lambda - \mu)\}\,e_{2,n} & \{\eta/(\lambda - \mu)\}\,e_{1,n} - e_{3,n} \\ \{\eta/(\lambda - \mu)\}\,e_{1,n} + e_{3,n} & I - \{\eta/(\lambda - \mu)\}\,e_{2,n} \end{pmatrix}.$$

(11.8)

Then the system of commuting Hamiltonians given by Lemma 11.3 and Lemma 11.5

can be expressed as algebraic functions with $N + 1$ coefficients of the following polynomial in λ:

$$\mathcal{P}_N(\lambda) = (\lambda - \mu)^N \operatorname{Tr}\{\mathcal{L}_N(\lambda) \ldots \mathcal{L}_1(\lambda)\} \qquad (11.9)$$

where $\operatorname{Tr}\{\,.\,\}$ over 2×2 matrix with elements from $U(g)$ is taken over $U(g)$. We put

$$\mathcal{P}_N(\lambda) = \sum_{j=0}^{N} (\lambda - \mu)^j H_j$$

where H_j is a Hamiltonian which is a polynomial in $e_{1,n}$, $e_{2,n}$ and $e_{3,n}$ (with $n = 1, \ldots, N$). The first two H_j are trivial: $H_N = 2I$, $H_{N-1} = 0$. The third Hamiltonian is already nontrivial:

$$H_{N-2} = \sum_{k_1,k_2=1; \, k_1 < k_2}^{N} \left\{ 2\, U_{11}^{(k_1)}\, U_{11}^{(k_2)} + U_{12}^{(k_1)}\, U_{21}^{(k_2)} + U_{21}^{(k_1)}\, U_{12}^{(k_2)} \right\}. \qquad (11.10)$$

This form of the Hamiltonian H_{N-2} can be simplified if we take into account the form of the Casimir operator for $QU(2)$, which is

$$e_{1,n}^2 + e_{2,n}^2 - e_{3,n}^2 = C_n I \qquad (11.11)$$

where C_n is a constant for every irreducible representation of $QU(2)$. In other words, in terms of the $e_{i,n}$ the Hamiltonian H_{N-2} can be taken as

$$H_{N-2} = \eta^2 \left\{ \left(\sum_n e_{1,n} \right)^2 + \left(\sum_n e_{2,n} \right)^2 - \left(\sum_n e_{3,n} \right)^2 - 2 \sum_n e_{2,n}^2 \right\} + C'_N I$$

where C'_N is a constant (for each irreducible representation of $QU(2)$). Now we rewrite the Hamiltonian H_{N-2} in Darboux coordinates in two different forms using parametrization.

One gets another representation from direct parametrization of $QU(2)$. In this case the formulae involve exponents in q and polynomials in p. It is easier to use $U_{ab}^{(n)}$ coordinates. In other words, we are working directly with the relations (11.5) and the parametrization is the following one

$$U_{12}^{(n)} = \eta \left\{ e^{q_n} \frac{d}{dq_n} - (l_n + \epsilon_n) e^{-q_n} \right\}, \quad U_{21}^{(n)} = \eta \left\{ -e^{q_n} \frac{d}{dq_n} - (l_n - \epsilon_n) e^{-q_n} \right\}, \qquad (11.12)$$

$$U_{11}^{(n)} = \eta \left\{ \epsilon_n - \frac{d}{dq_n} \right\}.$$

In this notation the value of the Casimir operator is

$$A^{(n)} = \eta^2 \left\{ l_n + l_n^2 \right\}$$

and is independent of ϵ_n. Consequently the Hamiltonian H_{N-2} has the following form

$$H_{N-2} = \eta^2 \sum_{n,m=1; n \neq m}^{N} \left\{ (p_n p_m - \epsilon_m p_n - \epsilon_n p_m + 2\epsilon_n \epsilon_m) \right.$$

$$\left. + e^{(q_n - q_m)} \left(-p_n p_m - (l_m + \epsilon_m)p_n + (l_n + \epsilon_n) p_m + (l_n + \epsilon_n)(l_m - \epsilon_m) \right) \right\}. \qquad (11.13)$$

This Hamiltonian belongs to a family of completely integrable Hamiltonians, the solutions of which are expressed in terms of hyperelliptic Abelian integrals. The Hamiltonian of Eq. (11.13) makes sense in the quantum case with the proper ordering of p and q. The Hamiltonian of Eq. (11.13) contains $2N$ parameters l_n and ϵ_m. For particular choices of these parameters the solutions of the Hamiltonian of Eq. (11.13) can be expressed in terms of elementary functions. In the quantization of this system for these special values of l_n and ϵ_m the eigenfunctions can also be expressed in terms of elementary functions using the Bethe Ansatz. It should be noted that the system of integrals given by Eq. (11.9) provides us with only $N - 1$ involutive first integrals of motion for a system with N degrees of freedom. There is an additional first integral which is in involution with all the other integrals H_i for $i = 0, 1, \ldots, (N-2)$. While all the integrals H_i: $i \leq (n-2)$ are nonlinear in p_i and q_i the additional integral is the usual integral of the center of mass

$$P = \sum_{n=1}^{N} p_n.$$

Without loss of generality one can assume that $P = 0$ and, moreover,

$$\sum_{n=1}^{N} q_n = 0.$$

The fact that the Hamiltonian (11.13) is completely integrable now follows simply from Liouville's Theorem. Explicit solutions of the classical equations of (11.13) can be obtained in terms of Θ-functions using results on the commutative algebra of differential operators [52, 53, 3].

We take the simplest example, when the value of the Casimir operator is zero and so we put $l_n = \epsilon_n = 0$ for all $n = 1, \ldots, N$. The Hamiltonian of Eq. (11.13) takes the form ($\eta = 1$)

$$H_{N-2} = 2 \sum_{n,m=1; n \neq m}^{N} \left\{ 1 + \text{ch}(q_n - q_m) \right\} p_n p_m.$$

Example 11.14: Let us take the simplest model example, when $N = 2$ and two commuting Hamiltonians are

$$H_o = \left\{ 1 + \text{ch}(q_1 - q_2) \right\} p_1 p_2, \quad \text{and} \quad P = p_1 + p_2.$$

In order to write down eigenfunctions for this quantum problem one can put $q_1 + q_2 = 0$ and we obtain a single Hamiltonian

$$H' = \left\{ 1 + \operatorname{ch}(2q) \right\} p^2 = 2 \operatorname{ch}^2 q \cdot p^2 .$$

The eigenfunctions $\psi = \psi(x)$ we are looking for satisfy the following second order linear differential equation

$$\operatorname{ch}^2 x \frac{d^2 \psi}{dx^2} = 2 E \psi .$$

This equation is solved using the substitution $x = \log\frac{x_1}{1 - x_1}$ which reduces the linear differential equation above to the hypergeometric form:

$$x_1 (x_1 - 1) \frac{d^2 \psi}{dx_1^2} + (2 x_1 - 1) \frac{d^2 \psi}{dx_1} + E \psi = 0$$

which means that $\psi(x)$ is expressed in terms of

$${}_2F_1\left(\alpha, \beta, 1; \frac{e^x}{1 + e^x}\right)$$

for $\alpha + \beta = 1$, $\alpha\beta = E$, and the second linearly independent solution of the hyper-geometric equation. It is important to note that for real x, the variable x_1 takes values inside the open interval $(0, 1)$ only. The spectrum of H' can be easily determined.

Hamiltonians which are generated by $\mathcal{L}_n(\lambda)$, as given by the expansion of $\mathcal{P}_N(\lambda)$, involve nonlinear interactions between all the neighbors in the lattice model, while the Toda lattice Hamiltonians are degenerate forms of the the above Hamiltonians and involve only nearest neighbor interactions. It is possible, however, to find Hamiltonians which commute with those of $\mathcal{P}_N(\lambda)$ and involve only nearest neighbor interactions. For this to happen one considers an expansion of $\log \mathcal{P}_N(\lambda)$ in the neighborhood of $\Theta = \eta$. Surprisingly, the first nontrivial Hamiltonian in the new sequence has the same form as H_{N-2} but with the summation extended only over nearest neighbors. For example, the Hamiltonian H_{N-2} has the following commuting counterpart:

$$H_{N-2} = 2 \sum_{n=1}^{N} \left\{ 1 + \operatorname{ch}(q_n - q_{n+1}) \right\} p_n \, p_{n+1}$$

with $p_{n+N} \equiv p_n$, $q_{n+N} \equiv q_n$.

A general statement describing the relationship between lattice models with nearest neighbor interactions and lattice models with nonlocal interactions is given at the end of this chapter.

The Hamiltonian H'_{N-2} and the others commuting with it which are generated from $\mathcal{P}_N(\lambda)$, are directly related to lattice versions of the operator nonlinear Schrodinger equation constructed in Section 8 (see 8.6). This relationship is based on the direct identification of local transfer matrices $\mathcal{L}_n(\lambda)$ with local transfer matrices $\mathcal{L}(\lambda)$ from Section 3 which define the Backlund transformation (see (3.3)). For example, one can notice that the local transfer matrix

$$\mathcal{L}(\lambda) = \begin{pmatrix} \lambda - \mu_n + A_n B_n & A_n \\ B_n & I \end{pmatrix}$$

from Section 8 (that defines the lattice version of the nonlinear Schrodinger equation) satisfies the statement of Baxter's Lemma 11.3 with the factorized S-matrix $R(\Theta)$ presented in (11.3) for $m = 2$. To verify this we must demand, however, that A_n and B_n satisfy an additional commutation relation $[A_n, B_n] = \eta \, I$, reflecting the quantum nature of the lattice Hamiltonian.

Moreover, the local transfer matrices $\mathcal{L}_n(\lambda)$, determining $\mathcal{P}_N(\lambda)$ and H_{N-2}, can be constructed from two successive applications of elementary BTs, described in Section 8. Hence, the local Hamiltonian H'_{N-2} can be considered as a particular subsystem of the lattice Hamiltonian (8.6) approximating the operator nonlinear Schrodinger equation.

12. GENERALIZED QUANTUM XYZ MODELS

We want to consider now the most general form of a completely integrable lattice system associated with $SO(3)$ (or $SU(2)$). By a lattice model we understand here a quantum Hamiltonian, expressed classically as a system of difference-differential equations (without constraints). Particular reductions of these Hamiltonians give models of statistical mechanics, where the Hamiltonian is an operator in a finite-dimensional space. We already presented examples of such lattice systems associated with $SO(3)$, including a generalization of the XXX-model. This time we consider general completely integrable lattice models associated with elliptic curves and having factorized S-matrices of Baxter form [51, 8]. In a sense, such lattice models can be considered as a proper lattice version of $SO(3)$ nonlinear σ-model.

The class of Hamiltonians under consideration associated with Baxter's S-matrix [51, 8] is given (by Baxter's Lemma) using local transfer matrices $\mathcal{L}_n(\lambda)$ and the expansion of

$$\mathrm{Tr}\{\mathcal{L}_N(\lambda)\ldots\mathcal{L}_1(\lambda)\} . \tag{12.1}$$

Here the $\mathcal{L}_n(\lambda)$ are 2×2 matrices from elements of $U(g)$. In the case of two-dimensional representations of $SO(3)$ (in terms of Pauli matrices), we suppose that the $\mathcal{L}_n(\lambda)$ turn into local transfer matrices of the eight-vertex model [51] or XYZ-model of statistical mechanics. As functions of λ, elements of $\mathcal{L}_n(\lambda)$ are to be elliptic functions. According to the Baxter Lemma (11.1) proper commutation relations between elements of $\mathcal{L}_n(\lambda)$ can be derived from star-triangle relation

$$R(\lambda-\mu)\left(\mathcal{L}_n(\lambda)\otimes\mathcal{L}_n(\mu)\right) = \left(\mathcal{L}_n(\mu)\otimes\mathcal{L}_n(\lambda)\right) R(\lambda-\mu) \tag{12.2}$$

for some nonsingular 4×4 matrix with elements depending on $(\lambda - \mu)$. Such an S-matrix $R(\lambda-\mu)$ should possess certain symmetries suggested by the structure of local transfer matrices. In the eight-vertex case, these symmetries are Z/Z_2 symmetries, which means according to [8], that this S-matrix is indeed a Baxter S-matrix having the following explicit form [51, 55]:

$$R = \begin{pmatrix} a & 0 & 0 & d \\ 0 & b & c & 0 \\ 0 & c & b & 0 \\ d & 0 & 0 & a \end{pmatrix} \tag{12.3}$$

with

$$a(\lambda-\mu) = \mathrm{sn}(\lambda-\mu+\eta; k) , \qquad b(\lambda-\mu) = \mathrm{sn}(\eta; k) ,$$

$$c(\lambda-\mu) = \mathrm{sn}(\lambda-\mu; k) , \qquad d(\lambda-\mu) = k\,\mathrm{sn}(\lambda-\mu; k)\,\mathrm{sn}(\eta; k)\,\mathrm{sn}(\lambda-\mu+\eta; k) .$$

Here $\mathrm{sn}(\Theta;k)$ is the Jacobi elliptic function with the modulus k.

In order to determine quantum models associated with a given S matrix $R(\lambda)$ we are going to solve Eq. (12.2) in terms of the unknown matrix $\mathcal{L}_n(\lambda)$. A solution $\mathcal{L}_n(\lambda)$ as it is shown in [10] generates a representation of "Zamolodchikov algebra" (see Section 1) of a square of the S-matrix. It was noted in [10] that vice versa, starting from a representation of "Zamolodchikov algebra" one can get a solution $\mathcal{L}_n(\lambda)$ of Eq. (12.2). We propose a representation of the corresponding "Zamolodchikov algebra" in the Baxter case (12.3) and for degenerate S-matrices as well. Let us clarify our terminology. We remind the reader that (see Section 1) to find a representation of the "Zamolodchikov algebra" with a given S-matrix $R(\lambda)$ means to find e operators $\vec{\mathcal{A}}(\lambda)$ $= (\mathcal{A}_1(\lambda), \ldots, \mathcal{A}_e(\lambda))$, generating the vector space $V(\lambda)$, satisfying the commutation relations:

$$\vec{\mathcal{A}}(\lambda)^\dagger \vec{\mathcal{A}}(\mu) = R(\lambda-\mu)\,\vec{\mathcal{A}}(\mu)^\dagger \vec{\mathcal{A}}(\lambda) . \tag{12.4}$$

We only consider those representations of the "Zamolodchikov algebra" that can be expressed in terms of the representations of the Weyl algebra.

One starts with the Baxter S-matrix formula Eq. (12.3), where $e = 2$ and operators $\mathcal{A}_1(\Theta)$ and $\mathcal{A}_2(\Theta)$ are expressed in terms of Θ-functions in p and q. One of the expressions depends on a point τ in the upper half-plane (defining an elliptic curve) and on the parameter η from $\mathscr{C}$ (mod $Z \oplus Z\tau$). Then one has [54, 55]

$$\mathscr{A}_1(\lambda) = \Theta_1(q+\lambda)\, e^{\eta p} \, , \qquad\qquad \mathscr{A}_2(\lambda) = \Theta_4(q+\lambda)\, e^{\eta p} \, . \qquad (12.5)$$

Here Θ_1 and Θ_4 are Jacobi Θ-functions defined as follows

$$\Theta_1(\lambda) = i \sum_{n \in Z} (-1)^n e^{i\pi\{\tau(n-1/2)^2 + 2\lambda(n-1/2)\}} \, ,$$

$$\Theta_4(\lambda) = \sum_{n \in Z} e^{i\pi\{\tau n^2 + 2n(\lambda+1/2)\}} \, .$$

The classical law of addition of Θ-functions [54,55] can be used in order to prove that the commutation relations of Eq. (12.4) for the S-matrix (12.3) are satisfied.

Unfortunately, in the case of the S-matrix (12.3) it is impossible to obtain directly from the representations of the "Zamolodchikov algebra" the solution of Eq. (12.2). However, in the special degenerate case, when $\tau \to \infty$, in order to construct a solution of Eq. (12.2) it is enough to construct a representation of the "Zamolodchikov algebra". Indeed, when $\tau \to \infty$, we have $k \to 0$ and $d(\lambda-\mu) = 0$ in Eq. (12.3). Under the condition $d = 0$ every column and every row of the matrix $\mathscr{L}_n(\lambda)$ which is a solution of Eq. (12.2) defines a representation of the "Zamolodchikov algebra" and satisfies Eq. (12.4). Hence, given a fixed representation (deduced from Eq. (12.5)) of Eq. (12.4) one can construct a 2×2 matrix $\mathscr{L}_n(\lambda)$ taking its rows and columns as $G \cdot \vec{\mathscr{A}}(\lambda)^t$ for scalar matrices G.

In the degenerate cases under consideration the expression (12.5) turns into the following

$$\mathscr{A}_1(\lambda) = \sin(\lambda+q)\, e^{\eta p} \, , \qquad\qquad \mathscr{A}_2(\lambda) = e^{\eta p}$$

and they satisfy the relations (12.4) for the S-matrix $R_2(\lambda)$ expressed in elementary functions:

$$R_2(\lambda) = \begin{pmatrix} 1 & 0 & 0 & 0 \\ 0 & b & c & 0 \\ 0 & c & b & 0 \\ 0 & 0 & 0 & 1 \end{pmatrix} \qquad (12.6)$$

with

$$b = \frac{\sin\eta}{\sin(\lambda+\eta)} \, , \qquad c = \frac{\sin\lambda}{\sin(\lambda+\eta)} \, .$$

In order to construct a solution of Eq. (12.2) of the S-matrix Eq. (12.6) we describe the most general solution of Eq. (12.2) with $R_2(\lambda)$ form Eq. (12.6) assuming $\mathscr{A}_i(\lambda)$ to be the simplest rational functions in e^λ (i.e. defined in the algebraic group G_m). Such a solution was given in [10] and has the form

$$\mathscr{A}_1(\lambda) = b \ , \qquad \mathscr{A}_2(\lambda) = a \ e^\lambda + a' \ e^{-\lambda} \ ,$$

where a, a' and b satisfy the following commutation relations

$$[a \ , a'] = 0 \ , \qquad ba = e^\eta \ ab \ , \qquad ba' = e^{-\eta} \ a'b \ . \tag{12.7}$$

These relations (12.7) are particular cases of Weyl relations and hence their solutions can be parametrized by exponents in p and q. A typical solution of Eq. (12.7) is the following:

$$b = e^{\eta p} \ , \qquad a = e^q \ , \qquad a' = \gamma \ e^{-q}$$

for a constant γ. Inspired by this solution and the simple commutation relations of Eq. (12.7) we consider a 2×2 matrix $\mathscr{L}_n(\lambda)$ being the solution of Eq. (12.2) rational on the variety G_m ($\cong \mathscr{C}^*$) with the smallest possible degree in e^λ. It is easy to show that this degree is two:

$$\mathscr{L}_n(\lambda) = \begin{pmatrix} a_n(\lambda) & b_n \\ c_n & d_n(\lambda) \end{pmatrix}$$

where b_n and c_n are λ-independent. We put:

$$a_n(\lambda) = a \ e^\lambda + a' \ e^{-\lambda} \ , \qquad b_n = b \ , \qquad c_n = c \ , \qquad d_n(\lambda) = d \ e^\lambda + d' \ e^{-\lambda} \ . \tag{12.8}$$

The commutation relations implied by Eq. (12.2) are the following

$$[a \ , a'] = [a \ , d] = [a \ , d'] = [a' \ , d] = [a' \ , d'] = [d \ , d'] = 0$$

$$\begin{array}{llll} ab = e^{-\eta} \ ba \ , & a'b = e^\eta \ ba' \ , & ca = e^{-\eta} \ ac \ , & ca' = e^\eta \ a'c \ , \\ db = e^\eta \ bd \ , & d'b = e^{-\eta} \ bd' \ , & dc = e^{-\eta} \ cd \ , & d'c = e^\eta \ cd' \ , \end{array} \tag{12.9}$$

$$[c \ , b] = 2 \ \text{sh}\eta \ (a'd - ad') \ .$$

The natural parametrization of solutions of Eq. (12.9) by elements of A_1 is the following one

$$a = A_0 \ e^{c_3 p} \ , \qquad d = A_3 \ e^{-c_3 p} \ , \qquad a' = A_1 \ e^{-c_3 p} \ , \qquad d' = A_2 \ e^{c_3 p} \ ,$$

$$b = C_0 \ e^{c_1 p} \ f(p) \ , \qquad\qquad\qquad c = C_2 \ e^{-c_1 p} \ g(p) \tag{12.10}$$

for constants $A_0, \ldots, A_3, C_0, \ldots, C_3$ and functions $f(x)$ and $g(x)$ satisfying the following equation:

$$g(x+C_1)\,f(x) - f(x-C_1)\,g(x) = \frac{2\,\text{sh}\,\eta}{C_0\,C_2}\left\{ A_1\,A_3\,e^{-2c_3 x} - A_0\,A_2\,e^{2c_3 x} \right\}, \qquad (12.11)$$

and the only relationship between the constants is the following:

$$e^{c_1 c_3} = e^{-\eta}.$$

The solutions to functional equations (12.11) (e. g., taking into account an additional symmetry $f(x) = g(x)$) can be given in the general case in terms of elliptic functions. Particular solutions are, however, expressed in terms of trigonometric functions.

Assuming additional symmetries (corresponding to $SO(3)$-symmetries of lattice models defined through transfer matrices satisfying Eq. (12.2) for the S-matrix (12.6)), we can represent the solution (12.10) in the following form:

$$a = A\,e^{Dp}, \qquad a' = A\,e^{-Dp}, \qquad d = A\,e^{-Dp}, \qquad d' = A\,e^{Dp},$$

$$b = e^{\{-(\eta q)/D\}}\,f(p), \qquad\qquad c = e^{\{(\eta q)/D\}}\,f(p), \qquad\qquad (12.12)$$

$$f(x+\eta/D)\,f(x) - f(x-\eta/D)\,f(x) = 4\,\text{sh}\eta\,A^2\,\text{sh}(2Dx).$$

Here p and q are canonical Heisenberg variables: $[p\,,\,q] = 1$.

A particular solution, expressed in terms of trigonometric functions, has the form

$$a = d' = e^{p}, \qquad a' = d = e^{-p}, \qquad b = 2\,e^{-\eta q}\,\text{ch}p, \qquad c = 2\,e^{\eta q}\,\text{ch}p \qquad (12.13)$$

so that $f(x) = \text{ch}(x)$.

We can now present families of commuting quantum (lattice) Hamiltonians associated with the factorized S-matrix $R_2(\lambda)$ of Eq. (12.6). These Hamiltonians, according to the Baxter Lemma 11.1 arise as an expansion (in powers of $(\lambda - \lambda_0)$ for a given λ_0, typically, $\lambda_0 = \eta$) of

$$\text{Tr}\{\mathcal{L}_N(\lambda) \ldots \mathcal{L}_1(\lambda)\} \qquad (12.14)$$

for $\mathcal{L}_n(\lambda)$ given above with its elements parametrized by p_n and q_n, and with $[p_n\,,\,q_m] = \delta_{nm}$. In particular $\mathcal{L}_n(\lambda)$ can be chosen as

$$\mathcal{L}_n(\lambda) = \begin{pmatrix} 2A_n\,\text{ch}(\lambda+Dp_n) & \exp\{-(\eta q_n/D_n)\}\,f(p_n) \\ \exp\{(\eta q_n/D_n)\}\,f(p_n) & 2A_n\,\text{ch}(\lambda-Dp_n) \end{pmatrix} \qquad (12.15)$$

where $f(x)$ satisfies the functional equation (12.12). For a particular trigonometric solution $f(x)$ we obtain:

Corollary 12.1: Let

$$\mathcal{L}_n(\lambda) = \begin{pmatrix} \mathrm{ch}(\lambda + p_n) & e^{-\eta q_n}\,\mathrm{ch}p_n \\ e^{\eta p_n}\,\mathrm{ch}p_n & \mathrm{ch}(\lambda - p_n) \end{pmatrix}$$

where p_n and q_n satisfy the Heisenberg relations:

$$[p_n\,,\,q_m] = \delta_{nm}\,, \qquad\qquad [p_n\,,\,p_m] = [q_n\,,\,q_m] = 0\,.$$

Then for every λ and μ the quantum Hamiltonians

$$H(\lambda) = \mathrm{Tr}\{\mathcal{L}_N(\lambda)\ldots\mathcal{L}_1(\lambda)\}\,,$$

$$H(\mu) = \mathrm{Tr}\{\mathcal{L}_N(\mu)\ldots\mathcal{L}_1(\mu)\}$$

commute.

In particular, expanding $H(\lambda)$ in powers of λ, one gets a family of commuting Hamiltonians, which can be considered as quantum generalizaions of the six vertex (or *XYZ*-) model of statistical mechanics.

The *XYZ*-model itself corresponds to the second term of the expansion of $\log H(\lambda)$ in powers of λ at $\lambda = \eta$, when additional constraints (of the Clifford algebra type) are imposed, making this Hamiltonian to be a matrix of finite size.

Lattice Hamiltonians generated by Eq. (12.14) also provide lattice (local) Hamiltonians, equivalent to and approximating the quantum sine-Gordon equation. For this we remind the reader that the quantum sine-Gordon equation has $R_2(\lambda)$ (Eq. (12.1)) as its *S*-matrix [55]. Local transfer matrices $\mathcal{L}_n(\lambda)$ (12.15) are, in fact, generated by *BTs* of the quantum sine-Gordon equation, taking into account additional (generalized gauge) symmetries of eigenfunctions of linear spectral problems associated with the sine-Gordon equation.

The degenerate version of the lattice version of the sine-Gordon equation turns into the lattice version of the nonlinear Schrodinger equation form Secton 8. We also remark that, on the other hand, the lattice version of the sine-Gordon equation can be represented as a subsystem of the lattice Hamiltonians of Section 8 (under relations such as $P_{2n} = Q_{2n}$ with restrictions on μ_n). However, this reduction leads to more complicated symplectic structure, as is obviously reflected in the difference between *S*-matrices. Indeed, the system of section 8 has an *S*-matrix (11.3) (for $m=2$) while the lattice version of the sine-Gordon equation has a more complicated *S*-matrix $R_2(\lambda)$, Eq. (12.6).

Similarly, to the expression Eq. (12.15), the local transfer matrices $\mathcal{L}_n(\lambda)$, satisfying Eq. (12.2) for Baxter's *S*-matrix Eq. (12.3), can be expressed in terms of Jacobi elliptic Θ-functions in p_n and q_n. The corresponding quantum (lattice) Hamiltonians (12.1) are generaliztions of the *XYZ*- or eight vertex models.

Recently we obtained Backlund transformation formulas for XYZ-models that can be used for construction of the generalized quantum XYZ-model. In particular, using the model of Section 8, one obtains a lattice version of the classical two dimensional XYZ-model known as the Landau-Lifshitz equation (E. Sklanin, LOMI Preprint E-3-1979; U. Bogdan and A. Kovalev, JETP Lett. 31 (1980), 453). The Landau-Lifshitz equation was presented by Sklanin in the form $\vec{S}_t = \vec{S} \times \vec{S}_{xx} + \vec{S} \times J\vec{S}$ for $\vec{S} \in C^3$, $\|\vec{S}\| = 1$, $J = diag(J_1, J_2, J_3)$ with the parametrization found by Bogdan and Kovalev: $S_1 = (f_1 g + f g_1)F^{-1}$, $S_2 = -i(f_1 g - f g_1)F^{-1}$, $S_3 = (ff_1 + gg_1)F^{-1}$, with $F = ff_1 + gg_1$. The elliptic curve, used by Baxter and Sklanin is determined by J and is parametrized by (w_1, w_2, w_3): $w_i^2 - w_j^2 = (J_i - J_j)/4$. The BT B_w of a solution of a Landau-Lifshitz equation corresponding to the spectral parameter $w = (w_1, w_2, w_3)$ is determined as in Section 8 by the following formula B_w: $(f, \ldots, g_1) \to (f^+, f_1^+, g^+, g_1^+)$, where

$$(f_x^+ + z_3 f^+)F - \tfrac{1}{2} f^+ F_x = z_3 (ff_1 - gg_1)f^+ + \{(z_1 - z_2)f_1 g + (z_1 + z_2)fg_1\}g^+ \ ;$$

$$(g_x^+ + z_3 g^+)F - \tfrac{1}{2} g^+ F_x = -z_3 (ff_1 - gg_1)g^+ + \{(z_1 - z_2)fg_1 + (z_1 + z_2)f_1 g\}f^+ \ .$$

Similar formulas can be presented for the BT $B_w{}^2$: $(f, \ldots, g_1) \to (f^-, f_1^-, g^-, g_1^-)$ (gauge equivalent to the inverse BT $B_w{}^{-1}$). These and similar formulas are useful because for different degenerations of the elliptic curve, the Landau-Lifshitz equation turns into the sine-Gordon and the nonlinear Schrödinger equations.

REFERENCES

[1]. D. V. Chudnovsky, G. V. Chudnovsky, Phys. Lett. 73A (1979) 292; Phys. Lett. 74A (1979) 185;

[2]. V. E. Zakharov, A. V. Mikhailov, ZhETP, 74 (1978) 1953.

[3]. D. V. Chudnovsky, Cargese lectures 1979 in *Bifurcation Phenomena in Mathematical Physics and Related Topics*, D. Reidel 1980, 385.

[4]. D. Levi, O. Ragnisco, and M. Bruschi, Nuovo Cimento 58A (1980) 56.

[5]. D. Levi, R. Benguria, Proc. Nat. Acad. Sci. USA 77 (1980) 5025.

[6]. D. Levi, J. Phys. A. Math. Gen. 14 (1981) 1083.

[7]. S. MacLane, *Categories for the Working Mathematician*, Springer, 1971.

[8]. A. B. Zamolodchikov, Comm. Math. Phys. 69 (1979) 165.

[9]. A. B. Zamolodchikov, Soviet Physics Reports, Vol. II (1981) 2.

[10]. D. V. Chudnovsky, G. V. Chudnovsky, Phys. Lett. 98B (1981) 83.

[11]. D. Iagolnitzer, Phys. Rev. 18 (1978) 1275.

[12]. E. K. Sklanin, L. A. Takhtadjan, and L. D. Faddeev, Theor. Math. Phys., 80 (1980) 688.

[13]. M. Jackel, M.-M. Maillard, Seminaire E. N. S. (J. L. Verdier) Exp. No. 2 (1979-1980) (to be published).

[14]. A. B. Zamolodchikov, Comm. Math. Phys. (1981)

[15]. D. V. Chudnovsky, G. V. Chudnovsky, Phys. Lett. 79A (1980) 36.

[16]. D. V. Chudnovsky, Phys. Lett. 81A (1981) 105.

[17]. L. Schlesinger, *Einfuhrung in die Theorie der Gewohnlichen Differential-gleichungen auf Functionentheoretischer grundlagen*, 3 aufl, Berlin-Leipzig, 1922.

[18]. R. Garnier, Circolo Math. Palermo, 43 (1919) 155.

[19]. J. A. Lappo-Danilevsky, *Mémoires sur la theórie des systemes des equations differentielles lineares.* Chelsea, 1953.

[20]. G. D. Birkhoff, Trans. of the Amer. Math. Soc. 10 (1909) 436; 11 (1910) 199.

[21]. F. D. Gakhov, Boundary value problems, Pergamon, 1966.

[22]. B. Riemann, Oeuvres mathematiques, Blanchard, Paris, 1968, 353

[23]. D. V. Chudnovsky, Les Houches Lectures, 1979, *Complex Analysis, Microlocal Calculus and Relativistic Quantum Theory*, Lecture Notes in Physics, Vol. 126, Springer, 1980, 352.

[24]. D. V. Chudnovsky, G. V. Chudnovsky, Lett. Math. Phys. 4 (1980) 373.

[25]. G. V. Chudnovsky, Lecture Notes in Physics, v. 120 (1980) 150.

[26]. M. Sato, T. Miwa, M. Jimbo, Publ. RIMS, Kyoto Univ. 15 (1979) 201.

[27]. M. Jimbo, T. Miwa, RIMS preprints 227, 319, 327 (1980) Kyoto, Japan (to appear).

[28]. M. J. Ablowitz, D. J. Kaup, A. C. Newell, and H. Segur, Studies Appl. Math. 53 (1974) 249.

[29]. F. Calogero, A. Degasperis, Nuovo Cimento 32B (1976) 201; 39B (1977) 1.

[30]. A. C. Newell, Proc. Royal Soc. London, A365 (1979) 283.

[31]. G. V. Chudnovsky, Cargese lectures 1979 in *Bifurcation Phenomena in Mathematical Physics and Related Topics*, D. Reidel 1980, 449.

[32]. G. V. Chudnovsky, Les Houches Lectures, 1979, *Complex Analysis, Microlocal Calculus and Relativistic Quantum Theory*, Lecture Notes in Physics, Vol. 126, Springer, 1980, 136.

[33]. H. Steudel, Ann. Physik 32 (1975) 205; R. Sasaki, Phys. Lett. 78A (1980) 7; B. G. Konopelchenko, Phys. Lett. 74A (1979) 189.

[34]. D. V. Chudnovsky and G. V. Chudnovsky, Phys. Lett. 82A (1981) 271; Phys. Lett. 74A (1981) 353;

[35]. R. Hirota, J. Phys. Soc. Japan 43 (1977) 2079.

[36]. *Backlund Transformations*, edited by R. Miura, Lecture Notes in Math, Vol. 515, Springer, 1976.

[37]. D. V. Chudnovsky and G. V. Chudnovsky, Phys. Lett. 87A (1982) 325.

[38]. D. V. Chudnovsky and G. V. Chudnovsky, J. Math. Pure. Appl. (1982)

[39]. D. V. Chudnovsky, Lecture Notes in Physics, V. 120, Springer 1980, 103.

[40]. V. E. Zakharov, A. B. Shabat, Funct. Anal. Appl. 13 (1979) 13

[41]. W. Wasow, *Asymptotic Expansions for Ordinary Differential Equations*, Interscience, 1965.

[42]. D. V. Chudnovsky and G. V. Chudnovsky, Phys. Rev. Lett. 47 (1981) 1093.

[43]. D. V. Chudnovsky and G. V. Chudnovsky, J. Math. Phys. 22 (1981) 2518.

[44]. P. Lax, Comm. Pure Appl. Math. 21 (1968) 467.

[45]. E. T. Whittaker, *A Treatise on the Analytical Dynamics of Particles and Rigid Bodies*, Cambridge University Press, Cambridge, 1927.

[46]. N. Ya. Vilenkin, *Special Functions and the Theory of Group Represen tations*, AMS, Rhode Island, 1968.

[47]. A. M. Polyakov, Phys. Lett. 82B (1979) 247.

[48]. R. Baxter, Trans. Royal Soc. London A289 (1978) 315.

[49]. R. Baxter, *Exactly Solved Models of Statistical Mechanics*, Academic Press, 1982

[50]. L. Onsager, Phys. Rev. 65 (1944) 117.

[51]. R. Baxter, Ann. Phys. 76 (1973) 1, 25, 48.

[52]. J. L. Burchnall, T. W. Chaundy, Proc. London Math. Soc. 21 (1922) 420. Proc. Royal Soc. London 118 (1928) 557.

[53]. H. F. Baker, Proc. Royal Soc. London, 118 (1928) 584.

[54]. I. Cherednick, Dokl. Acad. Sci. USSR 249 (1979) 1095.

[55]. L. A. Takhtadjan, L. D. Faddeev, Usp. Math. Nauk. 34 (1979) 13.

[56]. D. V. Chudnovsky, G. V. Chudnovsky, Lett. Math. Phys. 5 (1981) 43.

[57]. M. Luscher, K. Pohlmeyer, Nucl. Phys. B137 (1978) 46.

[58]. E. H. Lieb, Phys. Rev. 162 (1967) 162

UNFASHIONABLE PURSUITS

Freeman J. Dyson

Institute for Advanced Study
Princeton, New Jersey

1. HISTORY

When I first met Feza in London long, long ago, I think he was probably the first physicist I ever met. I had the idea all physicists were like that. That had a lot to do with my becoming a physicist myself; of course I found out afterwards I was under something of a misapprehension. The reason why I think it is important to talk about Unfashionable Pursuits is because people like Feza are so uncommon. Table 1.1 gives a list of things that Feza has done in his life, with no claim to completeness. They have this common thread, that in the short run they may not be terribly relevant, but in the long run they are. That's what characterizes Feza, I think.

Table 1-1: Feza's Unfashionable Pursuits

1. Conformal Group	6. Kerr Geometry
2. DeSitter Group	7. Quaternions
3. Classical Spinning Electron	8. Octonions
4. Non-relativistic $SU(6)$	9. Exceptional Lie Algebras
5. Machian Relativity	(Especially E_7)

Historically, the fruitful developments of mathematical physics have often been discovered for irrelevant reasons to solve problems which turned out not to be very interesting. Only many, many years later, typically fifty or a hundred years later, they turned out to be extraordinarily fruitful and interesting for reasons which the inventors would never have foreseen, and frequently would never have understood. That's why it's very important, if you want to be a mathematical physicist, to be unfashionable. To work on the borderline between physics and mathematics you had better be unfashionable. It is much more likely that you will do something that in the end will turn out to be important. So I have just collected a few historical examples to illustrate what I have in mind.

The first example is the well known fact that Gauss developed differential geometry in 1827 because he was interested in problems of geodesy, the representation of the earth on plane maps, a not very profound problem which certainly had nothing to do with general relativity. Much later, in 1854, in the hands of Riemann and Clifford, and finally Levi-Civita and Einstein it led to very great things, but the man who invented it was long ago dead and buried.

Another example concerns Felix Klein, who when he was a very young man--he was twenty-three and a full professor in Erlangen in 1873--gave an inaugural lecture which was one of the great documents in the history of mathematics, the so-called Erlanger Programm. This laid down the course which mathematics was supposed to follow for the next 100 years [Klein, 1872]. It's a wonderful description. He said right then and there that the function of mathematics is to classify geometries and dynamical systems by means of group theory. Nobody listened except Sophus Lie. It happened that Sophus Lie was also a radical young man. He had just been released from a French jail at that time. He was decidedly not an establishment type. He happened to be at Erlangen and helped Felix Klein write this lecture and they were both very enthusiastic about it. Later on, in about 1888, Lie invented Lie algebras and Lie groups. Lie thought they were great, but the mathematical profession did not agree. Ultimately they turned out to be absolutely essential to the development of physics. First, they were introduced into Quantum Mechanics by Weyl in 1928 and Wigner in 1931. In our own time, in the period 1960-1980, we have seen particle physics emerge as the playground of group theory with the work of Gell-Mann, Nambu, Salam, Weinberg and Gürsey. But let me read to you what the historian Rouse Ball wrote in the "Short History of Mathematics", Fourth Edition [Rouse Ball, 1908], describing how things looked in 1908: "Lie seemed to have been disappointed and soured by the absence of any general recognition of the value of his results. He brooded over what he deemed was the undue neglect of the past, and the happiness of the last decade of his life was much affected by it." So poor old Lie, he never knew that he was to be so honored and revered in the years to come. Thanks to you people, Feza didn't suffer that fate.

Let me give you another equally striking example. I think the example of Felix Klein and Sophus Lie was in some respects unusual in that they did know what they were doing. The problem that they were addressing was in fact the real problem; that was very unusual. It was certainly not true of Gauss and Riemann, and is certainly even less true of Hermann Weyl when he invented gauge. When Hermann Weyl invented gauge in 1919, he was trying to fix up Einstein's theory of gravitation and derive Maxwell's equations from the equations of gravitation. He somehow got the right formalism but completely misunderstood how it fitted in. As we all know, that was finally clarified only after we had quantum mechanics and understood that i was an important thing to put in front of the electromagnetic potentials (Table 1.2).

Table 1-2: The Gauge Principle

Gauge Theory of Gravitation
Weyl 1919

Quantum Electrodynamics
Fermi-Heisenberg-Pauli-Dirac
1930

Non-Abelian Gauge Theory
Yang-Mills 1954

Quantum Chromodynamics

And so the gauge became phase instead of scale. This gauge idea of Hermann Weyl has turned out to be another of the central ideas of physics in the last fifty years. I didn't put down who invented quantum chromodynamics because the list is too long.

Here is a fourth example of historical lag. The average time that it took from the idea to the application was fifty to a hundred years--this fourth example shows about the longest lag I can turn up. It begins with Hermann Grassmann who published a wonderful treatise "Ausdehnungslehre", which is a very difficult book to read; it's probably well worth while if you make the effort [Grassmann, 1844]. He was a high school teacher in Stettin, and nobody had heard of him when he published this extraordinary work in which he invented vectors and vector-spaces and anticommuting algebras, all sorts of wonderful things. I don't think anybody understood what it was all about. It was very largely ignored. And it led eventually, after many vicissitudes, long after Grassmann was in his grave, to all sorts of nice things, like supersymmetry and supergravity and superspace and so on.

The last example is not really from physics at all. It's something that I just happen to be very much interested in. Finite groups are now in a process of great flowering; mathematicians are doing great things with finite groups. Physics hasn't yet caught up; there's very little interest among physicists in finite groups. But I think this may be something for a hundred years from now; we really don't know. It is hopeless to try to predict what's going to be useful. The mathematicians sow the seed, some of it falls on stony soil and some doesn't. This one I have hopes for, but certainly without any evidence. Anyway, if you look at the papers of Émile Mathieu in 1861 and 1873, you see that he discovered these wonderful sporadic groups, that is finite groups which don't fit into any of the classical series of groups such as the classical orthogonal and permutation groups. He discovered the first of these sporadic groups which we call M_{12} and M_{24}. In the papers that Mathieu wrote the word group is never mentioned [Mathieu, 1861, 1873]. He talked about functions of several variables, and what he was interested in was some problem that we would consider totally irrelevant. But he found these things and he understood at least that they were beautiful, so eventually he got around to publishing them. He wasn't in any hurry--it took him twelve years to go from his first publication to his second, and he hadn't done anything in the meantime. And nobody was very much interested in it at the time. But now a hundred years later things begin to hum. The second stage in this particular development began in 1967 with the invention of the Leech lattice, which I shall have something to say about later. It turned out to be a key to what you might call the zoology of the finite sporadic groups. The Leech lattice is the beautiful lattice that exists in a space of 24 dimensions and which seems to have a very unique quality, [Leech, 1967]. The automorphism group of the Leech lattice is the Conway group which is another very elegant finite group, and out of that, just in the last ten years, has grown the monster which is the most exciting part of the subject. I'll have something to say about that later, too. This is an example of something that hasn't yet found its application, but otherwise it's very typical. The time-scale--because of our general slowness and general dumbness--doesn't seem to accelerate with the progress of material civilization or with the number of published journals or with the number of entries in the citation index, or with anything else. It seems to be just about as slow now as it always was.

2. SPHERE PACKING

I have talked so far about history, and what I'd like to do next is to discuss the things that are now going on that I happen to be interested in. Most of these may turn out to be quite trivial. You can't see the future. You can't tell. Most of the things that I shall talk about are very likely of no importance whatever. They are things that are now going on, things that mathematicians do that most physicists don't know about. I thought it would be interesting to go through a few of them--to give you a taste of mathematical developments that may or may not turn out to be important to your grandchildren. And they also are a little bit in the style and in the tradition of the things that Feza does, so I thought he might be amused by them too. The first thing on my list is the packing of spheres, which is an old problem--how many spheres can you pack together in a Euclidean space? It's a problem that many mathematicians have been interested in. And it has some interesting connections with other things-- that's the beauty of pure mathematics--one is always discovering new connections between things that looked unconnected. There is first of all a connection between packings of spheres and Lie algebras which has been discovered by Coxeter in the course of his geometrical interpretation of the Lie groups. Coxeter discovered that certain finite groups, which are the automorphism groups of the root diagrams of the Lie algebras, are connected very closely with packings of spheres. These groups are also symmetry groups of sphere packings.

Table 2-1: Maximum Density of Sphere Packings

ρ_n = Density of densest lattice packing of unit spheres in $\mathfrak{R}_n$.

$$\rho_{11} = \rho_{12} = \rho_{13} = 2^{-5}$$

$$\rho_{10} = \rho_{14} = 2^{-4}\, 3^{-1/2}$$

$$\rho_9 = \rho_{15} = 2^{-9/2}$$

$$\rho_7 = \rho_8 = \rho_{16} = \rho_{17} = 2^{-4}$$

$$\rho_6 = \rho_{18} = 2^{-3}\, 3^{-1/2}$$

$$\rho_5 = \rho_{19} = 2^{-7/2}$$

$$\rho_4 = \rho_{20} = 2^{-3}$$

$$\rho_3 = \rho_{21} = 2^{-5/2}$$

$$\rho_2 = \rho_{22} = 2^{-1}\, 3^{-1/2}$$

$$\rho_1 = \rho_{23} = 2^{-1}$$

$$\rho_{24} = 1$$

Theorem: $\rho_n = \rho_{24-n}$.

Table 2.1 comes from a paper by Leech and Sloane [1971]. You don't have to look at all this--the thing is to start at the bottom and work your way up. These are the densities of the densest packings of spheres on lattices in spaces of n dimensions. So this is Euclidean n-space. You take a lattice and around each lattice point you draw a unit sphere, and the rule of the game is that the spheres must not overlap. They may touch each other, but they may not overlap, and the trick is to find the densest lattice that gives you the highest density of spheres in the space. So you start at the bottom and work your way up. The first is ρ_1--that's the case of packing circles along a one-dimensional line--the diameter of a circle is two so obviously the density is one-half. That's a trivial case. It turns out that three points on a line make the root diagram of the group $SU(2)$, a trivial root diagram having three roots. The next case ρ_2 is packing of circles in two dimensions, where the density is one over the twice the square root of three, and that corresponds to the Coxeter group of $SU(3)$, that's the next Lie algebra A_2 in the classical series. Sure enough $SU(3)$ comes in there. The next one is ρ_3 where the density is one over four times square root of two, and that corresponds to a very nice lattice with the symmetry of the root diagram of the Lie algebra D_3 which is the same as A_3 since the groups $SU(4)$ and $O(6)$ are locally isomorphic. It jumps from the A series to the D series at that point. The next case is ρ_4, that's the density of the lattice packing in four-dimensions which has density one eighth, 1/8, and corresponds to the algebra D_4 or the group $O(8)$. The next case is D_5, but then it jumps from the D to the E series and goes on to E_6, E_7, E_8. There is a very beautiful one in eight dimensions which has density one-sixteenth. This lattice has an extremely high degree of symmetry which corresponds to the exceptional Lie algebra E_8, which some of us, including Feza, are very fond of; and so it goes.

The interesting question is what happens after 8 -- the sphere packings continue but the Lie algebras don't. There's no Lie algebra corresponding to the sphere packings in nine dimensions and beyond. Well, why not? What's going on? It's an interesting question. We don't know. There should be something there which the mathematicians have missed. But anyhow, there continue to be these interesting symmetrical configurations beyond eight dimensions, and you go on up the left-hand side of Table 2.1 and you find row 9, row 10, row 11 and row 12.

What is unexpected is that the thing comes to a minimum at row 12, and 12 dimensions is the point of return, so to speak, where the density is one-thirty-second, and that's the smallest density you ever get in any number of dimensions. After that it starts to go up again, and you find that it goes up at fourteen to the same density as ten. Very much to their surprise, when Leech and Sloane worked out these densities, at first it looked as if it might be symmetrical, then it looked unsymmetrical, but in the end it always turned out when it wasn't symmetrical somebody had made a mistake. So now after it's all cleaned up and we have the best lattices in all these dimensions, it turns out there is a precise duality going down the right-hand side between dimension d and dimension $(24-d)$. There is something magic about 24 dimensions, and for every close-packed sphere in d-dimensions there is an analogous one in $(24-d)$ dimensions with the identical density. They don't look alike, there's no obvious connection between them at all, but on a deep level there must be, and so that's a nice problem for somebody.

When you get back finally to 24 you have the Leech lattice which corresponds to zero dimension. The Leech lattice in 24 dimensions has density one, and is the most beautiful of all, with the highest symmetry. When you go on beyond 24 there is no possibility of a duality anymore, and beyond that point we know very little. We know there are sphere packings--the densities then go up beyond one--they are mostly greater than one when you go beyond 24, but we don't know anything about what happens there. So there is this beautiful theorem $\rho_n = \rho_{24-n}$, which is verified empirically, but there is no proof of the duality which leads to any sort of understanding. The only proof is just writing down the table of numbers and saying--Look!

3. POLYHEDRA AND CAGES

Here's another example of something nice which was discovered recently which may not have any connection with anything but it's beautiful. That is Coxeter's 11-faced polyhedron [Coxeter, 1981]. Plato discovered regular polyhedra, or maybe he didn't discover them, but anyway, he publicized the fact that there were five regular solids, the tetrahedron, the cube, the octahedron, the dodecahedron, and the icosahedron, which are aesthetically very satisfying--they are nice regular solids--they have beautiful faces which are regular polygons arranged symmetrically at each vertex. They are in every way as symmetrical as possibly could be. After Plato a thousand years went by, or two thousand, and in the nineteenth century people began looking again. They began looking in four dimensions, thinking they might find some polyhedra in four dimensions, and they did. They found six in four dimensions which have nice properties, but they all have familiar symmetries. The largest of them has 600 faces, and it has a symmetry group which is just the direct product of two icosahedra. It's a very beautiful solid, but it has nothing that one would say is radically new. And at that point everybody thought the game was finished; it was a nice problem for a graduate student or maybe for a high school kid, but there wasn't much more to be done.

Then just last year Donald Coxeter, the same Coxeter who discovered the Coxeter groups for the root diagrams of the Lie algebras, now 40 years later suddenly found a new regular solid, to everybody's great astonishment. Table 3.1 gives some of its properties. It has 11 faces, and it exists in four dimensions. Nobody would have believed such a thing is possible. How could something have 11 faces and be symmetrical? Anyhow, Fig. 3-1 shows a picture of it. There it is, there are its 11 faces,

Table 3-1:   Coxeter's New Regular Polyhedron

11 Vertices	55 Edges (Triangles)
11 3-Cells	55 2-Cells (Faces)

1. Self-dual configuration in hyperbolic 3-space.
2. Each face is a hemi-icosahedron. (An icosahedron with pairs of opposite vertices identified.)
3. Faces meet at each vertex in a hemi-dodecahedron arrangement.
4. The symmetry group of this object is $PSL(2,11)$ which has order 660.

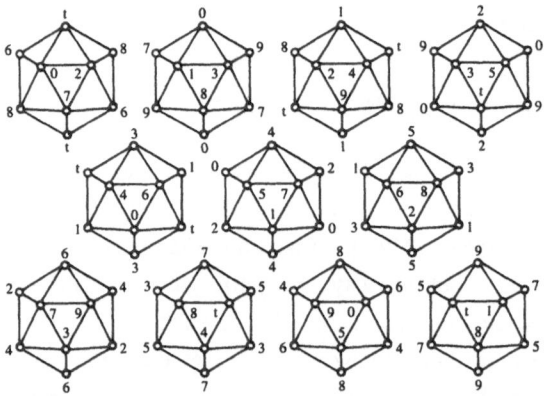

Figure 3-1:   Coxeter's New Regular Polyhedron.

and you see each face is a hemi-icosahedron. So instead of a standard icosahedron with 12 vertices you use a hemi-icosahedron which only has 6. If you identify opposite vertices you get something which is just as symmetrical as an icosahedron, but it has only six vertices and only ten triangular faces instead of twenty. Fig. 3-1 tells you how to stick them together and get your 11-faced polyhedron. It's a lovely thing, it's self-dual, that is to say, there is a duality symmetry between vertices and the faces. It has 11 vertices, 55 edges, 55 triangular 2-cells and 11 3-cells which are the faces. The symmetry group is $PSL(2,11)$. As you might expect, it is not a new group. It has order 660. It's a well-known group belonging to one of the classical families; it is a projective special linear group PSL over the Galois field of order 11. Coxeter's polyhedron is one of these nice sporadic objects. What attracts me about these objects is that they all have the property of being unique and individualistic.

There is another thing which Coxeter has been interested in which is connected with his polyhedron. That's the problem of the cage, which some pure mathematicians have been amusing themselves with. A cage is a graph which has the highest degree of tree-like quality combined with a high degree of symmetry. You try to construct a finite graph which locally behaves like a tree, that is, it doesn't have any closed cycles, but globally it has a very high degree of symmetry. Of course this is a contradiction in terms, because an actual tree has to have twigs and has to have a lot of endpoints which are not like the interior vertices, so a real tree has an exterior which is quite unsymmetrical as compared with the interior. When you are trying to make a symmetrical graph which is like a tree, you've got to make a compromise. You don't say there are no cycles, you say that all the cycles in the graph have to be as long as possible. You try to make a finite graph in which the length of the shortest cycle is given and you have to find the graph which has the smallest number of vertices with the given shortest closed cycle. That's the problem. And you can define the cage as being that particular graph. You can do it with any number of edges meeting at each vertex. The classification problem considers trivalent graphs which have just three edges meeting at each vertex. You try to construct a graph with not too many vertices and no short cycles. So it's a kind of approximation to a tree.

The case of an even number of vertices in the cycle is the simplest and in some ways the most interesting. The odd case has also some interesting twists to it--twists in the literal sense, but the even case is the one I'll talk about. The order of a graph means the total number of vertices, and the girth of a graph means the length of the shortest closed cycle. When the girth is an even number $2k$, you can easily prove that the order of the graph must be $2(2^k-1)$. For example, for the girth 8 you can prove that the cage has order 30. There is a very beautiful symmetric graph of order 30, with 30 vertices and girth 8, which was discovered by Tutte, and is in some ways the most symmetrical of all graphs [Tutte, 1966]. It has a very high symmetry group. So Tutte conjectured that for all even numbers you could construct a cage which had order precisely $2(2^k-1)$. Well, it turns out that that's not true. It was proved by a graduate student in Princeton called Singleton that you can only do it with graphs of order 2, 6, 14, 30 and 126. That's all there are. For all the other cases, there is no highly symmetrical graph which has all the properties we want it to have. And that's in a way a disappointment, but in some ways also nice. It's like the exceptional Lie algebras, it's nice that there's only a finite number of them, so that you can get to know them as individuals.

The case of girth 12 is particularly beautiful. That's the largest one, and for a time, after Singleton proved that you couldn't do it for any others, nobody found a good way of constructing the cage of girth 12. It was done with computer programs, verifying laboriously that you could find one, but the symmetry was not obvious. Then Clark Benson came along and showed how to construct it in an elegant fashion, and

I'll tell you the construction just to show you how beautiful something like that can be. This is the construction of the 12-cage, the graph of 126 vertices with girth 12, and it can be done using finite geometry in a very elegant style [Benson, 1966]. You use geometry over the Galois field of order 2, that's the field whose elements are just zero and one. You look at the six-dimensional projective space over this field, that is, points which are described by seven homogeneous coordinates; each coordinate just takes the value zero or one. It's a little bit like the Fermion dimensions in superspace. It's a space in which each coordinate only takes 2 values. The projective space has 127 points, because each coordinate can be zero or one but the point with all coordinates zero doesn't count. The nice thing is you can use physicists's notation--you call the first three coordinates an electric field, the next three coordinates components of a magnetic field, and the seventh one is just normalization, you can call it a scalar potential if you like. Benson then considers the quadric Q defined by the equation $E{\cdot}H + \Phi^2 = 0$. Just half the points of the space are on Q, 63 points altogether. These 63 points on Q are the even vertices of Benson's graph. Next, Benson constructs a family of lines of which there are also 63, by writing down equations which look superficially like the Maxwell equations, to set up a duality of the space between points and lines. These 63 lines are the odd vertices of his graph. Given any point on the quadric you can construct a set of lines which are solutions of these Maxwell equations, in fact there are three lines through each point, and those are precisely the three vertices of the graph which are incident on the vertex that you started from. So you get a very beautiful construction without any work at all, without any computation, you can see at a glance that this configuration has exactly the properties that you want. The fact that the cage exists in this particular number of dimensions is somehow connected with the fact that the Maxwell field has six components. Well, let me go ahead. This is what you might call mathematical salad.

4. MODULAR GROUP AND MACDONALD IDENTITIES

The last collection of things I'm going to talk about is concerned with the modular group, which is something the mathematicians, starting with Felix Klein himself, have always been obsessed with. Klein wrote innumerable papers about the modular group, and it must be important, but no physicist has yet been smart enough to use it. It's a nonlinear group, and, in addition to all that, it brings geometry into the center of the picture. Let me not talk generalities, let me talk about one or two things in detail. The modular group is in mathematician's notation $SL(2,Z)$. It's the group $SL(2)$ taken over the integers. It's the group of unimodular substitutions with integer coefficients and unit determinant. It can also be represented as a group of transformations of one complex variable in the upper half-plane. Many of you have probably seen pictures of these transformations of the upper half-plane into itself.

Here are some the things which were unexpected and interesting which turned up in connection with the modular group. One of the things which you can do when you have a modular group, is to invent things called modular forms. A modular form is a function of one variable which transforms not exactly into itself, but up to a scale factor it transforms into itself under the operations of the group. The genus of the form is just the degree of the scale factor that comes in when you make the transformation. Here is an example, the Dedekind modular function:

$$\eta(x) = x^{1/24} \prod_{n=1}^{\infty} (1 - x^n) .$$

It has the factor of $x^{1/24}$ at the beginning, and the 1/24 is very essential, nobody understands why. The rest of it is just the infinite product of $(1 - x^n)$. That's the Dedekind eta function, a modular form of genus one-half. In some way's it's the most fundamental of the modular forms. You can't have anything less than one-half. The genuses have to be half-integers or integers, so the eta function is a sort of unit in the space of modular forms, and in the theory of the modular group it is the basic quantity that you manipulate. It's a very essential covariant of the modular group.

Now Ian MacDonald, who is a very pure mathematician, a geometer who lives in Oxford, has discovered weird identities which connect the Dedekind eta function with Lie algebras in a very strange fashion [MacDonald, 1972]. You have one identity of this kind corresponding to each Lie algebra, or you may have several because you can extend the Lie algebra in various ways by adjoining commuting elements to it, but in any case to a first approximation you have one identity of this kind for each Lie algebra. For example, the Lie algebra corresponding to G_2 is a very nice one, and this is the corresponding identity:

MacDonald Identity for G_2

$$\eta^7(x) \, \eta^7(x^3) = \frac{1}{120} \sum_{(abc)} abc(a-b)(a-c)(b-c) \, x^{(a^2 + b^2 + c^2)/12}$$

$$a, b, c, \equiv 1, 2, 3 \ (\mathrm{mod}\ 6) , \qquad\qquad a + b + c = 0 .$$

On the left-hand side appears the seventh power of the eta function of x multiplied by the seventh power of the eta function of x^3, and that's connected with the way the root diagram looks. The root diagram for G_2 has six short roots and six long roots, and two more in the center which are zero. You have fourteen altogether, which is the dimension of the algebra. If you look carefully at this identity you see on the left side the powers of the eta function which is a nice modular form. On the right-hand side you have a power series, and the coefficients of the various powers of x are given simply as products of factors, each factor corresponding to one root of the algebra. They have this beautiful geometrical structure $(abc)(a-b)(b-c)(a-c)$, which corresponds exactly to the roots in the root diagram of the algebra. MacDonald could prove this, but he couldn't understand it, and nobody yet has really understood it on a deep level. There's this strange connection between the modular group on one side and the general structure of Lie algebras on the other.

I couldn't resist writing down the identity for E_7 since E_7 is Feza's favorite, and I happen to enjoy this kind of thing too. Here is the identity for E_7:

MacDonald Identity for E_7

$$\eta^{133}(x) = \frac{1}{2^{110}\,3^{22}\,5^{10}\,7^6\,11^3\,13^2\,17} \times \sum_{(abcdefg)} \left[x^{(a^2+b^2+c^2+d^2+e^2+f^2+g^2-s^2)/144} \right] \times$$

$$\prod_{7}(a+s) \prod_{21}{}'(a-b) \prod_{35}(a+b+c-s)]$$

$a, b, c, d, e, f, g \equiv 9,\ 11,\ 13,\ 15,\ 17,\ 19,\ 21 \pmod{36}$

$3s + a + b + c + d + e + f + g = 0$.

As you may know, the dimension of of E_7 is 133, so you get a beautiful formula for the 133rd power of the eta function; MacDonald also proved this one. You see on the right-hand side again this product of now 63 linear factors, each corresponding to one pair of roots of E_7. And it's all true, you can verify it—it really is so. But we still don't understand why. There's also what is called the strange formula, which is something that pure mathematicians have been intrigued by for a long time. The strange formula says, if you take all the positive roots and add them together and form a scalar product with the Killing metric for any Lie algebra, then that is equal to the dimension of the Lie algebra divided by 24, and it's only because that strange formula is true that these identities can hold. If you look at the way the $x^{1/24}$ appears on the left-hand side raised to the power 133, this means that $d/24$ is the power of x that sits outside on the left-hand side. On the right-hand side you have a certain quantity in the exponent which is a sum of squares of positive roots, and you see the identity can only be true because this strange formula holds. That's something which MacDonald discovered in the course of proving these identities. But there is no understanding of this from any deeper point of view. Again 24 seems to be very central in the whole business but nobody knows why.

Here is another thing, this is dimension 26 for a change. In particle physics the string model for some reason likes 26 dimensions, and so maybe we might have a glimmering of a connection with physics, but I doubt it. Anyway, 26 dimensions also turns out to be very interesting from the point of view of the modular forms. Oliver Atkin, who is a pure mathematician with a taste for numerology, who loves computers and loves to tinker around with modular forms on computers, started looking at the 26th power of the Dedekind eta function. He expanded it in a power-series and computed the coefficients.

$$\prod_{n=1}^{\infty} (1 - x^n)^{26} = \sum_{n=0}^{\infty} p_{26}(n) \, x^n .$$

And for some reason he started factorizing these coefficients, just for the hell of it, because he liked factorizing, he had written a clever program to factorize big numbers very fast. He found that these coefficients have very weird factors.

$p_{26}(100) = 3 \times 13 \times 23 \times 37^2 \times 47 \times 61 \times 71$

$p_{26}(101) = 2^2 \times 11 \times 13 \times 23^2 \times 37 \times 97 \times 107$

$p_{26}(102) = -3 \times 13 \times 23^2 \times 47 \times 59 \times 97 \times 109$

$p_{26}(103) = 2 \times 5^2 \times 13 \times 23 \times 37 \times 83 \times 97 \times 131$

$\cdots \cdots$

$p_{26}(1000) = 2^2 \times 3 \times 5^2 \times 11 \times 13 \times 17 \times 19 \times 23 \times 59 \times 157 \times 277 \times 863 .$

You see, big numbers are not supposed to have many prime factors. Hardy in fact proved the theorem that in a very precise sense, if you take a random very large number N, it has exactly log log N prime factors. So these numbers that Atkin factorized are not behaving as random large numbers should. They all seem to have a large number of comparatively small prime factors. And that intrigued him very much. He found that this goes on, as far as he went, but it's not always true. He found there are some of these coefficients with only a few big prime factors. Some in fact he wasn't able to factorize with his program. That intrigued him even more, because the strange behavior is sometimes true but not always. What could be the explanation?

Atkin went one step further and found some beautiful identities for the 26th power of the Dedekind eta function. Here's an example of one of Atkin's identities:

Atkin Identity

If $N = 12n + 13$ is prime, we have the unique representations

$$N = a^2 + b^2 = c^2 + 3d^2 , \quad a \equiv 0 \quad (mod\ 3)$$

$$p_{26}(n) = \frac{(-1)^a}{2^3\ 3^4\ 11^2\ 13}(a - c)(a + c)(b - c)(b + c)$$

$$\times(2a + c + 3d)(2a + c - 3d)(2a - c + 3d)(2a - c - 3d)$$

$$\times(2b + c + 3d)(2b + c - 3d)\ (2b - c + 3d)(2b - c - 3d) .$$

This one happens to be true if the number $N = (12n + 13)$ is prime. N is the exponent that you get if you put in the $x^{1/24}$ in the right place. If N happens to be prime, then you get a formula for the nth coefficient in the power series for the 26th power of the eta function. It looks just like the one we had before for E_7; it's a product of twelve linear factors, so that explains why you got so many different prime factors. And it also looks like the root diagram of a Lie algebra, but there is no Lie algebra of dimension 26. So that indicates again something special. What is it that happens in dimension 26? We don't know. It's not a Lie algebra, but it must be something. There it is, so again mystery. That's not the only formula of that kind. When N is not prime you get other formulas which sometimes give you beautiful factorizations. That's more than I can go into here. But there it is, you have this strange object which appears in dimension 26.

There is one other interesting object that exists in dimension 26, the exceptional Jordan algebra over the octonions, that's something which Feza certainly knows all about, maybe some of you do too. The octonions are the simplest nonassociative algebra, and they have many beautiful properties. People like Feza and me are always trying to find ways of dragging the octonions into physics. Even Feza really hasn't succeeded. But anyway the chances are it will work. And if you look for extensions of the octonion algebra, there is precisely one which is very beautiful; namely this Jordan nonassociative algebra, which is the algebra of 3×3 matrices which are self-adjoint with octonion elements, divided by the center which is simply the scalar algebra. The generators of the algebra are traceless 3×3 matrices, which form a manifold of dimension 26. Maybe that has something to do with Atkin's identities or maybe not.

5. ZETA-FUNCTION ZEROS

As many of you may know, the greatest unsolved problem in pure mathematics, or the most famous unsolved problem, is the Riemann hypothesis, the hypothesis that all the complex zeros of the zeta function have real parts equal to one-half. Well, suppose the Riemann hypothesis is true; you look at the distribution of zeros on the critical line, on the line real part equal to one-half, and see how they behave. You can look for things like the pair correlation function of zeta function zeros. It turns out that for reasons which again we don't understand, the pair correlation function for the zeros of the zeta function is equal to the very interesting quantity $1 - (\frac{\sin \pi x}{\pi x})^2$; that is, the quantity $P(E_1, E_2) = 1 - (\frac{\sin \pi x}{\pi x})^2$, where $x = (E_1 - E_2)/\Delta$ and Δ is the mean spacing. This is an elementary function, and it happens to be identical with the pair correlation function for the eigenvalues of a very large random Hermitian matrix. There is a large body of information about the statistics of eigenvalues of random matrices, which may or may not have something to do with physics [Mehta, 1967]. It's a remarkable fact that these two correlation functions happen to be the same. That of course has not been proved. First of all you have to prove the Riemann hypothesis and then maybe you might be able to prove this. So it's very far from being proved. But it's numerically verified to a high degree of accuracy by a mathematician called Andrew Odlyzko who has the good luck to be working at Bell Labs and has a Cray computer at his disposal. So he can run off zeros of the zeta function on the Cray by the tens of millions, and he has a nice little program that lets him do that. He's gone way beyond anybody else in computing zeros of the zeta function.

Figure 5-1 shows the pair correlation function that Odlyzko finds. You can see that it very beautifully follows the curve $1 - (\frac{\sin \pi x}{\pi x})^2$. Actually, if you look at it carefully, it overshoots slightly. The wiggles are very slightly but quite definitely outside the statistics. They are larger than they should be, so it slightly overshoots at each wiggle. So there's another problem that we don't understand. Why should the zeros of the zeta function behave like the eigenvalues of a random Hermitian matrix? Of course, the obvious answer is that they are the eigenvalues of some particular Hermitian matrix. That would be wonderful. If they are the eigenvalues of a random Hermitian matrix, then you are only left with the trivial exercise of finding out which one it is. If you found out which random Hermitian matrix these are the eigenvalues of, you would prove the Riemann hypothesis, and you'd go down in history, and you'd get the Fields medal and everything else.

You can also look at the spacing distribution, which is something that physicists have been more accustomed to. When physicists find a series of eigenvalues they tend

Freeman J. Dyson

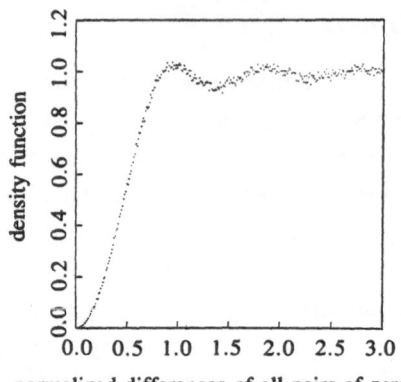

normalized differences of all pairs of zeros

Figure 5-1:   Pair Correlation Function of Zeta-Function Zeros.

to look at the statistics of the spacings, so Odlyzko thought, why not look at the distribution of the spacings of the zeros of zeta function?   Figure 5-2 shows Odlyzko's graph--you see it looks very much like curves that the physicists have been drawing. Madan Lal Mehta did this for the first time, and the Mehta curve for the spacing distribution of the eigenvalues of a random Hermitian matrix looks exactly like that.   It agrees beautifully and so you think you have a nice conjecture.

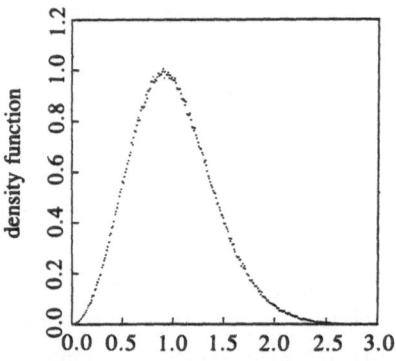

normalized differences of all pairs of zeros

Figure 5-2:   Spacing Distribution of Zeta-Function Zeros.

However, let's go one step further because you have the Fast Fourier Transform program which is trivial to do on the Cray.   So Odlyzko took a Fourier spectrum of the spacing distribution of the zeros of the zeta function.   You would expect to find a nice smooth curve, and in fact if you took $1 - (\frac{\sin \pi x}{\pi x})^2$, that has a very simple Fourier transform, it's just a triangle.   Figure 5-3 is the Fourier transform of the spacing distribution calculated by Odlyzko.   I should say that he has subtracted out the trivial part,

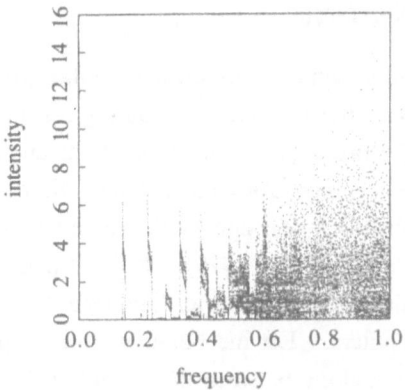

Figure 5-3: Fourier Transform of Spacing Distribution of Zeta Function Zeros.

the part that's given by the random matrix formula, which is just a triangle. Figure 5-3 is what's left over after you have subtracted the elementary conjecture. You might expect that you have nothing but noise after you subtracted that out. Figure 5-3 shows the Fourier transform--frequency plotted horizontally, intensity plotted vertically. And you see it's a weird and wonderful function. You can hardly see on that scale what it's doing. Figure 5-4 is an expanded scale.

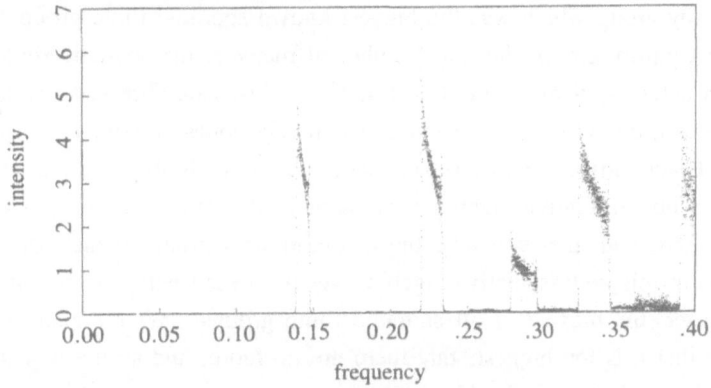

Figure 5-4: Fourier Transform of Spacing Distribution of Zeta Function Zeros.
Expanded Scale.

It's a very clean spectrum. Starting from zero frequency, it stays very close to zero, with practically no noise at all, until abruptly at frequency 0.14 it suddenly shoots up and then it comes down in rather a ragged fashion until about 0.149 where it suddenly drops down to zero. Then if you go a little further along the axis it does that three or four more times. It's a very strange spectrum. I couldn't say more about that without getting into fine points of number theory and analytic function theory.

6. MONSTROUS MOONSHINE

I come now to my last item, which goes back to finite groups and modular functions. I left this until the end, after the zeta function digression, although it is connected with what I was saying before about modular functions and Lie algebras and finite groups. The last item is called monstrous moonshine. Monstrous moonshine is the title of a paper that was published a year ago by the same Conway who discovered the Conway groups. [Conway and Norton, 1979]. He's also the inventor of the game "Life" which has been publicized by Scientific American. He's a very bright fellow who likes to play with computers. Let me begin with a couple of words about the monster. I want to draw an analogy between the exceptional Lie groups and some of the sporadic finite groups. Consider the two simplest exceptional Lie groups, G_2 and F_4. G_2 is the automorphism group of the octonion algebra. That's how the exceptional groups come to be linked to octonions. Then you look at the simplest nonassociative Jordan algebra of dimension 26. And the group of automorphisms of that turns out to be F_4. F_4 is the automorphism group of the exceptional Jordan algebra. So you can make the step from G_2 to F_4 by enlarging the algebra and taking the automorphism group. Well, now do the same thing with the Leech lattice which is the lattice of the densest sphere-packing in dimension 24, and the automorphism group of that is the finite Conway group which was the biggest known sporadic finite group until very recently. The Conway group, the grandmother of many of the sporadic finite groups, is the group of automorphisms of the Leech lattice. Now Bob Griess, a mathematician who is at Princeton this year, constructed a very lovely nonassociative algebra over the vectors of the Leech lattice [Griess, 1981]. You take these 24-dimensional vectors and you construct a nonassociative algebra over them in the same sort of way it is done over the octonions, and then you take the automorphism group of that, the group of automorphisms of this nonassociative algebra over the Leech lattice, and that turns out to be the monster, the biggest of all sporadic finite groups. What's also nice is that you can prove that it is the biggest, that there are no more, and so the monster stands there in grand isolated glory all by himself.

Here is the order of the monster group:

$$2^{46} \times 3^{20} \times 5^9 \times 7^6 \times 11^2 \times 13^3 \times 17 \times 19 \times 23 \times 29 \times 31 \times 41 \times 47 \times 59 \times 71 \ .$$

It's a number which is roughly equal to 8×10^{53}, and you might wonder whether that has anything to do with the size of the gravitational constant. It also has 194 irreducible representations, all of which have been worked out. The dimensions of the first four are

1	$= 1$
196883	$= 47 \times 59 \times 71$
21296876	$= 2^2 \times 31 \times 41 \times 59 \times 71$
842609326	$= 2 \times 13^2 \times 29 \times 31 \times 47 \times 59 \; .$

The first irreducible representation has dimension one and it's trivial, the next one is 196883 and it happens to be the product of the three largest primes that enter into the order of the group. For some reason the small representations seem to go with the big primes, it's not clear why. You can write down a table of 194 irreducible representations. That's all very nice, it's fun for people who are interested in finite groups, but the beauty of it, what Conway calls the "Moonshine", is the connection with modular forms and modular groups. This is extraordinary, totally not understood.

You can construct modular forms with positive or negative genus, and you can construct a modular form of genus zero by taking a quotient. The form of genus zero was discovered by Jacobi and is called a modular invariant. To construct it you take an Eisenstein series which is a modular form of genus 4, and you take the Dedekind eta function which is a modular form of genus one-half, and you take the quotient of the cube of the Eisenstein series divided by the 24th power of the Dedekind function. That gives you a power series in x which is a complete invariant. It's the only function over the modular group which has only one pole and is completely invariant over the group, so it is a unique analytic function characteristic of the group. It is:

$$J(x) = \frac{Q^3}{\eta^{24}}$$

$$Q(x) = 1 + 240 \sum_{1}^{\infty} \frac{m^3 x^m}{(1 - x^m)} \text{ (Eisenstein)}; \quad \eta(x) = x^{1/24} \prod_{1}^{\infty} (1 - x^m) \text{ (Dedekind)}.$$

It's called $J(x)$ and you can expand it in a power series, and these are the coefficients.

$$J(x) = x^{-1} + 744 + 196884 \, x + 21493760 \, x^2 + 864299970 \, x^3 + \ldots$$

1	$= 1$
196884	$= 196883 + 1$
21493760	$= 21296876 + 196883 + 1$
864299970	$= 842609326 + 21296876 + 2 \times 196883 + 2 \times 1$

. . . . .

Conway and various other mathematicians noticed the above decomposition by serendipity. They just happened to look at these numbers and remembered something--

didn't I see something like that before? If you take this coefficient 196,884, it's one plus 196,883; it's just the sum of the dimensions of the first two irreducible representations of the monster. Then you look at the next one. You see the next coefficient is the sum of the first three irreducible representations of the monster, and so it goes. The next time it's not just the sum, it's the 4th plus the 3rd plus twice the 2nd plus the 1st. But you can go on and it always works. Each of the coefficients in this power series corresponds to a linear combination of irreducible representations of the monster. Nobody understands why. They haven't the slightest inkling of why. But there it is.

What's much more interesting is to say, the dimension of a representation isn't its only characteristic number. The Monster group has 194 characters, each of which is defined on each of its 194 irreducible representations. The first character is just the dimension of the representation, the trace of the unit element. The other characters are the traces of other elements in the group. You can do the same trick with each of these. You take the character of the first few irreducible representations, and you add them together in the same linear combinations, and you form a power series with that. You have a look then at Jacobi's "Fundamenta Nova Theoriae Functionum Ellipticarum", Jacobi's treatise on elliptic functions, which I think was the last major scientific work written in Latin, and you find they are all there [Jacobi, 1829]. These other characters, each of them gives you a power series, and each of them is a well-known elliptic function. So you find that, given any particular character which corresponds to a class of elements in the monster group, you get a particular modular form which turns out to be not an invariant of the entire modular group, but an invariant of a subgroup of the modular group. So you get 193 different modular forms, each of them breaking the symmetry of the modular group in a different way, each of them belonging to a particular subgroup of the modular group. So you have a wonderful richness of cross connections between the modular group and its subgroups and the monster and its representations and characters and subgroups; almost all sporadic groups are subgroups of the monster in one way or another. You get this wonderful interconnection between broken symmetries of the monster and broken symmetries of the modular group. And there it stands. It's up to us physicists to find something useful to do with it.

REFERENCES

[1]. Benson, C. T. 1966. *Minimal Regular Graphs of Girths Eight and Twelve*, Can. J. Math., 18, 1091-1094.

[2]. Conway, J. H. and S. P. Norton, 1979. *Monstrous Moonshine,*, Bull. London Math. Soc. 11, 308-339.

[3]. Coxeter, H.S.M. 1981. *A Symmetrical Arrangement of Eleven Hemi-Icosahedra*, to be published.

[4]. Grassman, H. 1844. *Die Lineale Ausdehnungslehre*, (Otto Wigand, Leipzig).

[5]. Griess, R. L. 1981a. *A Construction of F_1 as Automorphisms of A 196883-dimensional Algebra*, Proc. Nat. Acad. Sci. USA, 78, 689-691. Griess, R. L. 1981b. *The friendly Giant*, to be published.

[6]. Jacobi, C. G. J. 1829. *Fundamenta Nova Theoriae Functionum Ellipticarum*, (Königsberg).

[7]. Klein, F. 1872. *Vergleichende Betrachtungen über neuere geometrische Forschungen*, (Erlangen, A. Deichert), reprinted in F. Klein, *Gesammelte Mathematische Abhandlungen*, (Berlin, Springer, 1921), Vol. 1., 460-497.

[8]. Leech, J. 1967. *Notes on Sphere Packings*, Can. J. Math. 19, 251-267.

[9]. Leech, J. and N. J. A. Sloane, 1971. *Sphere Packings and Error-Correcting Codes*, Can. J. Math. 23, 718-745.

[10]. MacDonald, I. G. 1972. *Affine Root Systems and Dedekind's Eta function*, Invent. Math. 15, 91-143.

[11]. Mathieu, E. L. 1861, 1873. *Mémoire sur l'étude des fonctions de plusieurs quantités*, J. de. Math. Pures et Appliquées, 6, 241-323, *Sur la fonction cinq fois transitive de 24 quantités*, J. de Math. Pures et Appliquées, 18, 25-46.

[12]. Mehta, M. L. 1967. *Random Matrices*, (New York, Academic Press), pp. 74-79.

[13]. Rouse Ball, W. W. 1908. *A Short Account of the History of Mathematics*, 4th edition, (MacMillan, London), page 478.

[14]. Tutte, W. T. 1966. *Connectivity in Graphs*, (University of Toronto Press).

EPILOGUE

Maurice Goldhaber

Brookhaven National Laboratory
Associated Universities, Inc.
Upton, Long Island, New York 11973

We have come here tonight for a very happy occasion: to celebrate Feza Gürsey's 60th birthday. Yoichiro Nambu told us this morning that someone started this universe by pulling strings. We all agree that this was a good beginning considering that it finally led to Feza. Freeman Dyson in the last talk today told us that there is a long time gap before the work of a man of such depth as Feza is appreciated; clearly we shall have to come back here for his 120th birthday. When Feza makes a grant application to one of the agencies he could say in his modest way: "This may show its value in fifty or a hundred years." Feza has been honored many times before, but to be honored right at home is a very special honor indeed, and I hope he will remember that. We all enjoy Feza's success, especially because it disproves Leo Durocher's famous conjecture that nice guys finish last.

As all of you have heard today, group theory is Feza's real love. The physicist uses group theory to describe beautiful symmetries found in Nature. He also finds that these symmetries are sometimes broken, e.g., the important symmetries of parity conservation and time reversal invariance. The man in the street knows this. He knew it before the physicists did. If you ask the man in the street if left is different from right, he says "Of course." If you ask him, is the past different from the future, he says "Of course." But the physicist took a while to find that out.

There are two extreme types of scientists, with a broad spectrum in between: you might call the one extreme the "Bloodhound" who follows every clue from experiments as well as from theoretical developments till he gets a coherent picture; the other extreme you might call the "Olympian" who is guided by general principles to arrive at theories of predictive value. Of course, no one is a pure example of one or the other extreme, but if anyone comes close to the Olympian it is Feza, who early realized the power of group theoretical methods to guide us in the understanding of the properties of fundamental particles. I do not have to enumerate his many achievements; they were mentioned often today.

We are proud at Brookhaven that Feza was at one time a member of our Laboratory and that he did some of the important work with Luigi Radicati on $SU(6)$ at that time; this work has had a great impact on the modern quark model. In more recent years he has done important work on exceptional groups which, unlike some of the other groups used in grand unified theories, allow the possibility of a stable proton. Though I am personally engaged in a search for the decay of the proton, in a collaboration with Irvine and Michigan physicists, I have kept an open mind, as you can see from a toast I gave the other day to the proton (which some of you may have heard before): "The proton, may it live forever, but if it dies, may it die in our arms." Experimenters are a bit selfish.

I once heard Murph Goldberger say that group theory is very easy. He has learned it many times. But if you have a real difficulty in group theory there is nothing better that to go to Dr. Feza for a little group therapy.

The organizers of this conference must have had a hard time deciding on a speaker for tonight. What we clearly need is a Michelin guide to after dinner speakers. Such a guide would be very popular and useful; I suggest that it should show by some symbol beside the names of potential speakers how desirable they are. For instance, it might show the number of cups of coffee which are needed to keep the audience awake. In the absence of such a guide, the organizers had to look around for physicists at a certain point in their career.

In the beginning of his career a physicist gives contributed papers at meetings, 10 minutes or so. Then, as he advances, he is asked to give invited papers. A little later he takes part in round-table discussions. Then he is asked to summarize conferences, then to introduce conferences, and finally on his way down he is asked to give after dinner speeches. One easy way to choose him: The oldest person present gives the after dinner speech. I remember still, when I held a certain Fellowship at Magdalene College in Cambridge, I had an interesting quite difficult duty. In fact, that Fellowship had only a single explicit duty attached to it, and that duty was to "pour the wine when the youngest Fellow present." Since the other Fellows, who were charged by the glass, didn't like it if I didn't fill the glass completely, nor did they like it if I overdid it, I had to learn to fill it exactly full and I still have this ability. For anyone who wants to see it I'll demonstrate later.

The health of a business is often measured by its backlog; and if you want to measure the health of physics, you can ask yourself how many good questions there are around to be answered. In this sense physics today is a very healthy subject: We have a big backlog. At present we have no dearth of good, or big, questions and one reason they stay around is that they are also such hard questions, either theoretical or experimental. It is very difficult these days for a researcher who has raised a question to see the final answer; but when it is all over it may also be an anti-climax. I remember the case of Louis Michel who introduced the famous Michel parameter ρ; he predicted a value of $\rho = 0.75$. The first measurement was 0.25 and then, as the experiments improved, it went up to something like 0.30, 0.40, and finally 0.75, 0.76. That was a little too much and it came back to 0.75. To this Michel reacted: "Alas, they will speak of it no more."

Today we have many good questions; what are some of these questions? I'm not going to try to give a complete catalogue. Many of you know these questions. One of the most important questions today is: Are there any free quarks? The leading theories of the day tell us that quarks are confined in hadrons. Experiments can never prove such a statement; they could only disprove it if the theories should turn out to be wrong. However, as the astronomers are fond of saying, the absence of knowledge is not the knowledge of absence. So far experiments have only shown, as a wit has pointed out, that quarks are confined to niobium. There are many other good questions around, good, hard, deep questions which will stay around for a while, e.g., questions of whether the neutrino has a finite mass, whether there are neutrino oscillations, etc., etc.. I recently noticed that a new disease has hit some in the physics community. I call it the "good question syndrome": If you work on a good question, never mind what the answer is! We have to beware of this syndrome.

I was very interested to learn a few years ago from a paper of Julia and Tom Gaisser (this is a wife and husband team, one a classical scholar and the other a physicist whom some of you know) that the Greeks who invented the atom had also something to say about partons or quarks. In his famous poem, *"De Rerum Natura"*, Lucretius tells us of the idea of the Greek philosopher Epicurus, who lived in the third century B. C. Epicurus talked of "minimal parts" or "the least or smallest things". He adds that since they cannot exist by themselves they must necessarily cohere so that they can never be torn apart. Very clever foresight! In the spirit of what we have heard today it is only to be expected that somebody has said it long ago.

Today we know of twelve flavors of particles, or three generations of quarks and leptons. It is worth noting, especially at a dinner like this, that physiologists have found that our tongue can distinguish twelve fundamental (pure) flavors. Most of us would agree that these pure flavors are best confined to cookbooks! It is certainly a good question to ask tonight whether our mind can distinguish more than that number of flavors among the particles. We shall see.

You might say that man is Nature's way to find out about itself. Nature is probably self-conscious and would like to know about itself. The fact that we exist shows that the constants of Nature have about the right strength for us to be able to exist. In fact some people have tried to see what kind of a world this would be if we would change the constants by a little. Small changes could have big effects; the world we live in is remarkably sensitive to the values of the constants. It may even be that this is not only the best of all possible worlds but the only one of all possible worlds. This is one of the big questions which are around, one of the great questions, and another one of the big questions, one which Dirac has raised and which still has no answer, is "Are the constants really constant?" So I could go on. There are many, many of these good questions around, and I am sure that Feza will remain interested in the answers and often contribute to the solutions.

I would like to drink a toast to Feza, to Suha, and to their son, Yussef, and wish them a long and happy active life.

Publications of Feza Gürsey

1. *A Symbolic Treatment of Special Relativity*, The Scientific Journal of the Royal College of Science **26**, 128-137 (1947).

2. *Classical Statistical Mechanics of Rectilinear Assembly*, Proc. Cambridge Phi. Soc. **46**, 182-194 (1950).

3. *On Two-Component Wave Equations*, Phys. Rev. **77**, 844-845 (1950).

4. *On the Virial Coefficients of an Imperfect Gas*, Proc. 8th. Congress of Pure and Applied Mechanics, 459-460 (1952).

5. *Gravitation and Cosmic Expansion in Conformal Space-Time*, Proc. Cambridge Phil. Soc. **49**, 285-291 (1953).

6. *Dual Invariance of Maxwell's Tensor*, Rev. Fac. Sci. Univ. Istanbul A **19**, 154-160 (1954).

7. *Connection Between Dirac's Equation and a Classical Spinning Particle*, Phys. Rev. **97**, 1712-1713 (1955).

8. *Contribution to the Quaternion Formulation of Special Relativity*, Rev. Fac. Sci. Istanbul A **20**, 149-164 (1955).

9. *Correspondence Between Quaternions and Four Spinors*, Rev. Fac. Sci. Univ. Istanbul A **21**, 33-54 (1956).

10. *New Algebraic Identities and Divergence Equations for the Dirac Electron*, Rev. Fac. Sci. Univ. Istanbul A **21**, 85-95 (1956).

11. *On Some Conformally Invariant World-Lines*, Rev. Fac. Sci. Univ. Istanbul A **21**, 129-143 (1956).

12. *On a Conformally Invariant Spinor Wave Equation*, Nuovo Cimento **3**, 988-1006 (1956).

13. *General Relativistic Interpretation of Some Spinor Wave Equations*, Nuovo Cimento **5**, 154-171 (1957).

14. *Relativistic Kinematics of Classical Spinning Particle in Spinor Form*, Nuovo Cimento **5**, 784-809 (1957).

15. *Coulomb Scattering of Polarized Electrons*, Phys. Rev. **107**, 1734-1735 (1957).

16. *Relation of Charge Independence and Baryon Conservation to Pauli's Transformations*, Nuovo Cimento **7**, 411-415 (1958).

17. *On the Group Structure of Elementary Particles*, Nuclear Physics **8**, 675-691 (1958).

18. *Possible Connection Between Strangeness and Parity*, Phys. Rev. Lett. **1**, 98-100 (1958).

19. *Space Time Properties and Internal Symmetries of Strong Interactions*, with G. Feinberg, Phys. Rev. **114**, 1153-1170 (1959).

20. *On the Symmetries of Strong and Weak Interactions*, Nuovo Cimento **16**, 230-240 (1960).

21. *On the Structure and Parity of Weak Interaction Currents*, Proc. 10th International High Energy Conference, 572-577 (1960).

22. *On the Structure and Parity of Weak Interaction Currents*, Annals of Physics **12**, 91-117 (1961).

23. *Lepton Pairings in the Two-Neutrino Theory*, with G. Feinberg and A. Pais, Phys. Rev. Lett. **7**, 208-210 (1961).

24. *Approximate Symmetries in the 2-Neutrino Theory*, with G. Feinberg, Phys. Rev. **128**, 378-385 (1962).

25. *Spin 1/2 Wave Equation in the de Sitter Space*, with T. D. Lee, Proc. Nat. Acad. Sci. **49**, 179-186 (1963).

26. *Introduction to the de Sitter Group*, Proceedings of the Istanbul Summer School of 1962, 365-389, Gordon-Breach Co. (New York, 1964).

27. *Reformulation of General Relativity in Accordance with Mach's Principle*, Annals of Phys. **24**, 211-242 (1963).

28. *Introduction to Group Theory*, Summer School (1963) on Relativity, Groups and Topology, 91-161, Gordon and Breach Co. (New York, 1964).

29. *Implications of Approximate SU(3) Symmetry and Mass Formalae for the Mesons*, with T. D. Lee and M. Nauenberg, Phys. Rev. **135B**, 467-477 (1964).

30. *Spin and Unitary Spin Independence of Strong Interactions*, with L. A. Radicati, Phys. Rev. Lett. **13**, 173-175 (1964).

31. *Spin and Unitary Spin Independence of Strong Interactions II*, with A. Pais and L. A. Radicati Phys. Rev. Lett. **13**, 299-301 (1964).

32. *Equivalent Formulations of the SU(6) Group for Quarks*, Phys. Letters **14**, 330-331 (1965).

33. *Remarks on Relativistic Formulations of SU(6)*, High Energy and Elementary Particles IAEA Trieste Seminar 1965, 696-707 (Vienna 1965).

34. *Groups Combining Internal Symmetries and Spin*, High Energy Physics, Les Houches Lectures 1965, ed. M. Jacob and C. de Witt, 55-87 (Gordon and Breach, New York 1965).

35. *Approximate Higher Symmetries of Strong Interactions*, Proceedings of Fourth Yeshiva Annual Science Conference 1965, 341-59 (New York 1967).

36. *Some Non Compact Groups Associated with the Vertex Operator*, Non Compact Groups in Particle Physics, Milwaukee Conference ed. by Chew, 181-201 (Benjamin, New York 1966).

37. *The Algebra of Spin Currents and the Relativistic Formulation of SU(6)*, with K. Bitar, Phys. Rev. **164**, 1805 (1967).

38. *Unified Formulation of Effective Non-Linear Pion-Nucleon Lagrangians*, with P. Chang, Phys. Rev. **164**, 1752-1761 (1967).

39. *Tensorial Current Operators as Representations of the Extended Poincaré Group*, Theoretical Physics, Boulder Lectures 1967, Edited by W. Brittin and A. Barut, 303-323 (1968).

40. *Non-Linear Lagrangians Invariant Under the Generalized Chiral Groups SL(4,C) and SL(12,C)* with P. Chang, Phys. Letters **26B**, 520-523 (1968).

41. *Effective Lagrangians in Particle Physics*, Acta. Phys. Austriaca, Suppl. V., 185-225 (1968).

42. *Generalized Chiral Symmetry Groups and the Classification of Hadrons*, Acta. Phys. Hung. **26**, 127-137 (1969).

43. *Some Nonlinear Lagrangian Models Combining Chiral Symmetry and Spin Independence*, with P. Chang, Nuovo Cimento **63A**, 617-688 (1969).

44. *SU(6) Symmetry and Its Relativistic Generalizations*, Contemporary Physics, Trieste Symposium, ed. A. Salam, vol. **2**, 193-210 (1969).

45. *Effective Lagrangians at High Energy*, with M. Koca, Lett. Nuovo Cimento **1**, 228-232 (1969).

46. *Empirical Mass Formulae, Schwinger's Quantum Numbers and Non Leptonic Decays*, with Meral Serdaroglu, Lett. Nuovo Cimento **1**, 233-236 (1969).

47. *Meson Trajectories with Increasing Multiplicities and Trilocal Fields*, with Mehmet Koca, Nuovo Cimento **1A**, 429-444 (1971).

48. *Duality and the Lorentz Group*, with I. Bars, Phys. Rev. **4D**, 1669-1776 (1971).

49. *Nonlinear Realizations of Broken SU(3)xSU(3) for Pseudoscalar Mesons*, with M. Serdaroglu, Nuovo Cimento **7A**, 584-604 (1972) and **9A**, 263-264 (1972).

50. *Operator Treatment of the Gelfand-Naimark Basis for SL(2,C)*, with I. Bars, J. Math Phys. **13**, 131-143 (1972).

51. *Extended Hadrons, Scaling Variables and the Poincaré Group*, with S. Orfanidis, Nuovo Cimento **11A**, 225-278 (1972).

52. *An Octonionic Representation of the Poincaré Group*, with Murat Günaydin, Nuovo Cimento Letters **6**, 401-406 (1973).

53. *Conformal Invariance and Field Thory in Two Dimensions*, with S. Orfanidis, Phys. Rev. **D7**, 2414-2437 (1973).

54. *Quark Structure and Octonions*, with Murat Günaydin, J. Math. Phys. **14**, 1651-1667 (1973).

55. *General Chiral SU(2)xSU(2) Lagrangian and Representations Mixing*, with Abdul Ebrahim, Nuovo Cimento Letters **9**, 9-14 (1974).

56. *Quark Statistics and Octonions*, with Murat Günaydin, Phys. Rev. **D9**, 3387-3391 (1974).

57. *Color Quarks and Octonions*, The Johns Hopkins University Workshop on Current Problems in High Energy Particle Theory, 15-42 (1974).

58. *Derivation of the String Equation of Motion in General Relativity*, with Metin Gürses, Phys. Rev. **D11**, 967-969 (1975).

59. *Lorentz Covariant Treatment of the Kerr-Schild Geometry*, with Metin Gürses, J. Math. Phys. **16**, 2385-2390 (1975).

60. *Algebraic Methods and Quark Structure*, International Symposium on Mathematical Problems in Theoretical Physics, ed. H. Araki, 189-195 (Springer, 1975).

61. *A Six Quark Model for the Suppression of $\Delta S=1$ Neutral Currents*, with P. Ramond and P. Sikivie, Phys. Rev. **D12**, 2166-2168 (1975).

62. *A Universal Gauge Theory Model Based on E_6*, with P. Ramond and P. Sikivie, Phys. Lett. **60B**, 177-180 (1976).

63. *Exceptional Groups and Elementary Particles*, Proc. of the 4th International Colloquium on Group Theoretical Methods in Physics, ed. A. Jenner, 225-233 (Springer, 1976).

64. *Supersymmetric Ansatz for Spontaneously Broken Gauge Field Theories*, Proc. of the Conference on Gauge Theories and Modern Field Theory, ed. P. Nath and R. Arnowitt, 369-376 (MIT Press, 1976).

65. *E_7 as a Universal Gauge Group*, with P. Sikivie, Phys. Rev. Lett. **36**, 775-778 (1976).

66. *Charge Space, Exceptional Observables and Groups*, New Pathways in High Energy Physics I, 231-248 (Plenum Press, 1976)

67. *Some Special Kerr-Schild Metrics*, with Metin Gürses, Nuovo Cimento **3B**, 226 (1977).

68. *Of Trials and Errors*, The Significance of Nonlinearity in the Natural Sciences, ed. A. Perlmutter and I. Scott, 349-407 (Plenum Press, 1977)

69. *The Graded Lie Groups SU(2,2/1) and OSp(1/4)*, with Louis Marchildon, J. Math. Phys. **19**, 942-951 (1978).

70. *Spontaneous Symmetry Breaking and Nonlinear Invariant Lagrangians: Applications to SU(2)xU(1) and OSp(1/4)*, with Louis Marchildon, Phys. Rev. **D17**, 2038-2047 (1978).

71. *Symmetry Laws (Physics)*, McGraw-Hill 1978 Yearbook of Science and Technology, 355-358 (1978).

72. *Basic Fermion Masses and Mixings in the E_6 Model*, with M. Serdaroglu, Lett. Nuovo Cimento **21**, 28-32 (1977).

73. *Quark and Lepton Assignments in the E_7 Model*, with P. Sikivie, Phys. Rev. **D16**, 816-834 (1977).

74. *Some Algebraic Structures in Particle Theory*, Second Workshop on Current Problems in High Energy Theory, ed. G. Domokos, 3-25 (The Johns Hopkins University, 1978).

75. *Quaternion Analyticity in Field Theory*, Second Workshop on Current Problems in High Energy Theory, ed. G. Domokos, 179-221 (The Johns Hopkins University, 1978).

76. *Symmetries of Quarks and Leptons*, The Ways of Subnuclear Physics, ed. A. Zichichi, 1059-1164 (Plenum Press, 1979).

77. *Octonionic Structures in Particle Physics*, Group Theoretical Methods in Physics, Proceedings Austin, 1978, ed. W. Beiglbock, A. Bohm and E. Takasugi, 508-521 (Springer-Verlag, 1979).

78. *Quaternionic Multi $S^4 = HP(1)$ Gravitation and Chiral Instantons*, with H. C. Tze and M. A. Jafarizadeh, Phys. Lett. **88B**, 282 (1979).

79. *Invariance Principles in Physics*, American Institute of Physics Handbook (1978).

80. *Geometrization of Unified Fields*, To Fulfill a Vision (Jerusalem Einstein Centennial Symposium (1979)), ed. Y. Neeman, 22-36 (Addison-Wesley, 1981).

81. *Complex and Quaternionic Analyticity in Chiral and Gauge Theories I*, with H. C. Tze, Annals of Physics **128**, 29-130 (1980).

82. *Quaternion Methods in Field Theories*, Fourth Workshop on Current Problems in Particle Theory, Bad Honnef, Bonn, ed. G. Domokos and S. Kövesi-Domokos, 255-268 (Johns Hopkins, 1980)

83. *Symmetry Breaking Patterns in E_6*, New Hampshire Workshop on Grand Unified Theories, ed. P. Frampton, S. L. Glashow and A. Yildiz, 39-55 (Math. Sci. Press, 1980).

84. *Exceptional Groups for GUT's and Remarks on Octonions*, Proceedings of Workshop on Weak Interactions as Probes of Unification, ed. G. B. Collins, L. N. Chang and J. R. Ficenec, 635-646 (AIP, 1981).

85. *E_6 Gauge Field Theory Model Revisited*, with M. Serdaroglu Nuovo Cimento **65A**, 337-354 (1981).

86. *Introduction to the Completely Integrable Systems in Physics*, Lecture Notes in Proceedings of the Istanbul Workshop on Completely Integrable Systems, ed. M. Hortacsu and M. Serdaroglu (Turkish Physical Society, 1982).

87. *A Class of Supersymmetric Effective Actions*, with S. Catto, Lett. Nuovo Cimento **35**, 241-248 (1982).

88. *Potential Scattering, Transfer Matrix and Group Theory*, with Y. Alhassid and F. Iachello, Phys. Rev. Lett. **50**, 873 (1983).

89. *A Dirac Algebra Approach to Supersymmetry*, Foundations of Physics **13**, 289-296 (1983).

90. *Scattering and Transfer in Some Group Theoretical Potentials*, Proceedings of XIth International Colloquium on Group Theoretical Methods in Physics, ed. M. Serdaroglu and E. Inönü, 106-122 (1983).

91. *Group Theory Approach to Scattering*, with Y. Alhassid and F. Iachello, Annals of Physics **148**, 346-380 (1983).

92. *Group Theory of the Morse Oscillator*, with Y. Alhassid and F. Iachello, Chem. Physics Letters **99**, 27-30 (1983).

93. *Octonionic Torsion on S^7 and Englert's Compactification of Supergravity*, with H. C. Tze, Phys. Letters **B127**, 191-196 (1983).

94. *A Three-Dimensional Geometric Model with a Kink and a Higgs Mechanism*, with H. C. Tze, Phys. Letters **B129**, 205-208 (1983).

95. *General Supersymmetric Equations Involving Unconstrained Chiral Superfields*, with S. Catto, Phys. Rev. **D29**, 653-657 (1984).

96. *Dyson Representation of Su(3) in Terms of 5 Boson Operators*, with R. Dündarer, J. Math. Phys. **25**, 431-432 (1984).

97. *Generalized Vector Products, Duality and Octonionic Identities in D=8 Geometry*, with R. Dündarer and H. C. Tze, J. Math. Phys. **25**, (1984).

98. *Quaternion Analyticity and Conformally Kählerian Structure in Euclidean Gravity*, with H. C. Tze, Lett. Math. Phys. (1984).

Curriculum Vitae of Feza Gürsey

Feza Gürsey was born in Istanbul, Turkey on April 7, 1921. He attended Istanbul University and received his Ph.D. from Imperial College, England in 1950, after which he spent one year at Cambridge University as a postdoctoral fellow. In 1951 he returned to Turkey and stayed at the University of Istanbul until 1956 (becoming "Docent" in 1953). The period 1957-61 saw him as a visiting scientist, first at the Brookhaven National Laboratory, then at the Institute for Advanced Study at Princeton, and finally as a visiting Associate Professor at Columbia University. In 1961 he became Professor of Physics at the Middle East Technical University in Ankara where (except for short stays at the I. A. S. and at Yale) he remained until 1968. In 1969 he left Ankara to take up a professorship of physics at Yale University where he has held the J. Willard Gibbs Chair since 1977. Among the many honors bestowed upon him we mention only the J. R. Oppenheimer Prize in 1977, the Einstein Medal in 1979 and the A. Cressey Morrison Award in Natural Sciences, the Medaille du College de France, and the Medal of the University of Istanbul in 1981.

Hermann Weyl would have called Feza a physicist with the soul of a mathematician. Feza has long believed that certain beautiful mathematical systems have a correspondence in physical reality. Through the years he has diligently strived to construct the deeper theories using his marvellous, unique mathematical and physical insights. The fruitful marriage of mathematics and physics is evident in many of his papers, but especially in his masterly use of symmetries, group theory, quaternions and octonions in a wide variety of topics in particle physics. As Dyson observes in this volume, Feza's physics pursuits span a broad range of topics. They include 1) the conformal and DeSitter groups; 2) Machian relativity; 3) the classical spinning electron; 4) pioneering works on nonlinear realizations particularly the introduction of the $SU(2) \times SU(2)$ chiral group and the formulation of the first nonlinear sigma model; 5) the celebrated $SU(6)$ symmetry of hadrons; 6) the exceptional E_6 grand unified theory; 7) the applications of group representation theory to exactly solvable scattering problems; and 8) the relevance of quaternionic and octonionic structures to chiral and gauge field theories.

Through the years Feza has built at Yale, the home of Willard Gibbs, a unique school of theoretical physics. He has gladly shared the fruits of his discoveries and erudition with his many students, friends and colleagues. We wish him well and look forward to enjoying his company and the poetry of his physics for many, many years to come.

The Editors

List of Contributors

T. Appelquist J. W. Gibbs Physics Laboratory, Yale University, New Haven, CT 06511

D. V. Chudnovsky Department of Mathematics, Columbia University, New York, NY 10027

G. Chudnovsky Columbia University, New York, NY 10027

G. Domokos Department of Physics, Johns Hopkins University, Baltimore, Maryland, 21218

Freeman J. Dyson Institute for Advanced Study, Princeton, New Jersey

Peter G. O. Freund The Enrico Fermi Institute and the Department of Physics, The University of Chicago, Chicago, Illinois 60637

Maurice Goldhaber Brookhaven National Laboratory, Associated Universities, Inc. Upton, Long Island, NY 11973

Gerard 't Hooft University of Utrecht, Institute for Theoretical Physics, Postbox 80.006, 3508 TA Utrecht, The Netherlands

F. Iachello Wright Nuclear Structure Laboratory, Yale University, New Haven, CT 06520 and Kernfysisch Versneller Instituut, Rijksuniversiteit Groningen, The Netherlands

S. Kövesi-Domokos Department of Physics, Johns Hopkins University, Baltimore, Maryland, 21218

T. D. Lee Department of Physics, Columbia University, New York, NY 10027

S. W. MacDowell J. W. Gibbs Physics Laboratory, Yale University, New Haven, CT 06511

Louis Michel Institut des Hautes Etudes, Scientifiques, 35, route de Chartres, 91440 Bures-Sur-Yvette, France

Yoichiro Nambu The University of Chicago, The Enrico Fermi Institute, 5630 Ellis Avenue, Chicago, Illinois 60637

V. I. Ogievetsky Joint Institute for Nuclear Research, Laboratory of Theoretical Physics, Dubna, USSR

Luigi A. Radicati di Brozolo Scuola Normale Superiore, Pisa, Italy

Ralph Roskies Department of Physics and Astronomy, University of Pittsburgh, Pittsburgh, PA 15260

Abdus Salam International Center for Theoretical Physics, Trieste, Italy and Imperial College, London, England

Charles M. Sommerfield J. W. Gibbs Physics Laboratory, Yale University, New Haven, CT 06511

J. Strathdee International Center for Theoretical Physics, Trieste, Italy

INDEX

Abelian algebras and groups
  Backlund transformations, 229, 232, 234
  and dimensional reduction, 195
  fermions, 111
  gauge field, 116–117
  lattices, 226, 229
  quarks, 22–23
  quartic Hamiltonians, 67
  renormalization group equations, 67
Action, and equations of motion, 185
Action principle, *see* Invariant action
Atkin, Oliver, work of, 277

Backlund transformations (*see also* Lattice systems, completely integrable; Linear differential equations)
  compositions of
    Abelian groups, 234
    boundary conditions, 237
    and Backlund transformation space, 236
    commutativity conditions, 234
    consistency of linear problems, 236–237
    difference-differential equations, 235
    independence of integrals, 237
    inductive definition, 234
    and lattice models, 234
    law of, 234
    linear differential operator, 236
    linear fractional transformation, 236
    nonlocal conservation laws, 236–237
    pair representation, 235
    quadratic difference equations, 234–235
    scattering matrix, 235–236, 237
  defined, 226–227
  examples
    one-dimensional operator, 231
    of $\phi(\lambda)$, 230
    projection operator, 231
  exponential matrices, 228
  Jordan normal form, 228
  λ-plane, 229–230
  linear transformation, 227–228
  matrix for, 227
  monodromy, 227, 228
  multipliers, 227
  scattering interpretation
    Abelians, 232
    block matrices, 232
    eigenfunctions, 233
    first integrals, 232
    inverse scattering method, 231
    Korteweg de Vries equation, 232
    Lie groups, 233
    $n \times n$ matrix, 231
    two-dimensional system, 232
  singularities, 228, 229

Bosons
  mixing with fermions, 179
  in nuclei, 49
  quadrupole, 53
  scalar, 53
  vector, 53

Cage problem
  Benson's graph, 274
  closed cycles, 273
  even number of matrices, 273
  Galois field, 272, 274
  girth of graph, 273–274
  girth 12, 273–274
  Maxwell fields, 274
  order of graph, 273
  physical uses, 274
  Singleton's proof, 273
  as tree-like graphs with symmetry, 272–273
  12-cage, with 126 vertices, 274
Casimir operators
  dynamic symmetries, 48, 50
  infinitesimal operators, 250
  lattice models, 246–247
  quantum lattice models, 253
Chiral functions and fields, *see* Fermions; Quantum lattice models; Supergravity
Color confinement (*see also entries under* Quantum chromodynamics, Quarks)
  compared to superconductivity, 94, 95
  diagram, 94
  gluon pairs in QCD, 94
  in vacuum, 94
Compactification, *see* Dimensional reduction
Confinement of quarks (*see also entries under* Quantum chromodynamics; Quarks), 27–28
Conway group, 268
Coxeter, Donald, work of, *see* Sphere packing; Polyhedra

De Rerum Natura, 289
Decoupling theorems, *see* Goldstone theorem
Dedekind eta function
  monster group, 283
  26-space, 277
Dimensional reduction
  charge conjugation, geometrization of
    C-parity, 196
    C-trivial, 196
    in D dimensions, 196
    in 11-space, 197
    interpretation, 196
    Majorana fields, 196
    reflections of coordinates, 197
    and scalar fields, 197

Dimensional reduction (*cont.*)
  curl radius of rolled-up dimensions, 191
  forces acting on point particles, geometric
    derivation
    connection coefficients, 195
    equations of motion, 195
    Killing/Minkowski metric, 194
    Lorentz expressions, 194
    non-Abelian generalization, 195
    one-parameter Abelian group, 195
  for gravity
    coordinate basis, 193–194
    geodesics, 193
    metric, in $D - d$ dimensions, 192
    scalar fields, 193
  higher dimensions, physicality of, 198
  and Planck length, 191
  power of, 191–192
Diophantine approximations, new
  four-term linear recurrence, 217
  initial conditions, 216
  irrationality measures, 216
  $\pi/\sqrt{3}$ approximation, 218, 219
  π^2, 220
  properties of approximations, 218–219
  rational numbers, 216–217
  singularities, 216, 217
Dirac
  free particle equation, 134
  lattice models equation, 241–242
  matrices for spinors, 133
  operator, 133
  quark confinement, 27
  spinors, 119, 133
Dynamic symmetries
  in atoms
    chains, 58–59
    Coulomb potential, 58
    eigenvalue problem, 59
    Hamiltonians, 58
    helium, excited states of, 58
  baryon decuplet, 49
  and Casimir operators, 48
  group chain, 48
  Hamiltonians, 47–48
  in molecules
    CO_2, diagram, 57
    diagrams, 54
    energy formulas, 54
    neutron/proton pairs, Hamiltonians of, 56
    $O(4)$ spectra, 55
    phase diagram, 56
    quadrupole bosons, 53
    rotation group, 53–54
    scalar boson, 53
    triatomic molecules, Hamiltonian of, 56
    $U(4)/U(6)$, 54
    vector boson, 53
    vibron model, 53

Dynamic symmetries (*cont.*)
  in nuclei
    bosons, 49
    Casimir operator, 50
    chains, 49
    chain energies, 50
    coherent state, 52
    coset variables, ellipsoid, 52–53
    creation/annihilation pairs, 49
    eigenvalue problem, 50
    $O(6)$, 51–52
    phase diagram, 52
    $SU(3)$, spectrum, 51
    $U(5)$, spectrum, 51
  $SU(3)$, Gell-Mann–Ne'eman, 48
  supersymmetry, 60

Effective actions
  Bethe–Salpeter kernel, 150
  connected Green's functions, 147–148
  dual of chiral superfield, 148
  first kind, Legendre transformations, 148
  generator formula, 148
  inverse two-point Green's functions, 148
  second kind, 149

Fermions, determinants of in massless two-
    dimensional QCD
  A_μ, gauge conditions on
    Abelian, 111
    complete, 112
    derivatives, 110–111
    non-Abelian, 111
  Euclidean formulation, generating functional,
    107
  fermion determinant, 106, 109–110
  fermion point of integration, 107
  gauge choice, and fermion determinant
    axial transformations, 109
    decoupling from gluons, 109, 110
    evaluation, 110
    infinitesimal variables, 109
  gauge choice, mathematics of
    complex unit, 108
    Hermitian matrix, 108
    and Lie algebras, 108
  gauge field, 106
  gluon fields, 106
  non-Abelian, 106
  Schwinger model, two-dimensional, 106
Fermions, dynamic mass generation for
  algebra simplifications, 118
  amplitude, 117
  axial gluons, 118
  chiral projector, 121
  chiral triplets, 116
  color group, 119
  computability, 118
  Dirac spinors, 119

Fermions, dynamic mass generation (*cont.*)
  Dyson expression for fermion propagator, 121
  Dyson–Schwinger, 117
  equation decoupling, 124–125
  fermion propagator, complete, 121
  gauge propagators, 119
  gauge theory, 115
  graphical summation, 115
  Higgs fields, 115, 116
  Higgs mechanisms, 119
  interacting Lagrangian, 121
  Landau gauge, 120
  mass matrix, 120, 121–122
  massless, 116
  mixing angles, 118
  off-diagonal components, 123–124
  propagator, diagram, 117
  reduction of matrix, 123
  rotated coupling matrices, 122–123
  self-energy, 117
  self-mass calculation, 116
  ultraviolet diagonal terms, 121
  ultraviolet convergence, 116
  vector masses, 123, 214
  Wick rotation, 122
*Fundamenta Nova Theoriae Functionum Ellipti-
  carum,* 284

Gluons
  axial, 118
  fields, 106
  gauge choice, 109–110
  QCD pairs, 94
  and topological gauge artifacts, 24
Gaisser, Julia and Tom, 289
Gauge, *see entries under* Fermions
Gauss, work of, 266
Goldstone theorem, connection to decoupling
    theorems
  Ansatz, 152
  decomposition, 152
  decoupling theorems, 154–155
  fields identities, 154–155
  Goldstone fermion, 154
  infrared singularities, 154
  inverses of Green's functions, 152
  propagators, 153
  supersymmetry, spontaneously broken, 153
  vacuum expectation values, 153
Grassmann analyticity
  coordinates, 143
  nature of, 188
  variables, 179–180
Grassmann, Hermann, 267
  and *Ausdehnungslehre,* 267
  functions of
    effective action, 147, 148
    inverse of, 152

Gravity, *see* Dimensional reduction
Group theory
  and cosmos from chaos, problem of, 35
  convection onset, 36–37
  early work in physics, 33–34
  excitation states, 34
  gases, mixing of, 35, 36
  Hesiod, on Void, 35
  and Klein, Felix, 33
  and Lie, Sophus, 33
  loaded rod problem, 36
  magnetic dipoles, 35, 36
  rotating fluid mass
    bi- to triaxial ellipsoid transition, 36
    ellipsoids to pears transition, 36
  spontaneous order, 35
  symmetry breaking, 34

Hadron strings (*see also entries under* Fermions;
    Quarks; QCD)
  Abelian gauge field
    bag surface, two static point sources, 16–17
    and lattice gauge theory, 16, 17
    and QCD, 16
  fields associated with, Kalb–Ramond theory, 13
  glueballs, 13, 14
  and hydrodynamic vortices, 13
  mathematics about
    electromagnetic formula, 15
    Hamilton–Jacobi, 14
    internal clocks, 14
    mass points, 14, 15
    string formulas, 15
    surface family, 14–15
    times in mass point, 14
    world sheet, evolution toward, 14
  and older views of Void, 13–14
  open strings, 14
    and cosmology, 14
  and vector curl, 13
Hamiltonians
  Backlund transformation, 238–240
  Baxter method, 251
  commuting, 253
  dynamic symmetries, 47–48, 58
  factorized S-matrix, 260–261
  hadron strings, 15
  lattice models, 243, 246, 248
  molecules, 56
  neutron/proton pairs, 56
  quartic, 67
  renormalization, 65–66
  scalar field, 131
  σ-polynomial, 89
  spinor field, 133
  stable fixed points, 74, 89
Heavy ion collisions, relativistic
  A-dependence of, 97–98

Heavy ion collisions (*cont.*)
  antinuclei, 98
  binding energy, 96
  center-of-mass equations, 99
  coherent processes, defined, 97
  collision problems, 98
  Coulomb repulsion of, 96
  cross-sectional equation, 96–97
  cylinders of particles, facing, effects, 97
  diagram, of collision, 97
  energy density of nuclei, 96
  extreme energy, 99
  impact parameter, 97
  Lorentz contraction, 98
  metastable extended object, properties of, 100
  nucleons, energy density of, 96
  quark–antiquark and gluon plasma
    absence of eruption, signs of, 101
    collisions of, 99, 100
    core of, 99
    correlation functions, 102
    decay diagram, 101
    diagram, cross-sectional, 100
    energy limits, 101
    γ, emission of, 102
    hadron bags, disappearance of, 99
    leptons, pairs of, 102
    radius, finding of, 103
    soft pions, 102
    τ detection, 101–102
    volcanoes, 100–101
Hermitians
  matrix, for fermions, 108
  quantized fields rotations, 137
  zeta function zeros, 279
Hessians
  covariant symmetric algebra, 78, 84
  eigenvalues, 84, 85
  stable fixed points, 76
  2-plane, 82
Higgs effects (*see also* Fermions, dynamic mass
      generation for)
  electroweak theory, 6
  strongly interacting, 6
  and symmetry breakdown, 5–6

Infinitesimal operators
  Casimir operators, 250
  generators, choice of, 249
  Lie group, 249
  Lorentz group, 250
  parameterization, 249
  Schur lemma, 249–250
  Weyl group, 250
Infrared singularities, 154
Invariant action, and the action principle
  action form, formula, 167
  action interval, requirements for, 168

Invariant action (*cont.*)
  Euler–Lagrange equations, 169–171
  horizontal form, 167
  integration by parts, 168–169
  invariance under transformations, 168
  Lagrangian, 169
  vierbein, 167, 168
Inverse N ($1/N$) expansions, asymptotically free
  expansion of SU theories, 30
  and g^2N theory, 30
  and limits of N, 30
  planar Feynman diagrams, 30
Isotropy groups, and symmetry and
      renormalization groups
  group families, 86
  n-space, tensor product, 86
  orthogonal matrix, 85
  permutation matrices, 87
  polynomials, 85
  quadratic invariants, 85–86
  subgroups, irreducible, 86–87
  Sylow subgroups, 87

Jacobians
  in supergravity, 165
  θ-functions, 258
Jordan
  algebras, 282
  forms, 228
  octonions, 278

Kaluza–Klein theories, *see* Dimensional reduction
Killing matrix, 163, 194
  and supergravity, 163
Klein, Felix, 266
  and *Erlanger Programm*, 266

Lagrangians
  interactive form, for fermions, 121
  pionic, 2
Legendre transformations
  and effective action, 148
  polynomials, 211
Lattice models, completely integrable, with sym-
      plectic structures and associated with classi-
      cal algebras
  arbitrary classic group, 248–249
  Casimir operator, 246–247
  commutation relations, two-dimensional
      manifold, 247–248
  Darboux coordinates, 246, 247
  foliation into orbit family, 247
  Hamiltonians, 248
  Lie algebra, 246
  Poisson brackets, 246, 247
  Toda lattice systems, 248
  vector bundles, 248
  and XYZ-model, Hamiltonian, 246

Lattice models, nonlinear Schrödinger equations and related species
  Backlund transformation constant, 241
  continuous n-limit, 244
  Dirac equation, 241–242
  and eigenfunctions, 244
  Hamiltonians
    equation system, 243
    integrable, 243
    quantum, 243
  iteration, 242
  linear transformation, 241
  matrices, 240
  negative numbers, 242
  scattering matrix, 240
  sine-Gordon equation, 245
  time evolution equations, 243
  Toda lattice, 245
Lattice systems, completely integrable, and Backlund transformations
  Abelian symmetries, 226
  associated with Backlund transformations
    BT sequences, 238
    commutativity, 239
    Hamiltonians, 238, 239, 240
    Heisenberg ferromagnet, 239, 240
    integer labels, 238
    projector for, 238
    symplectic structure, 239
    XXX-models, 239–240
  coherence theorem, 225
  commutativity, 223–224, 225, 226
  diagrams of, 223, 224
  discussion, 221–223
  equations, defined, 226
  factorization axioms for S-matrices, 223–226
  functoriality, 224–226
  hexagon, 225
  isomorphisms, 223, 224
  monoidal categories, 223
  rectangle, 225, 226
  scattering, 224
  3-string interaction, 226
  triangle, 224–225
  vector spaces, 224
Leech lattices, 268
Levy superform
  defined, 187
  hypersurface, 185–186
  plane tangent, 186
  superplane, 186
Lie, Sophus, work of, 266
Lie groups and algebra
  and Backlund transformation, 233
  infinitesimal groups, 249
  lattice models, 246
  MacDonald identities, 273
  sphere packing, 269–270
  supergravity, 162

Linear differential equations, Backlund transformations and (*see also* Lattice systems)
  complex numbers, 206
  contiguous relations, 203
  discussion, 201–202
  lemma, 203–204
  multiplicities, 203
  $n \times n$ matrix, 202
  Padé approximations, 204
  Poincaré lemma, 207–208
  projector, 204
  rational approximations, sequences of
    dense, nature of, 205
    irrationality, 205
    Padé approximations, 205
  rational integers, 207
  Stokes multipliers, 202–203
Lorentz boosts
  matrix, 139
  spinor field, 139
  tomographic transformation, 139
Lucretius, 289

MacDonald identities
  algebra in, 276
  connection of Dedekind eta function with Lie algebras, 275
  E_7 identity, 276
  G_2 case, 275
    and Lie algebra, 276
    133-dimensional space, 276
    strange formula, 276
Majorana fields
  dimensional reduction, 196
  in supergravity, 162–163
Mathieu, Émile, 268
  and sporadic groups, 268
Micro- *vs* macrocosm, quantum problems, 95
Modular groups
  Dedekind modular function, 275
  eta function, 275
  Klein, work on, 274
  nature, 274, 275
  operation of, 275
  SL group, 274, 275
Monopoles
  gauge principle, 12
  phantom sheets, 25
  quantized fields, 138
Monster group
  characters of, 284
  Conway, work on, 282
  decomposition, 283–284
  Dedekind eta function, 283
  Eisenstein series, 283
  elliptic functions, 284
  genuses of modular forms, 283
  irreducible representations, 282, 283
  Jacobi, 284

Monster group (*cont.*)
  Jordan algebras, 282
  Leech lattice, 282
  linear combinations *vs* coefficients, 284
  modular groups, subgroups of, 284
  nonassociative algebra, 282
  and octonions, 282
  symmetry breaking, 284
  24-dimensional space, 282
  26-dimensional space, 282

Nonlinear sigma model
  counterterms, one-loop
    and Goldstone bosons, 4
    loop expansion parameters, 4
    nonlinear constraint, Honerkamp on, 5
    nonlinear sigma model, 4
    off-shell structures, 5
    shells, 4–5
  counterterms in loop expansion, 3
  linear sigma model, advantages of, 3
  pion-nucleon interactions, 1
  proton-neutron doublet, 1
  and QCD, 3
  $SU(2)_L \times SU(2)_R$, 1–3
    fields, representation of by 2×2 matrix, 2
    Goldstone bosons, 2,3
    Lagrangian, pionic, 2
    nonlinear constraint, 2
    π, transformations of, 2
    σ-field, linearization of, 2,
  symmetry breaking, 2, 3
Nonlinearity
  absolute symmetry of chaos, 42
  bifurcation points, 38
  dynamo problem, 41
  ellipsoids, response diagram, 39, 41
  forbidden isotropy groups, 41
  G-invariance, breaking of, 41–42
  geometry, power of, 42
  λ control parameters, 37–38
  multiple eigenvalue branch points, 38
  nonlinear response, 39, 40
  symmetry breaking, 36, 37, 38
  trivial symmetries, linear, 39

$O(N)$ covariant symmetric algebra
  attraction basin, 81
  eigenspaces, 85
  eigenvalues of Hessians, 84–85
  and eigenvectors, 80
  extremum, 79
  Hessians, 78, 84
  inequalities, 83
  irreducible subgroups, 81
  lemma, 80–81
  linear map, 77
  matrix, fourth-degree homogeneous
    polynomials, 78

$O(N)$ covariant symmetric algebra (*cont.*)
  number of in 2-plane, 83
  operators, 77
  orthogonal subspaces, 83
  pair, 81–82
  $\Phi^{(1)}$, extremum of, 80
  polynomial invariant, 76–77
  renormalization, 81
  stable criteria, 78–79
  theorem, 79
  third-degree polynomials, 80
  trilinear form, 76, 77
  two-plane of Hessians, 82
Octonions, 278
  and Jordan algebra, 278
  3×3 matrices, 278

Padé approximations, and contiguous relations
  approximation problem, 210
  asymptotics, 209
  coefficients, 210
  cubic irrationalities, 214
  diophantine approximations, 211, 212
  defined, 208
  discussion, 208
  distinct complex numbers, 209
  Gaussian functions, 209–210
  integers *vs* near-integers, 213
  irrationality measure, 212
  Legendre polynomials, 211
  logarithmic functions, 213
  modular functions, 215
  π, irrationality measure, 213
  three-term recurrence, 214, 215
  Thue–Siegel approximation, 213–214
Phantom sheets
  domain walls, 25–26
  functioning of as bubbles, 26–27
  ghost mass, 25
  monopole singularities, 25
  sine-Gordon equation, 26
Polyhedra
  Coxeter's 11-sided figure in 4-space, 271–272
    diagram, 272
    table, 272
  4-space, symmetric, 271
  3-space, Platonic, 271

Quantized fields, tomographic representation of
  discussion, overview, 128
  Radon transformation, 128
  rotations, 136–140
    Hermitian, 137
    monopole, 138
    scalar field, 136–37
    and singularity, 138
    spin-differential operator, 137
    spinor field, 137

Quantized fields (*cont.*)
  scalar field, canonical transformation
    commutation relations, 130–131
    derivatives, 129–130
    dimensions, 130
    and Hamiltonians, 131
    inverses, 129, 130
    motion equation, 131
    one-dimensional, 131
    proof, 129
    relations, 129
    spin, 131
  spinor field
    anticommutation relations, 133
    Dirac matrices, 133
    Dirac operator, 133
    eigenfunctions, 132
    eigenspinors, 132, 133
    four-component, 132
    free particle, Dirac equation for, 134
    general equation, 131–132
    Hamiltonians, 133
    one-dimensional, 134
    orthogonality, 132
    three-dimensional, 134
  under Poincaré group, 136
    translations, 136
  vector field
    4-vector, 134
    gauge choice, 136
    inverse formulas, 134–135
    and Maxwell equations, 135
    polarization, 134
    3-vector, 134
    tomographic transformations, 135–136
Quantum chromodynamics (*see also* Quarks)
  and freedom, 29
  problems with, 28–29
  and quantum gravity, 29
Quantum lattice models, Baxter method
  Casimir operator, 253
  coadjoint representation method, 251
  commutation algebra, 252
  commuting Hamiltonians, 253
  Darboux coordinates, 253
  Hamiltonians, 251
  hypergeometry, 255
  integrals, 254
  Liouville's theorem, 254
  local transfer matrices, 250–251, 256
  nonlinear Schrödinger equation, 255–256
  parameterization, 253–254
  $QU(2)$, 252, 253
  S-matrix in, 250, 251
  Toda lattice, 253
  trivial *vs* nontrivial, 253
  Weyl algebras, 253
Quantum XYZ models, generalized
  Backlund transformation formulas, 262

Quantum XYZ models (*cont.*)
  canonical Heisenberg variables, 260, 261
  commutation relations, 259
  degenerate matrices, 257, 258
  degenerate version of sine-Gordon equation, 261
  8-vertex case, symmetries in, 257
  elliptic curve, 262
  Hamiltonians associated with factorized S-matrix, 260, 261
  Hamiltonians, discussion, 256
  Jacobi θ-functions, 258
  Landau–Lifshitz equation, 262
  lattice symbols, $SO(3)$, 260
  local transfer matrices, 256–257
  parameterization, 259–260
  quantum sine-Gordon equation, 261
  S-matrix, Baxter type, 257
  S-matrix, solution for, 258–259
  and $SO(3)$, 256
  symplectic structures, 261
  trigonometric solutions, 260–261
  Weyl relations, 259
  XYZ model, nature, 261
  Zamolodchikov algebra, 257, 258
Quarks, and confinement problem in QCD (*see also* Color confinement; Confinement; Heavy ion collisions)
  difficulties with, 19–20
  instantons, problems with, 20
  renormalization groups, and bare coupling constant, 20
  topological conditions, 20
  topology in gauge condition, 21–23
  2-loop perturbation theorem, 20
Quartic Hamiltonians, symmetry and renormalization group fixed points of
  group actions
    centralizer, 67
    compact groups, 68–69
    decomposition of quadratic forms, 69
    and dimensions, 73
    equalities, 68
    harmonic polynomials, 69
    hypercube, 72
    inequalities, 68
    invariant elements, 67
    invariant subgroups, 68
    irreducible representations, 70
    irreducibility, 72–73
    isotropy groups, 71–72
    n-dimensional orthogonal group, 69–70
    normalizer, 68
    orbits of isotropy groups, 67
    scalar product, invariant, 70–71
    stratums, defined, 67
    summary, subgroups, 73
    surjective linear map, 70
  idempotents of "d" algebras, 64
  in Landau theory, 64

Quartic Hamiltonians (*cont.*)
  n-vector model, Hamiltonian density of, 63–64
  renormalization group equations
    Abelian algebras, 67
    bonds, expressions of, 65–66
    constraints, 65
    equations, 65
    extrema, 66
    irreducibility constant, 65
    stable fixed points, equation, 66
    symmetry of coupling constants, 65
    stable fixed points, with large ν in Hamiltonian, 64

Relativistic effects, *see* Heavy ion collisions
Renormalization
  covariant symmetric algebra, $O(N)$, 81
  group equations, quartic Hamiltonians, 65–66
  and quark confinement, 20

S-matrices, *see* Quantum XYZ models
Scalar fields, *see* Quantized fields
Sine-Gordon equations
  lattice models, 245
  phantom sheets, 26
  quantum, 261
Singularities
  diophantine, 216, 217
  infrared, 154
  monopoles, 25
  quantized fields, 138
Sphere packing
  and Coxeter, work on Lie groups, 269
  8-space, stop of Lie algebras at, 270
  essential point, 269–270
  jump to D series, 270
  and Leech lattice, 24-space, 271
  maximum density, in 1- to 24-dimensional space, 269
  and SU groups, 270
  24-space, uniqueness of, 270–271
Spinor fields, *see* Quantized fields
Stable fixed points of σ-polynomial
  eigenvalues, 89
  Hamiltonian with quartic polynomial, 89
  Hessian, 88
  idempotents, 88
  isotropy groups, 91
  n-cubes, 91
  notation, 87–88
  orthogonality, 88
  in 3-plane, 90
Stable fixed points, defined
  attraction basin, 76
  bifurcation, 75
  ϵ-expansion, 75
  Hamiltonians, 74
  Hessian, 76
  isotropy groups, 74

Stable fixed points (*cont.*)
  normalization, 74–75
  orbit of solutions, 75
Superfields, chiral, variational calculus on
  auxiliary components, 146
  constraints, 145
  dynamic components, 146
  field equations, 146–147
  functional derivatives, 145
  infinitesimal variation, 146
  naive derivations, 145–146
  right-handed functions, 145
Supergravity
  action in superspace of $SO(N)$ variety
    advantages of model, 174
    base supermanifold, 163–164
    basis of vectors, 161
    Bianchi identities, 165, 171, 173
    bundle of frames for, 160–167
    bundle space, 161–162
    composition law, 167
    connection ω, 165, 166
    coordinate transformation, general, 167
    coset spaces, 162–163
    curvature, 160, 161, 164, 166
    Euler–Lagrange equations, 173, 174
    fibers, 160–161
    general relativity, 160–161
    generators, 162
    horizontal one-forms, 161
    infinitesimal transformation, 166–167
    invariance, 160
    invariant integral, 171
    Jacobi identities, 165
    Killing metric, 163
    Lie algebra, 162
    Lorentz invariance, 160
    Majorana representation, 162, 163
    on-shell constraints, 172, 173
    one-form ϕ, 164
    solder form, 164, 166
    supergroups, 162
    topological invariance, 173
    torsion, 161, 164, 166, 171, 172
    two-form ϕ, 164
    vertical vectors, 161
  cases, remarks about
    chiral superfield, 187
    curved case, 187
    Grassmann analyticity, 188
    $N = 1$, 187
    supersymmetry, 188–189
    torsion constraints, 188
  intrinsic geometry of
    analytic transformations, 181
    Berezinian, 182, 183
    boson–fermion mixing, 179
    chiral superfield for, 181
    complex superspace, 180–181

Supergravity (*cont.*)
  intrinsic geometry of (*cont.*)
    curved tensor components, 184
    degrees of freedom, 180
    Einstein supergravity group, 181–182
    embedding, 182–183
    fields/coordinates equivalent, 177–178
    general coordinate transformation group, 180
    Grassmann variables, 179–180
    and gravity, 181
    group covariant restriction, 183
    hypersurfaces, 178
    mass shell, 182
    $N = 1$ model, 178
    $N = 2$ model, 178
    Riemann tensor, 184
    Siegel–Gates formulation, 182
    spinor derivatives, 183
    superalgebra, 179
    superfields, 179–180, 184
    supersymmetry, 179
    torsion, 178, 184
    and unification theory, 177
Supersymmetry, spontaneously broken
  chiral superfield, decomposition of, 143–144
  Fermi generators, 142
  Grassmann coordinates, 143
  invariance, Poincaré–Lorentz, 142
  notation, points to remember, 144
  stable group of superspace, 142
  superfields, 143
  supersymmetry fields, 144
  vacuum, properties of, 142–143
    choices about, 143
    expectation values, 142, 143
  Wess–Zumino theory, 142
Symmetry quantum numbers, missing,
      problem of, 93

Toda lattices
  in Backlund transformations, 245, 248
  and quantum lattice, 253
Topological gauge artifacts, and QCD
  eigenvalues, 24
  gauge quantum numbers, 23
  gluons, 24
  photons, 24
  $SU(3)$, 24
  $U(1)$ subgroup, 23–24
  X-vector, 23
Turbulent aether
  gauge principle, 12

Turbulent aether (*cont.*)
  and Gürsey, work on, 9
  Lagrangian in, chiral, 9
  nature
    aether, as pseudomechanical "substance," 10
    and basic law, 10
    Einstein, field equation of, 11
    and electromagnetism, 11
    mass as stumbling block, 11
    massless fields, 11
    particles, 10
    phase transitions of vacuum, 11
    symmetry breakdown, 11
  24-dimensional space, *see* Sphere packing
  26-dimensional space, connection of with strings
      (*see also* Sphere packing)
    Atkin identity, 278
    coefficients, factors in, 277
    and Dedekind eta function, 277
    Hardy, proof of, 277
    prime numbers as factors in large numbers,
      277

Ultraviolet
  convergence, 116
  diagonal terms, 121

Vacuum (*see also* Turbulent aether)
  properties of, 142–143
  superconducting, 28

Ward identities
  derivation of, 150
  integration by parts, 150–151
  internal symmetries, linearly related, 151
  second kind, 151
  and Volterra series, 151
Weyl, Hermann
  development of theory, steps, 267
  gauge theory invention by, 266–267
Weyl algebras, of quantum lattices, 253

Zeta-function zeros
  fast Fourier transform, discussion, 280–281
  and Hermitian eigenvalues, 279
  and large random Hermitian matrices, 279
  Mehta curve, 280
  Odlyzko, Andrew, work of, 279
  pair correlation function, 279, 280, 281
  problem of, 279
  spacing distribution of zeros, 280
  triangle subtraction, 280, 281